Springer Theses

Recognizing Outstanding Ph.D. Research

Aims and Scope

The series "Springer Theses" brings together a selection of the very best Ph.D. theses from around the world and across the physical sciences. Nominated and endorsed by two recognized specialists, each published volume has been selected for its scientific excellence and the high impact of its contents for the pertinent field of research. For greater accessibility to non-specialists, the published versions include an extended introduction, as well as a foreword by the student's supervisor explaining the special relevance of the work for the field. As a whole, the series will provide a valuable resource both for newcomers to the research fields described, and for other scientists seeking detailed background information on special questions. Finally, it provides an accredited documentation of the valuable contributions made by today's younger generation of scientists.

Theses are accepted into the series by invited nomination only and must fulfill all of the following criteria

- They must be written in good English.
- The topic should fall within the confines of Chemistry, Physics, Earth Sciences, Engineering and related interdisciplinary fields such as Materials, Nanoscience, Chemical Engineering, Complex Systems and Biophysics.
- The work reported in the thesis must represent a significant scientific advance.
- If the thesis includes previously published material, permission to reproduce this must be gained from the respective copyright holder.
- They must have been examined and passed during the 12 months prior to nomination.
- Each thesis should include a foreword by the supervisor outlining the significance of its content.
- The theses should have a clearly defined structure including an introduction accessible to scientists not expert in that particular field.

More information about this series at http://www.springer.com/series/8790

Paulo André Dias Gonçalves

Plasmonics and Light–Matter Interactions in Two-Dimensional Materials and in Metal Nanostructures

Classical and Quantum Considerations

Doctoral Thesis accepted by
Technical University of Denmark,
Kongens Lyngby, Denmark

Springer

Author
Dr. Paulo André Dias Gonçalves
Center for Nano Optics
University of Southern Denmark
Odense M, Denmark

Supervisors
Prof. N. Asger Mortensen
Center for Nano Optics
University of Southern Denmark
Odense M, Denmark

Prof. Antti-Pekka Jauho
Department of Physics
Technical University of Denmark
Kongens Lyngby, Denmark

Prof. Nuno M. R. Peres
Department of Physics
University of Minho
Braga, Portugal

ISSN 2190-5053 ISSN 2190-5061 (electronic)
Springer Theses
ISBN 978-3-030-38290-2 ISBN 978-3-030-38291-9 (eBook)
https://doi.org/10.1007/978-3-030-38291-9

This Springer imprint is published by the registered company Springer Nature Switzerland AG
The registered company address is: Gewerbestrasse 11, 6330 Cham, Switzerland

Supervisors' Foreword

Nature is not always intrinsically friendly towards a scientist looking for applications. An example is light–matter coupling: it follows from the numerical values of natural constants that this coupling is inherently weak. This is unfortunate because charges and electromagnetic waves have their own respective highly attractive characteristics exploited massively in computer technologies and optical communication, and a possibility to exploit both of these features simultaneously would open the way towards many superior technologies. A possible remedy for this weakness is provided by the field of plasmonics: collective oscillations of charges in materials couple much more strongly to electromagnetic waves than individual charges do. The coupling can be made particularly strong in modern nanostructures and two-dimensional systems which have become available in recent years. But a quantitative assessment of this coupling is an arduous task: the geometries are complicated, experimentally relevant structures consist of many different materials, quantum mechanical features must be taken into account, and the required calculations become extremely tedious and time-consuming. This is the task that Dr. Gonçalves' Ph.D. thesis addresses. The thesis begins with a summary of classical electrodynamics and the way it is applied in conventional plasmonics. This allows the reader to fully grasp and appreciate the complications that follow when applying these concepts to two-dimensional (2D) geometries. Several examples are treated in great detail, including 2D nanoslits, ultrathin metallic films, and anisotropic materials. The next level of complication consists of nonplanar structures, and the thesis shows how grooved or wedged structures can demonstrate extreme light localization, and, consequently, strong light–matter coupling. In later chapters systems going beyond classical plasmonics are analyzed with a variety of methods. Effects discussed include the hydrodynamical model—which allows a discussion of nonlocal effects—and, as a subsequent development, the role of quantum nonlocal effects is analyzed. Finally, mesoscopic electrodynamics is given a unified treatment which allows a smooth transition between the classical and quantum regimes.

The thesis gives a comprehensive view of state-of-the-art nanoplasmonics. It derives a number of effective, simple methods which may become the standard tools for analyzing many emerging experiments. The thesis is self-contained and complete: a serious student of the text will be able to enter this fascinating field of research which holds many promises for the future.

Kongens Lyngby, Denmark
October 2019

Prof. N. Asger Mortensen
Prof. Antti-Pekka Jauho
Prof. Nuno M. R. Peres

Abstract

This thesis presents a comprehensive theoretical description of classical and quantum plasmonics in three and two dimensions and also in hybrid systems containing elements with different dimensionalities. It focuses on the theoretical understanding of the salient features of plasmons in nanosystems as well as on the multifaceted aspects of plasmon-enhanced light–matter interactions at the nanometer scale, with special emphasis on the modeling of nonclassical behavior across the transition between classical and quantum domains.

We have divided the content of this thesis into two main parts. In Part, I we present the results obtained using classical theories of plasmonics, and in Part II we go beyond classical electrodynamics by extending our theoretical considerations to the nonclassical, quantum regime.

In the first part of the thesis, following an introduction to the foundational concepts behind the theory of classical electrodynamics and to the core elements of classical plasmonics in three and two dimensions, we develop a general theoretical formalism for calculating plasmons in various two-dimensional (2D) geometries. We have then applied that framework to study plasmon coupling and hybridization in 2D nanoslits, and have fully characterized the properties of the resulting two hybrid plasmon modes and how these can be tuned upon varying the slit's width. Our method is valid for any 2D plasmon-supporting material (including ultrathin metallic films); both the case where the material's response is isotropic and the case where it is anisotropic have been considered. Next, after having dealt with strictly planar 2D configurations, we expand our investigations to nonplanar structures based on 2D materials. Concretely, we consider one-dimensional channels formed by engineering the 2D material into either a V- or a Λ-shape (i.e., resulting in a groove or in a wedge, respectively). We have developed two distinct theoretical models for computing the features of the 2D channel plasmons supported by such structures. Our results show that these modes exhibit levels of light localization that are deeply subwavelength, even larger than what could possibly be obtained by exploiting plasmons in the planar, continuous host 2D medium.

In the second part of the thesis, we identify the main shortcoming associated with classical treatments of plasmonics, and then propose a number of different theoretical approaches for overcoming those shortcomings. We start by reviewing the hydrodynamic model for the homogeneous electron gas in three-dimensions, which includes nonlocality to lowest-order; both the planar and the spherical configuration are analyzed. We then briefly discuss the so-called specular reflection model which includes the full nonlocal dielectric function of the homogeneous electron gas. Next, we investigate how quantum nonlocal effects influence the dispersion of acoustic-like graphene plasmons, which are ultraconfined graphene plasmons that can be excited when a graphene sheet lies in close proximity to a metal substrate. We find significant deviations from classicality pertaining to the plasmonic response of graphene, and then exploit the remarkable confinement attained by this kind of graphene plasmons to probe nonlocal effects in the metal's optical response. Lastly, we present a unified theoretical treatment of mesoscopic electrodynamics—rooted on the d-parameter formalism—whose applicability encompasses both the classical and quantum regimes, and, crucially, spans the challenging transition region where classical and quantum effects can coexist. In particular, our approach allows a simultaneous account of nonlocality, electronic spill-out, and surface-enabled Landau damping, while also including retardation. We derive analytical expressions for the nonclassical scattering coefficients in selected geometries, from which we determine the systems' plasmonic excitations. Our results show that, for a broad range of experimentally-relevant parameters, the latter incur in substantial nonclassical resonant shifts (broadenings) parameterized by a geometry-dependent factor times the real (imaginary) part of the $d_\perp$-parameter. Furthermore, we extend and apply our mesoscopic formalism to a plethora of plasmon–emitter interactions. Specifically, we investigate the role of quantum surface corrections to the plasmonic Purcell enhancement along with their influence in enhancing dipole-forbidden transitions, plasmon-mediated energy transfer between two emitters, as well as two-photon emission. Our findings underscore the importance of incorporating nonclassical corrections in quantum nanoplasmonics; this becomes increasingly important as current state-of-the-art nanoplasmonic studies continue to probe ever-smaller nanostructures and/or emitter–metal separations.

Publications Related to this Thesis

The list of papers that have been published by the author in international peer-reviewed journals during the course of the author's Ph.D. studies is presented below (in inverse chronological order). Publications A, F, J, K, and N are discussed in this thesis.

Papers published in peer-reviewed journals during the author's Ph.D. studies:

A. **P. A. D. Gonçalves**, T. Christensen, N. Rivera, A.-P. Jauho, N. A. Mortensen, M. Soljačić, *Plasmon–Emitter Interactions at the Nanoscale*, Nat. Commun. **11**, 366 (2020)

B. M. Geisler, X. Cui, J. Wang, T. Rindzevicius, L. Gammelgaard, B. S. Jessen, **P. A. D. Gonçalves**, F. Todisco, P. Bøggild, A. Boisen, M. Wubs, N. A. Mortensen, S. Xiao, N. Stenger, *Single-Crystalline Gold Nanodisks on WS_2 Mono- and Multilayers for Strong Coupling at Room Temperature*, ACS Photonics **6** (4), 994–1001 (2019).

C. Y. You, **P. A. D. Gonçalves**, L. Shen, M. Wubs, X. Deng, S. Xiao, *Magnetoplasmons in monolayer black phosphorus structures*, Opt. Lett. **44**, 554–557 (2019).

D. C. Tserkezis, **P. A. D. Gonçalves**, C. Wolff, F. Todisco, K. Busch, N. A. Mortensen, *Mie excitons: Understanding strong coupling in dielectric nanoparticles*, Phys. Rev. B **98**, 155439 (2018).

E. N. A. Mortensen, **P. A. D. Gonçalves**, M. Khajavikhan, D. N. Christodoulides, C. Tserkezis, C. Wolff, *Fluctuations and noise-limited sensing near the exceptional point of parity-time-symmetric resonator systems*, Optica **5**, 1342–1346 (2018).

F. E. J. C. Dias, D. A. Iranzo, **P. A. D. Gonçalves**, Y. Hajati, Y. V. Bludov, A.-P. Jauho, N. A. Mortensen, F. H. L. Koppens, N. M. R. Peres, *Probing non-local effects in metals with graphene plasmons*, Phys. Rev. B **97**, 245405 (2018).

G. A. J. Chaves, B. Amorim, Y. V. Bludov, **P. A. D. Gonçalves**, N. M. R. Peres, *Scattering of graphene plasmons at abrupt interfaces: An analytic and numeric study*, Phys. Rev. B **97**, 035434 (2018).

H. **P. A. D. Gonçalves**, L. P. Bertelsen, S. Xiao, N. A. Mortensen, *Plasmon-exciton polaritons in two-dimensional semiconductor/metal interfaces*, Phys. Rev. B **97**, 041402(R) (2018).

I. B. Amorim, **P. A. D. Gonçalves**, M. I. Vasilevskiy, N. M. R. Peres, *Impact of Graphene on the Polarizability of a Neighbour Nanoparticle: A Dyadic Green's Function Study*, Appl. Sci. **7** (11), 1158 (2017).

J. **P. A. D. Gonçalves**, S. Xiao, N. M. R. Peres, N. A. Mortensen, *Hybridized Plasmons in 2D Nanoslits: From Graphene to Anisotropic 2D Materials*, ACS Photonics **4** (12), 3045–3054 (2017).

K. **P. A. D. Gonçalves**, S. I. Bozhevolnyi, N. A. Mortensen, N. M. R. Peres, *Universal description of channel plasmons in two-dimensional materials*, Optica **4**, 595–600 (2017).

L. U. Levy, M. Grajower, **P. A. D. Gonçalves**, N. A. Mortensen, J. B. Khurgin, *Plasmonic silicon Schottky photodetectors: The physics behind graphene enhanced internal photoemission*, APL Photonics **2**, 026103 (2017).

M. **P. A. D. Gonçalves**, E. J. C. Dias, Y. V. Bludov, N. M. R. Peres, *Modeling the excitation of graphene plasmons in periodic grids of graphene ribbons: An analytical approach*, Phys. Rev. B **94**, 195421 (2016).

N. **P. A. D. Gonçalves**, E. J. C. Dias, S. Xiao, M. I. Vasilevskiy, N. A. Mortensen, N. M. R. Peres, *Graphene Plasmons in Triangular Wedges and Grooves*, ACS Photonics **3** (11), 2176–2183 (2016).

Acknowledgements

First and foremost, I must thank my Ph.D. supervisors, Prof. N. Asger Mortensen and Prof. Antti-Pekka Jauho, and my co-supervisor Prof. Nuno M. R. Peres, for guiding me throughout my Ph.D. project. To Asger in particular, I am profoundly thankful for his continuous support, guidance, invaluable counsel, and for many inspiring scientific (and non-scientific) discussions. I could not have asked for a better Ph.D. supervisor. To Antti, I wish to express my gratitude for many insightful discussions and for his inestimable sagacity and wit. To Nuno, for his dedication and for accompanying me during yet another chapter of my academic education.

I would also like to extend my thanks to all current and former members of the Structured Electromagnetic Materials group at DTU, with whom I had the pleasure to interact. I acknowledge Assoc. Profs. Nicolas Stenger, Sanshui Xiao, and Martijn Wubs for various collaborations and fruitful scientific discussions; I am grateful to Nicolas for valuable insights from an experimental viewpoint and many interesting conversations. Moreover, special thanks to my Ph.D.-mates, namely, Moritz Fischer, Emil C. André, Mathias Geisler, Johan Maack, Kåre Wedel, and Christian Frydendahl; thank you all for making such an amazing working environment and for countless cheerful discussions about pretty much everything.

In addition, I would like to express my gratitude towards all the members of the Quantum Plasmonics group at SDU from whom I have learned much. In particular, I am grateful to Christos Tserkezis, Christian Wolff, and Francesco Todisco for many stimulating discussions and rewarding collaborations.

Furthermore, I wish to thank Prof. Marin Soljačić for hosting me at the Massachusetts Institute of Technology within the scope of my Ph.D. external research stay. I must also thank Thomas Christensen for tutoring me during my almost six month stay at MIT and for introducing me to the Feibelman d-parameter formalism. In addition, I thank Nicholas Rivera for many excellent interactions.

I further thank the members of my Ph.D. thesis committee, Prof. Peter Nordlander, Prof. Javier García de Abajo, and Prof. Jesper Mørk for taking time out of their busy agendas to consider this thesis and to serve on my thesis assessment committee.

On a more personal level, I would like to wholeheartedly thank my parents and brother for all their affection and for always being there for me. Finally, my deepest thanks to Inês for her unwavering encouragement and support, for her kindness, and for her unconditional love.

Contents

Acronyms

1D	One-dimensional
2D	Two-dimensional
2DEG	Two-dimensional electron gas
3D	Three-dimensional
3DEG	Three-dimensional electron gas
AGP	Acoustic-like graphene plasmon
BEM	Boundary element method
BP	Black phosphorus
BPP	Bulk plasmon polariton
CPP	Channel plasmon polariton
DDA	Discrete-dipole approximation
DFT	Density functional theory
DOS	Density of states
EIM	Effective-index method
ET	Energy transfer
FDTD	Finite-different time-domain
FEM	Finite-element method
GFIEM	Green's function integral equation method
GGP	Groove graphene plasmon
GNOR	Generalized nonlocal optical response
GP	Graphene plasmon
HDM	Hydrodynamic model
IR	Infrared
LDOS	Local density of states
LRA	Local-response approximation
LSP	Localized surface plasmon
QED	Quantum electrodynamics
RPA	Random-phase approximation
RTA	Relaxation-time approximation
SERS	Surface-enhanced Raman spectroscopy

SNOM	Scanning near-field optical microscope
SPP	Surface plasmon polariton
SRM	Specular reflection model
TDDFT	Time-dependent density functional theory
TE	Transverse electric
TM	Transverse magnetic
TMDC	Transition metal dichalcogenide
TMM	Transfer-matrix method
WGP	Wedge graphene plasmon
xc	Exchange-correlation

Part I
Classical Plasmonics

Chapter 1
Introduction

Ever since the dawn of humankind our species has been continuously pursuing new ways to tailor materials and to tame light to its advantage. In early human history, the ages of civilization were named after the prevailing tool-making material. Even today, our technological evolution is intimately associated with advances in materials science, involving both the synthesis of new materials and the purposeful processing and combination of them [1]. Along with materials innovation, since the early nineteenth century that the understanding and ensuing exploitation of electromagnetic phenomena has been contributing enormously—in ways that the pioneers of the field could not possibly have fathomed at the time—to our technological and scientific revolution [2] as well as to our general well-being.

Plasmonics—the central topic of this thesis—lies precisely at the intersection between materials science and electromagnetism, or else between condensed matter physics and photonics. Broadly speaking, plasmonics [3–7] is a sub-branch of physics that focuses on the study of plasmons and plasmon-enabled phenomena. In classical terms, plasmons are self-sustained collective excitations of the free-electron plasma mediated by the Coulomb interaction between its charge carriers. In a quantum mechanical picture, plasmons are often viewed as coherent intraband excitations of electron-hole pairs around the ground-state's Fermi level; because a plasmon is a *collective* excitation, brought about by the Coulomb interactions, its energy is considerably higher than the one associated with each individual electronic transition. The first theoretical description of plasmons in a quantum mechanical setting is generally attributed to Bohm and Pines [8–12], who have developed a microscopic formalism for the dynamic response of the homogeneous electron gas that became known as the random-phase approximation (RPA) [13–15].

At metal surfaces or in finite-sized metal structures plasmons can couple to free-space electromagnetic waves, thereby giving rise to surface plasmon polaritons (SPPs) [3, 4, 7]. Here, the polariton epithet is indicative of the mixing between electronic and photonic degrees of freedom. The relative fraction of "light-like" and "matter-like" that characterizes the SPP mode varies along its energy-wavevector dispersion curve.

P. A. D. Gonçalves, *Plasmonics and Light–Matter Interactions in Two-Dimensional Materials and in Metal Nanostructures*, Springer Theses,
https://doi.org/10.1007/978-3-030-38291-9_1

One of the most tantalizing properties of surface plasmons lies in their ability to provide a mean to achieve subwavelength light confinement beyond the diffraction limit of conventional optics [16, 17]. This feature has remarkable consequences, such as, for instance, in delivering extreme field enhancements [18, 19], subwavelength waveguiding [17, 20], enhanced spectroscopy [21–23], control over the dynamics of emitters [24–26], ultrasensitive biochemical sensing [22, 27–31], medical theragnostics [32–34], and metamaterials [35, 36] and antenna design [37, 38] for shaping wavefronts. Although all of these feats have now been realized, some of them have somewhat fallen short of the community's earnest expectations [39], in particular those concerning applications for which a delicate control of losses is imperative [40] (curiously, for other applications, plasmon-related losses can actually act as performance-improvers [41]). In spite of this, new research avenues and exciting opportunities have been recently recognized [42], including the potential of plasmonics for enhancing nonlinearities [43, 44], producing ink-free plasmonic colors [45–48], boosting light–matter interactions up to the strong-coupling regime [49–52], tailoring spontaneous decay rates beyond the dipole-approximation [53–56], and for plasmon-driven hot-electron physics and catalysis [57–61], just to name a few (Fig. 1.1). In parallel, recent advances in nanofabrication and characterization techniques have brought the possibility to investigate plasmons and plasmon-empowered phenomena in systems characterized by length scales approaching the intrinsic quantum mechanical scales of the plasmon-supporting electron gas [62–68]. In this context, classical descriptions of plasmonics must be either abandoned [69, 70] or augmented with nonclassical corrections [56, 71–73], which may be of semi-classical [74–81] or quantum mechanical origin [56, 67, 73, 82–86]. The study of plasmonics beyond classical electrodynamics is what motivates and defines the vibrant field of quantum nanoplasmonics [52, 69, 87–89].

Another enticing development in the field of plasmonics is the one concerned with the investigation of plasmons in two-dimensional (2D) materials [7, 90–92].

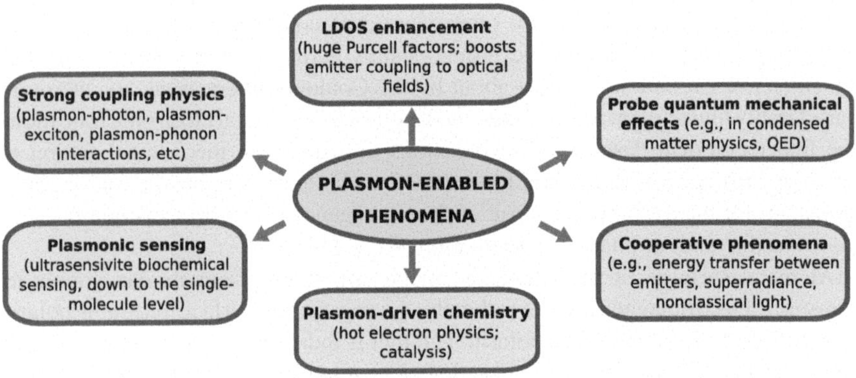

Fig. 1.1 Examples of applications where surface plasmons can be used to enhance, tailor, and/or probe tailor light–matter interactions across various sub-branches of the physical sciences

The recently renewed interest in 2D plasmonics is deeply intertwined with the advent of graphene, a one-atom-thick carbon allotrope whose atoms are arranged in a planar honeycomb lattice, first isolated in 2004(–2005) [93–96]. Graphene possesses remarkable electronic properties [97], spearheaded by its peculiar low-energy linear spectrum where charge carriers behave as massless particles governed by an effective relativistic massless Dirac Hamiltonian [94, 97, 98], together with extraordinary mechanical [99] and thermal properties [100]. Besides these, graphene also exhibits outstanding optical properties [7, 90, 98, 101], namely, a nearly constant absorption of $\approx 2.3\%$ in the visible region [102] of the electromagnetic spectrum (which is unarguably surprisingly large for a *single* atomic layer). When doped, graphene supports 2D plasmons spanning the terahertz (THz) and mid-infrared (mid-IR) spectral regions [7, 90–92]. Strikingly, graphene plasmons are capable of rendering light confinements that can be extremely subwavelength [7, 90–92, 103–108]. Furthermore, and in stark contrast with SPPs in three-dimensional (3D) metals, the frequency of graphene plasmons (GPs) can be actively tuned either by electrostatic gating [103, 104, 109, 110] or by chemical means [111]. Such versatility is pivotal from a technological, application-oriented perspective. At the time of writing, the subject of graphene plasmonics has matured [7], but it continues to expand considerably into numerous directions at an incredibly rapid pace.

The research presented throughout this thesis focuses on the theoretical description of surface plasmons in a plethora of plasmon-supporting nanostructures, encompassing not only plasmonics in three- and two-dimensional systems, but also an in-between, transdimensional regime. The latter spans an intermediate regime between the "strictly 3D" and the "strictly 2D". This definition, albeit somewhat loose, has recently been used [112] to classify systems comprised of a few-atomic layers, but could, in practice, also be extended to compound structures which simultaneously contain 3D and 2D elements.

1.1 Structure and Scope of This Thesis

The present thesis is divided into two parts, Part I and Part II, where it is discussed, respectively, the salient features pertaining to classical and quantum (or nonclassical) theoretical treatments of plasmonics and plasmon-related nanophotonic phenomena. In the first part, after reviewing the fundamentals of plasmonics in three- and two-dimensions (in the latter, with special emphasis on graphene), we theoretically describe plasmonic excitations in planar 2D nanostructures using a universal framework that is both scale- and material-independent, being fully specified by the structure's geometry (i.e., its shape). This analysis shall set the stage for our subsequent investigations pertaining to (i) the treatment of plasmon coupling and hybridization in 2D nanostructures; and (ii) the study of plasmons in nonplanar geometries, namely, of 2D channel plasmons propagating along a triangularly-shaped wedge or groove. In the second part of the thesis, we start by summarizing the main shortcomings of the classical approach, chiefly, the local-response approximation, incompleteness and the omission of Landau damping, and the assumption of abruptly-defined material interfaces (i.e., discontinuity). Evidently, these are not all the shortcomings associated with classical treatments of plasmonics, but they are arguably the most important in nearly all relevant scenarios encountered in nanoplasmonics. Next, we address the above-noted deficiencies and develop different models to overcome them—again, both in three- and two-dimensions—, ranging from a semiclassical account based on hydrodynamics to more sophisticated formalisms, including the nonlocal RPA and a novel framework for electrodynamics in the mesoscopic regime. The latter framework, in particular, is rooted on the microscopic formalism of Feibelman d-parameters [56, 71, 73, 82] and is capable of simultaneously remedy all the three aforementioned shortcomings; specifically, it enables a leading-order-accurate incorporation of nonlocality, electroninc spill-out/spill-in, and surface-enabled Landau damping. Crucially, we extend the applicability of the d-parameter formalism to an unprecedented range of length scales, from the classical, macroscopic domain down to the *truly* few-nanometer regime. Finally, we further develop our mesoscopic framework and investigate the impact of quantum surface corrections on the plasmon-enhanced light–matter interaction: we consider a broad array of plasmon–emitter interactions ranging from dipolar and multipolar spontaneous emission enhancement, to plasmon-assisted energy transfer between two emitters and plasmon-enhanced two-photon emission. Our formalism for mesoscopic electrodynamics gives a complete account of both plasmons and plasmon–emitter interactions at the nanoscale, constituting a simple yet rigorous and general platform to incorporate nonclassical effects in plasmon-empowered nanophotonic phenomena.

1.1.1 Outline

This thesis is structured into two parts, each addressing multifaceted aspects of classical and quantum nanoplasmonics and light–matter interactions. In both parts, we cover three-dimensional metal nanostructures as well as two-dimensional nanosystems. Below, we provide a succinct description of the content of each chapter.

Chapter 2 | **Classical Electrodynamics of Solids**
Reviews the foundations of classical electrodynamics including Maxwell's equations and corresponding boundary conditions together with the macroscopic constitutive relations. Next, we introduce the core elements of plasmonics in numerous settings and establish the fundamental properties of surface plasmon polaritons and localized surface plasmons. Finally, we provide an overview of the electromagnetic Green's dyadics and use it to introduce the concept of local density of states and describe the Purcell effect.

Chapter 3 | Electronic and Optical Properties of Graphene
Gives a crash course in the physics of graphene, focusing on its electronic and optical properties. It discusses the predominant theoretical models for describing the optoelectronic properties of graphene and provides their derivations along the way.

Chapter 4 | Fundamentals of Graphene Plasmonics
This chapter starts by introducing the field of graphene plasmonics in a self-contained fashion. The salient features of graphene plasmons in both continuous and nanopatterned graphene are reviewed. Finally, plasmon hybridization in 2D nanoslits made from different atomically-thin materials is investigated.

Chapter 5 | Two-dimensional Channel Plasmons in Nonplanar Geometries
Extends the previous treatments of plasmons (and other polaritons) in 2D systems to nonplanar structures, i.e., configurations where the 2D material is either folded or conformed to a pre-engineered substrate. It focuses on nonplanar geometries based on one-dimensional, triangularly-shaped channels.

Chapter 6 | Electrodynamics of Metals Beyond the Local-Response Approximation: Nonlocal Effects
Identifies the main shortcomings associated with classical descriptions of plasmonics and introduces the most common models for addressing some of those shortcomings. Here, we center our attention in three-dimensional metal nanostructures and investigate the impact of nonlocality in the large-wavevector (for planar systems) or small-size (for metal nanospheres) limits.

Chapter 7 | Quantum Nonlocal Effects Probed by Ultraconfined Graphene Plasmons
Reviews the outstanding properties of acoustic-like graphene plasmons arising from the electromagnetic interaction between plasmons in graphene and the screening exerted by a nearby metal. It is shown that this type of graphene plasmons can not only be exploited for confining electromagnetic fields to nanometer scales but can also be used as probes for revealing quantum nonlocal effects in condensed matter systems. Whenever possible, we compare our calculations against experimental data that has been reported in the literature. After exploring the consequences of nonlocality and many-body interactions in the graphene sheet alone, we then simultaneously account for nonlocal effects in the graphene and in the metal.

Chapter 8 | Quantum Corrections in Plasmonics and Plasmon–Emitter Interactions
In this chapter, we develop and apply a new, mesoscopic treatment of plasmons and light–matter interactions in nanoplasmonics, whose applicability encompasses an unprecedentedly wide range of length scales, and, in particular, conveniently bridges the gap between "fully quantum" (microscopic, atomistic) and "fully classical" (macroscopic) descriptions. This framework, rooted on the so-called Feibelman d-parameters, facilitates a simultaneous incorporation of electronic spill-out, nonlocality, and surface-assisted Landau damping—all intrinsically quantum mechanical mechanisms—through the derivation of *analytical* mesoscopic scattering coef-

ficients, thereby enabling the calculation of plasmon-mediated light–matter interactions in the mesoscopic regime.

Chapter 9 | **Conclusions and Outlook**
Revisits and summarizes the main content of the thesis as a whole and discusses its implications in a broader context, including potentially new opportunities and perspectives for future investigations.

References

1. N.R. Council (1975) Materials and man's needs: materials science and engineering—volume i, the history, scope, and nature of materials science and engineering. The National Academies Press, Washington, DC. https://www.nap.edu/catalog/10436/materials-and-mans-needs-materials-science-and-engineering-volume-i, https://doi.org/10.17226/10436
2. Darrigol O (2003) Electrodynamics from Ampère to Einstein. Oxford University Press
3. Maradudin AA, Barnes WL, Sambles JR (eds) (2014) Modern plasmonics, 1st edn. Elsevier
4. Maier SA (2007) Plasmonics: fundamentals and applications. Springer
5. Pelton M, Bryant GW (2013) Introduction to metal-nanoparticle plasmonics. Wiley, New York
6. Pitarke JM, Silkin VM, Chulkov EV, Echenique PM (2006) Rep Prog Phys 70(1):1. https://iopscience.iop.org/article/10.1088/0034-4885/70/1/R01/meta, https://doi.org/10.1088/0034-4885/70/1/r01
7. Gonçalves PAD, Peres NMR (2016) An introduction to graphene plasmonics, 1st edn. World Scientific, Singapore. http://www.worldscientific.com/worldscibooks/10.1142/9948, https://doi.org/10.1142/9948
8. Bohm D, Pines D (1951) Phys Rev 82:625. https://doi.org/10.1103/PhysRev.82.625
9. Pines D, Bohm D (1952) Phys Rev 85:338. https://doi.org/10.1103/PhysRev.85.338
10. Bohm D, Pines D (1953) Phys Rev 92:609. https://doi.org/10.1103/PhysRev.92.609
11. Pines D (1953) Phys Rev 92:626. https://doi.org/10.1103/PhysRev.92.626
12. Pines D (1956) Rev Mod Phys 28:184. https://doi.org/10.1103/RevModPhys.28.184
13. Giuliani G, Vignale G (2005) Quantum theory of the electron liquid. Cambridge University Press. https://doi.org/10.1017/CBO9780511619915
14. Mahan GD (2000) Many-particle physics, 3rd edn. Springer, New York
15. Bruus H, Flensberg K (2004) Many-body quantum theory in condensed matter physics: an introduction. Oxford graduate texts. Oxford University Press
16. Barnes WL, Dereux A, Ebbesen TW (2003) Nature 424:824. https://www.nature.com/articles/nature01937, https://doi.org/10.1038/nature01937
17. Gramotnev DK, Bozhevolnyi SI (2010) Nat Photon 4(2):83. https://www.nature.com/articles/nphoton.2009.282, https://doi.org/10.1038/nphoton.2010.282
18. Schuller JA, Barnard ES, Cai W, Jun YC, White JS, Brongersma ML (2010) Nat Mater 9:193. https://doi.org/10.1038/nmat2630
19. Hao E, Schatz GC (2004) J Chem Phys 120(1):357. https://doi.org/10.1063/1.1629280
20. Bozhevolnyi SI, Volkov VS, Devaux E, Laluet JY, Ebbesen TW (2006) Nature 440:508. https://doi.org/10.1038/nature04594
21. Kneipp K, Wang Y, Kneipp H, Perelman LT, Itzkan I, Dasari RR, Feld MS (1997) Phys Rev Lett 78:1667. https://doi.org/10.1103/PhysRevLett.78.1667
22. Haes AJ, Haynes CL, McFarland AD, Schatz GC, Van Duyne RP, Zou S (2005) MRS Bull 30(5):368–375. https://doi.org/10.1557/mrs2005.100
23. Perney NMB, Baumberg JJ, Zoorob ME, Charlton MDB, Mahnkopf S, Netti CM (2006) Opt Express 14(2):847. http://www.opticsexpress.org/abstract.cfm?URI=oe-14-2-847, https://doi.org/10.1364/OPEX.14.000847

24. Anger P, Bharadwaj P, Novotny L (2006) Phys Rev Lett 96:113002. https://doi.org/10.1103/PhysRevLett.96.113002
25. Russell KJ, Liu TL, Cui S, Hu EL (2012) Nat Photon 6:459. https://www.nature.com/articles/nphoton.2012.112, https://doi.org/10.1038/nphoton.2012.112
26. Akselrod GM, Argyropoulos C, Hoang TB, Ciracì C, Fang C, Huang J, Smith DR, Mikkelsen MH (2014) Nat Photon 8:835. https://www.nature.com/articles/nphoton.2014.228, https://doi.org/10.1038/nphoton.2014.228
27. Sönnichsen C, Reinhard BM, Liphardt J, Alivisatos AP (2005) Nat Biotechnol 23(6):741. https://www.nature.com/articles/nbt1100, https://doi.org/10.1038/nbt1100
28. Aćimović SS, Kreuzer MP, González MU, Quidant R (2009) ACS Nano 3(5):1231. https://doi.org/10.1021/nn900102j
29. Liu N, Mesch M, Weiss T, Hentschel M, Giessen H (2010) Nano Lett 10(7):2342. https://doi.org/10.1021/nl9041033
30. Mayer KM, Hafner JH (2011) Chem Rev 111(6):3828. https://doi.org/10.1021/cr100313v
31. Brolo AG (2012) Nat Photon 6:709. https://doi.org/10.1038/nphoton.2012.266
32. Hirsch LR, Stafford RJ, Bankson JA, Sershen SR, Rivera B, Price RE, Hazle JD, Halas NJ, West JL (2003) Proc Natl Acad Sci USA 100(23):13549. https://www.pnas.org/content/100/23/13549, https://doi.org/10.1073/pnas.2232479100
33. Lal S, Clare SE, Halas NJ (2008) Acc Chem Res 41(12):1842. https://doi.org/10.1021/ar800150g
34. Im H, Shao H, Park YI, Peterson VM, Castro CM, Weissleder R, Lee H (2014) Nat Biotechnol 32:490. https://doi.org/10.1038/nbt.2886
35. Engheta N, Ziolkowski RW (eds) (2006) Metamaterials: physics and engineering explorations. Wiley
36. Shvets G, Tsukerman I (eds) (2011) Plasmonics and plasmonic metamaterials. World Scientific
37. Fischer H, Martin OJF (2008) Opt Express 16(12):9144. https://doi.org/10.1364/OE.16.009144
38. Novotny L, van Hulst N (2011) Nat Photonics 5:83. https://doi.org/10.1038/nphoton.2010.237
39. Atwater HA (2007) Sci Am 296:56. https://www.scientificamerican.com/article/the-promise-of-plasmonics/, https://doi.org/10.1038/scientificamerican0407-56
40. Khurgin JB (2015) Nat Nanotechnol 10:2. https://doi.org/10.1038/s41565-017-0035-5
41. Boriskina SV, Cooper TA, Zeng L, Ni G, Tong JK, Tsurimaki Y, Huang Y, Meroueh L, Mahan G, Chen G (2017) Adv Opt Photon 9(4):775. http://aop.osa.org/abstract.cfm?URI=aop-9-4-775, https://doi.org/10.1364/AOP.9.000775
42. Fernández-Domínguez AI, García-Vidal FJ, Martín-Moreno L (2017) Nat Photon 11(1):8. https://www.nature.com/articles/nphoton.2016.258
43. Kauranen M, Zayats AV (2012) Nat Photon 6:737–748. https://doi.org/10.1038/nphoton.2012.244
44. Panoiu NC, Sha WEI, Lei DY, Li GC (2018) J Opt 20(8):083001. https://iopscience.iop.org/article/10.1088/2040-8986/aac8ed, https://doi.org/10.1088/2040-8986/aac8ed
45. Kristensen A, Yang JKW, Bozhevolnyi SI, Link S, Nordlander P, Halas NJ, Mortensen NA (2016) Nat Rev Mater 2:16088. https://doi.org/10.1038/natrevmats.2016.88
46. Clausen JS, Højlund-Nielsen E, Christiansen AB, Yazdi S, Grajower M, Taha H, Levy U, Kristensen A, Mortensen NA (2014) Nano Lett 14(8):4499. https://doi.org/10.1021/nl5014986
47. Zhu X, Vannahme C, Højlund-Nielsen E, Mortensen NA, Kristensen A (2016) Nat Nanotechnol 11:325. https://doi.org/10.1038/nnano.2015.285
48. Yu R, Mazumder P, Borrelli NF, Carrilero A, Ghosh DS, Maniyara RA, Baker D, García de Abajo FJ, Pruneri V (2016) ACS Photon 3(7):1194. https://doi.org/10.1021/acsphotonics.6b00090
49. Vasa P, Wang W, Pomraenke R, Lammers M, Maiuri M, Manzoni C, Cerullo G, Lienau C (2013) Nat Photon 7(2):128. https://www.nature.com/articles/nphoton.2012.340, https://doi.org/10.1038/nphoton.2012.340

50. Törmä P, Barnes WL (2014) Rep Prog Phys 78(1):013901. https://iopscience.iop.org/article/10.1088/0034-4885/78/1/013901/meta, https://doi.org/10.1088/0034-4885/78/1/013901
51. Chikkaraddy R, de Nijs B, Benz F, Barrow SJ, Scherman OA, Rosta E, Demetriadou A, Fox P, Hess O, Baumberg JJ (2016) Nature 535(7610):127. https://www.nature.com/articles/nature17974, https://doi.org/10.1038/nature17974
52. Flick J, Rivera N, Narang P (2018) Nanophotonics 7(9):1479. https://doi.org/10.1515/nanoph-2018-0067
53. Andersen ML, Stobbe S, Sørensen AS, Lodahl P (2011) Nat Phys 7(3):215. https://www.nature.com/articles/nphys1870, https://doi.org/10.1038/nphys1870
54. Rivera N, Kaminer I, Zhen B, Joannopoulos JD, Soljačić M (2016) Science 353(6296):263. http://science.sciencemag.org/content/353/6296/263, https://doi.org/10.1126/science.aaf6308
55. Cuartero-González A, Fernández-Domínguez AI (2018) ACS Photon 5(8):3415. https://doi.org/10.1021/acsphotonics.8b00678
56. Gonçalves PAD, Christensen T, Rivera N, Jauho, AP, Mortensen NA, Soljačić, M (2020) Plasmon-emitter interactions at the nanoscale. Nat Commun 11:366. https://doi.org/10.1038/s41467-019-13820-z
57. Brongersma ML, Halas NJ, Nordlander P (2015) Nat Nanotechnol 10(1):25. https://www.nature.com/articles/nnano.2014.311, https://doi.org/10.1038/nnano.2014.311
58. Mukherjee S, Libisch F, Large N, Neumann O, Brown LV, Cheng J, Lassiter JB, Carter EA, Nordlander P, Halas NJ (2013) Nano Lett 13(1):240. https://doi.org/10.1021/nl303940z
59. Zhou L, Swearer DF, Zhang C, Robatjazi H, Zhao H, Henderson L, Dong L, Christopher P, Carter EA, Nordlander P et al (2018) Science 362(6410):69. http://science.sciencemag.org/content/362/6410/69, https://doi.org/10.1126/science.aat696
60. Baumberg JJ (2019) Faraday Discuss 214:501. https://doi.org/10.1039/C9FD00027E
61. Seemala B, Therrien AJ, Lou M, Li K, Finzel JP, Qi J, Nordlander P, Christopher P (2019) ACS Energy Lett 4:1803. https://doi.org/10.1021/acsenergylett.9b00990
62. Scholl JA, Koh AL, Dionne JA (2012) Nature 483(7390):421. https://www.nature.com/articles/nature10904, https://doi.org/10.1038/nature10904
63. Kern J, Großmann S, Tarakina NV, Häckel T, Emmerling M, Kamp M, Huang JS, Biagioni P, Prangsma JC, Hecht B (2012) Nano Lett 12(11):5504. https://doi.org/10.1021/nl302315g
64. Chen X, Park HR, Pelton M, Piao X, Lindquist NC, Im H, Kim YJ, Ahn JS, Ahn KJ, Park N, Kim DS, Oh SH (2013) Nat Commun 4:2361. https://doi.org/10.1038/ncomms3361
65. Raza S, Kadkhodazadeh S, Christensen T, Di Vece M, Wubs M, Mortensen NA, Stenger N (2015) Nat Commun 6:8788. https://www.nature.com/articles/ncomms9788, https://doi.org/10.1038/ncomms9788
66. Campos A, Troc N, Cottancin E, Pellarin M, Weissker HC, Lermé J, Kociakand M, Hillenkamp M (2018) Nat Phys https://doi.org/10.1038/s41567-018-0345-z
67. Yang Y, Di Z, Yan W, Agarwal A, Zheng M, Joannopoulos JD, Lalanne P, Christensen T, Berggren KK, Soljačić M (2019) A general theoretical and experimental framework for nanoscale electromagnetism. Nature 576(7786):248–252. https://doi.org/10.1038/s41586-019-1803-1
68. Albert Polman MK, García de Abajo FJ (2019) Nat Mater. https://doi.org/10.1038/s41563-019-0409-1
69. Varas A, García-González P, Feist J, García-Vidal FJ, Rubio A (2016) Nanophotonics 5(3):409. https://doi.org/10.1515/nanoph-2015-0141
70. Zhang P, Feist J, Rubio A, García-González P, García-Vidal FJ (2014) Phys Rev B 90:161407. https://doi.org/10.1103/PhysRevB.90.161407
71. Liebsch A (1997) Electronic excitations at metal surfaces. Springer, New York
72. Zhu W, Esteban R, Borisov AG, Baumberg JJ, Nordlander P, Lezec HJ, Aizpurua J, Crozier KB (2016) Nat Commun 7:11495. https://doi.org/10.1038/ncomms11495
73. Christensen T, Yan W, Jauho AP, Soljačić M, Mortensen NA (2017) Phys Rev Lett 118:157402. https://doi.org/10.1103/PhysRevLett.118.157402
74. Boardman AD (1982) Electromagnetic surface modes. Wiley

75. Raza S, Bozhevolnyi SI, Wubs M, Mortensen NA (2015) J Phys Condens Matter 27(18):183204. https://iopscience.iop.org/article/10.1088/0953-8984/27/18/183204/meta, https://doi.org/10.1088/0953-8984/27/18/183204
76. Ruppin R (1973) Phys Rev Lett 31:1434. https://doi.org/10.1103/PhysRevLett.31.1434
77. Boardman AD, Paranjape BV (1977) J Phys F Met Phys 7(9):1935. https://iopscience.iop.org/article/10.1088/0305-4608/7/9/036, https://doi.org/10.1088/0305-4608/7/9/036
78. Christensen T, Yan W, Raza S, Jauho AP, Mortensen NA, Wubs M (2014) ACS Nano 8(2):1745. https://doi.org/10.1021/nn406153k
79. Ginzburg P, Zayats AV (2013) ACS Nano 7(5):4334. https://doi.org/10.1021/nn400842m
80. Mortensen NA (2013) Photon Nanostruct Fundam Appl 11(4):303. http://www.sciencedirect.com/science/article/pii/S1569441013000370, https://doi.org/10.1016/j.photonics.2013.06.002
81. Mortensen NA, Raza S, Wubs M, Søndergaard T, Bozhevolnyi SI (2014) Nat Commun 5:3809. https://doi.org/10.1038/ncomms4809
82. Feibelman PJ (1982) Prog Surf Sci 12(4):287. https://doi.org/10.1016/0079-6816(82)90001-6
83. Liebsch A (1993) Phys Rev B 48:11317. https://doi.org/10.1103/PhysRevB.48.11317
84. Teperik TV, Nordlander P, Aizpurua J, Borisov AG (2013) Phys Rev Lett 110:263901. https://doi.org/10.1103/PhysRevLett.110.263901
85. Yan W, Wubs M, Asger Mortensen N (2015) Phys Rev Lett 115:137403. https://doi.org/10.1103/PhysRevLett.115.137403
86. Jin D, Hu Q, Neuhauser D, von Cube F, Yang Y, Sachan R, Luk TS, Bell DC, Fang NX (2015) Phys Rev Lett 115:193901. https://doi.org/10.1103/PhysRevLett.115.193901
87. Bozhevolnyi SI, Martin-Moreno L, García-Vidal F (2017) Quantum plasmonics. Springer
88. Tame MS, McEnery KR, Özdemir K, Lee J, Maier SA, Kim MS (2013) Nat Phys 9:329–340. https://doi.org/10.1038/nphys2615
89. Bozhevolnyi SI, Mortensen NA (2017) Nanophotonics 6:1185. https://doi.org/10.1515/nanoph-2016-0179
90. Grigorenko AN, Polini M, Novoselov KS (2012) Nat Photon 6:749 https://doi.org/nphoton.2012.262
91. Low T, Avouris P (2014) ACS Nano 8(2):1086. https://doi.org/10.1021/nn406627u
92. García de Abajo FJ (2014) ACS Photon 1(3):135. https://doi.org/10.1021/ph400147y
93. Novoselov KS, Geim AK, Morozov SV, Jiang D, Zhang Y, Dubonos SV, Grigorieva IV, Firsov AA (2004) Science 306(5696):666. https://science.sciencemag.org/content/306/5696/666, https://doi.org/10.1126/science.1102896
94. Novoselov KS, Geim AK, Morozov SV, Jiang D, Katsnelson MI, Grigorieva IV, Dubonos SV, Firsov AA (2005) Nature 438:197. https://www.nature.com/articles/nature04233, https://doi.org/10.1038/nature04233
95. Zhang Y, Tan YW, Stormer HL, Kim P (2005) Nature 438:201. https://www.nature.com/articles/nature04235, https://doi.org/10.1038/nature04235
96. Novoselov KS, Jiang D, Schedin F, Booth TJ, Khotkevich VV, Morozov SV, Geim AK (2005) Proc Natl Acad Sci USA 102(30):10451. https://www.pnas.org/content/102/30/10451, https://doi.org/10.1073/pnas.0502848102
97. Castro Neto AH, Guinea F, Peres NMR, Novoselov KS, Geim AK (2009) Rev Mod Phys 81:109. https://doi.org/10.1103/RevModPhys.81.109
98. Katsnelson MI (2012) Graphene: carbon in two dimensions. Cambridge University Press. https://doi.org/10.1017/CBO9781139031080
99. Lee C, Wei X, Kysar JW, Hone J (2008) Science 321(5887):385. https://science.sciencemag.org/content/321/5887/385, https://doi.org/10.1126/science.1157996
100. Balandin AA, Ghosh S, Bao W, Calizo I, Teweldebrhan D, Miao F, Lau CN (2008) Nano Lett 8(3):902. https://doi.org/10.1021/nl0731872
101. Bonaccorso F, Sun Z, Hasan T, Ferrari AC (2010) Nat Photon 4:611. https://doi.org/10.1038/nphoton.2010.186

102. Nair RR, Blake P, Grigorenko AN, Novoselov KS, Booth TJ, Stauber T, Peres NMR, Geim AK (2008) Science 320(5881):1308. https://science.sciencemag.org/content/320/5881/1308, https://doi.org/10.1126/science.1156965
103. Fei Z, Rodin AS, Andreev GO, Bao W, McLeod AS, Wagner M, Zhang LM, Zhao Z, Thiemens M, Dominguez G, Fogler MM, Neto AHC, Lau CN, Keilmann F, Basov DN (2012) Nature 487:82. https://doi.org/10.1038/nature11253
104. Chen J, Badioli M, Alonso-González P, Thongrattanasiri S, Huth F, Osmond J, Spasenović M, Centeno A, Pesquera A, Godignon P, Elorza AZ, Camara N, García de Abajo FJ, Hillenbrand R, Koppens FHL (2012) Nature 487:77. https://doi.org/10.1038/nature11254
105. Woessner A, Lundeberg MB, Gao Y, Principi A, Alonso-González P, Carrega M, Watanabe K, Taniguchi T, Vignale G, Polini M, Hone J, Hillenbrand R, Koppens FHL (2015) Nat Mater 14:421. https://www.nature.com/articles/nmat4169, https://doi.org/10.1038/nmat4169
106. Lundeberg MB, Gao Y, Asgari R, Tan C, Van Duppen B, Autore M, Alonso-González P, Woessner A, Watanabe K, Taniguchi T, Hillenbrand R, Hone J, Polini M, Koppens FHL (2017) Science 357(6347):187. http://science.sciencemag.org/content/357/6347/187, https://doi.org/10.1126/science.aan2735
107. Alcaraz Iranzo D, Nanot S, Dias EJC, Epstein I, Peng C, Efetov DK, Lundeberg MB, Parret R, Osmond J, Hong JY, Kong J, Englund DR, Peres NMR, Koppens FHL (2018) Science 360(6386):291. https://science.sciencemag.org/content/360/6386/291, https://doi.org/10.1126/science.aar8438
108. Ni GX, McLeod AS, Sun Z, Wang L, Xiong L, Post KW, Sunku SS, Jiang BY, Hone J, Dean CR, Fogler MM, Basov DN (2018) Nature 557:530. https://doi.org/10.1038/s41586-018-0136-9
109. Ju L, Geng B, Horng J, Girit C, Martin M, Hao Z, Bechtel HA, Liang X, Zettl A, Shen YR, Wang F (2011) Nat Nanotechnol 6:630. https://www.nature.com/articles/nnano.2011.146, https://doi.org/10.1038/nnano.2011.146
110. Rodrigo D, Limaj O, Janner D, Etezadi D, García de Abajo FJ, Pruneri V, Altug H (2015) Science 349(6244):165. https://science.sciencemag.org/content/349/6244/165, https://doi.org/10.1126/science.aab2051
111. Liu H, Liu Y, Zhu D (2011) J Mater Chem 21:3335. https://doi.org/10.1039/C0JM02922J
112. Boltasseva A, Shalaev VM (2019) ACS Photon 6(1):1. https://doi.org/10.1021/acsphotonics.8b01570

Chapter 2
Classical Electrodynamics of Solids

The main goal of the present chapter is to introduce the elementary concepts governing the interaction of electromagnetic fields with matter, which together form the branch of physics known as classical electrodynamics of solids (or continuous media) [1–4]. As the name suggests, it describes electromagnetic phenomena in terms of classical physics. Nonetheless, the theory of classical electrodynamics has been remarkably successful in predicting and explaining empiric observations in various settings, and, even today, its impact in the field of nanophotonics remains superlative [5–7].

The theoretical framework set forth in this chapter shall provide context and establish the foundational basis for the theories developed in the subsequent chapters. The motivation for reviewing the fundamentals of classical electrodynamics here is two-fold: (i) to make this thesis as self-contained as possible, by providing a summary of the theoretical building blocks of macroscopic electrodynamics; and, (ii) to establish a standard that can serve as a platform for comparison and that will facilitate the interpretation of the nonclassical formalisms developed in Part II. It should be emphasized though that this succinct introduction is by no means a substitute for an all-encompassing treatment of classical electrodynamics and condensed matter physics, for which many authoritative works entirely devoted to these topics exist [1–3, 8, 9].

In what follows, we introduce Maxwell's equations—the workhorse of classical electromagnetism—and elucidate how the electromagnetic response of solids is accounted for via the so-called constitutive relations. We then enumerate the boundary conditions associated with Maxwell's equations at an arbitrary interface between two different media. Next, we discuss the optical response of solids and introduce the local-response approximation, followed by a derivation of the Drude model that describes the response of the free-electron gas driven by an external electromagnetic field. Finally, we provide a cursory overview on the fundamentals of plasmonics in three dimensions, and introduce the dyadic Green's function formalism and use it to describe the plasmon-enhanced Purcell effect.

P. A. D. Gonçalves, *Plasmonics and Light–Matter Interactions in Two-Dimensional Materials and in Metal Nanostructures*, Springer Theses,
https://doi.org/10.1007/978-3-030-38291-9_2

2.1 Foundations of Classical Electrodynamics

2.1.1 Maxwell's Equations and Constitutive Relations

The macroscopic Maxwell's equations governing classical electrodynamics read [1]

$$\nabla \cdot \mathbf{D}(\mathbf{r}, t) = \rho_{\mathrm{f}}(\mathbf{r}, t)\,, \qquad \text{(Gauss' law)} \tag{2.1a}$$

$$\nabla \cdot \mathbf{B}(\mathbf{r}, t) = 0\,, \qquad \text{(magnetic Gauss' law)} \tag{2.1b}$$

$$\nabla \times \mathbf{E}(\mathbf{r}, t) = -\frac{\partial}{\partial t}\mathbf{B}(\mathbf{r}, t)\,, \qquad \text{(Faraday's law)} \tag{2.1c}$$

$$\nabla \times \mathbf{H}(\mathbf{r}, t) = \frac{\partial}{\partial t}\mathbf{D}(\mathbf{r}, t) + \mathbf{J}_{\mathrm{f}}(\mathbf{r}, t)\,. \qquad \text{(Maxwell–Ampère's law)} \tag{2.1d}$$

In the above, **E** and **H** constitute the fundamental physical fields,[1] denoting, respectively, the electric and magnetic fields. The vectors **D** and **B** represent the electric displacement and the magnetic induction, respectively. Lastly, $\mathbf{J}_{\mathrm{f}}$ stands for the (free) current density, and ρ_{f} for the corresponding (free) charge density; these are interlinked by the continuity equation,

$$\nabla \cdot \mathbf{J}_{\nu}(\mathbf{r}, t) + \frac{\partial}{\partial t}\rho_{\nu}(\mathbf{r}, t) = 0\,, \tag{2.2}$$

where, in general, $\nu \in \{\mathrm{f}, \mathrm{ind}, \mathrm{ext}\}$ (denoting free, induced, and external, respectively). Equation (2.2), which essentially embodies charge conservation, is implicitly contained in Maxwell's equations (2.1) [1].

Considering monochromatic fields harmonic in time, i.e., evolving as $e^{-i\omega t}$, the previous equations become (or after a Fourier transform in time)

$$\nabla \cdot \mathbf{D}(\mathbf{r}, \omega) = \rho_{\mathrm{f}}(\mathbf{r}, \omega)\,, \tag{2.3a}$$

$$\nabla \cdot \mathbf{H}(\mathbf{r}, \omega) = 0\,, \tag{2.3b}$$

$$\nabla \times \mathbf{E}(\mathbf{r}, \omega) = i\omega\mathbf{B}(\mathbf{r}, \omega)\,, \tag{2.3c}$$

$$\nabla \times \mathbf{H}(\mathbf{r}, \omega) = -i\omega\mathbf{D}(\mathbf{r}, \omega) + \mathbf{J}_{\mathrm{f}}(\mathbf{r}, \omega)\,, \tag{2.3d}$$

for Maxwell's equations, and

$$\nabla \cdot \mathbf{J}_{\nu}(\mathbf{r}, \omega) = i\omega\rho_{\nu}(\mathbf{r}, \omega)\,. \tag{2.4}$$

for the continuity equation in the frequency domain.

---

[1] In macroscopic media, the fields **E** and **H** are assumed to be local macroscopic averages over volumes that are large compared to the volume of the unit-cell of the crystal (or the cube of the equivalent characteristic length).

We further note that in the preceding equations the *free* current density, can be decomposed into two contributions, namely, $\mathbf{J}_\mathrm{f} = \mathbf{J}_\mathrm{ind} + \mathbf{J}_\mathrm{ext}$. Here, $\mathbf{J}_\mathrm{ind}$ represents the *induced* current density driven by the electric field in the medium, whereas $\mathbf{J}_\mathrm{ext}$ accommodates the possible presence of external current densities in the system. Naturally, the same holds for the associated *free* charge density, $\rho_\mathrm{f} = \rho_\mathrm{ind} + \rho_\mathrm{ext}$.

In order to describe the medium's electromagnetic response one has to invoke the appropriate constitutive relations. These formally establish a connection between **D** and **E**, and between **B** and **H**; such interrelations depend on the bound electric polarization, **P**, and magnetization, **M**, according to $\mathbf{D} = \epsilon_0\mathbf{E} + \mathbf{P}$ and $\mathbf{B} = \mu_0\mathbf{H} + \mathbf{M}$, respectively, and where the quantities ϵ_0 and μ_0 denote the permittivity and the permeability of vacuum, correspondingly.

Concretely, the constitutive relations for homogeneous, nonmagnetic, and linear media, read [1, 9]

$$\mathbf{B}(\mathbf{r},t) = \mu_0\mathbf{H}(\mathbf{r},t)\,, \qquad \text{(nonmagnetic medium)} \tag{2.5a}$$

$$\mathbf{D}(\mathbf{r},t) = \epsilon_0 \int\!\!\int \epsilon(\mathbf{r},\mathbf{r}';t-t')\mathbf{E}(\mathbf{r}',t')d\mathbf{r}'dt'\,, \tag{2.5b}$$

$$\mathbf{J}_\mathrm{ind}(\mathbf{r},t) = \int\!\!\int \sigma(\mathbf{r},\mathbf{r}';t-t')\mathbf{E}(\mathbf{r}',t')d\mathbf{r}'dt'\,, \quad \text{(generalized Ohm's law)} \tag{2.5c}$$

where ϵ and σ are the material's response functions, dubbed as dielectric function and free-carrier conductivity, respectively. In writing of Eqs. (2.5), isotropy has been implicitly assumed. Nevertheless, in anisotropic media these equations possess the same mathematical structure[2] upon replacing the response functions ϵ and σ by their appropriate second-rank tensors, i.e., $\overleftrightarrow{\boldsymbol{\epsilon}}$ and $\overleftrightarrow{\boldsymbol{\sigma}}$. Moreover, in Eqs. (2.5b) and (2.5c) it is also implicitly assumed that the system's response is causal [i.e., $\epsilon(\mathbf{r},\mathbf{r}';t-t') = 0$ for $t' > t$ (and similarly for the conductivity)].

Furthermore, for a medium with translational invariance one may write $\epsilon(\mathbf{r},\mathbf{r}';t-t') = \epsilon(\mathbf{r}-\mathbf{r}';t-t')$ (and similarly for the conductivity), and the constitutive relations (2.5b) and (2.5c) acquire a particularly elegant form after carrying out a Fourier transform both in space and in time,[3] leading to

---

[2]Explicitly, the Cartesian components of the vectors **D** and $\mathbf{J}_\mathrm{ind}$ follow from $\mathbf{D}_\alpha(\mathbf{r},t) = \epsilon_0 \sum_\beta \int\int \epsilon_{\alpha\beta}(\mathbf{r},\mathbf{r}';t-t')\mathbf{E}_\beta(\mathbf{r}',t')d\mathbf{r}'dt'$ and $\mathbf{J}_{\alpha,\mathrm{ind}}(\mathbf{r},t) = \sum_\beta \int\int \sigma_{\alpha\beta}(\mathbf{r},\mathbf{r}';t-t')\mathbf{E}_\beta(\mathbf{r}',t')$ $d\mathbf{r}'dt'$, respectively.

[3]Throughout this work, we adopt the following convention for the Fourier transforms and corresponding inverse Fourier transforms:

$$g(\omega) = \int dt\, g(t)\, e^{i\omega t}; \qquad g(t) = \int \frac{d\omega}{2\pi}\, g(\omega)\, e^{-i\omega t}\,,$$

$$h(\mathbf{k}) = \int d\mathbf{r}\, h(\mathbf{r})\, e^{-i\mathbf{k}\cdot\mathbf{r}}; \qquad h(\mathbf{r}) = \int \frac{d\mathbf{k}}{(2\pi)^3}\, g(\mathbf{k})\, e^{i\mathbf{k}\cdot\mathbf{r}}\,.$$

$$\mathbf{D}(\mathbf{k}, \omega) = \epsilon_0 \epsilon(\mathbf{k}, \omega) \mathbf{E}(\mathbf{k}, \omega) , \tag{2.6a}$$

$$\mathbf{J}(\mathbf{k}, \omega) = \sigma(\mathbf{k}, \omega) \mathbf{E}(\mathbf{k}, \omega) , \tag{2.6b}$$

where the subscript has been dropped, for the reasons explained in the next paragraph. Clearly, the constitutive relations in **k**-space (or momentum representation) amount to simple algebraic multiplications, courtesy of the convolution theorem. Although this representation is clearly advantageous to describe electromagnetic phenomena in infinite, homogeneous media, the simplicity of Eqs. (2.6a) and (2.6b) comes at a price in the presence of interfaces between different media, as the application of the suitable boundary conditions is inherently natural in a real-space formulation.

At this stage it is pertinent to stress that, in principle, one may work with either the dielectric function or the conductivity alone, as each can be defined in such a way that they are simply two alternative approaches to describe the same fundamental physics. For instance, when both **D** and $\mathbf{J}_{\text{ind}}$ are used simultaneously, $\mathbf{J}_{\text{ind}}$ typically describes the part of the response due to free-carriers (via σ) while **D** is left to quantify the response arising from bound polarization charges (through $\epsilon \equiv \epsilon_\infty$). In insulators, the former is identically zero, whereas in metals it describes the contribution of the conduction electrons (or holes), with the latter accounting for the background polarization (e.g., arising from the core electrons). Yet, the partition of the material's response into these two parts—one accounted for via **D**, and another part through $\mathbf{J}_{\text{ind}}$—can be burdensome in many cases. Therefore, it is often more convenient in nanophotonics to combine both the effects of bound polarization and free-carriers into a single, *total* dielectric function of the form

$$\epsilon(\mathbf{r}, \mathbf{r}'; \omega) = \epsilon_\infty(\mathbf{r}, \mathbf{r}'; \omega) + \frac{i\sigma(\mathbf{r}, \mathbf{r}'; \omega)}{\omega \epsilon_0} , \tag{2.7}$$

which can be defined via $\mathbf{D}(\mathbf{r}, \omega) = \epsilon_0 \int \epsilon(\mathbf{r}, \mathbf{r}'; \omega) \mathbf{E}(\mathbf{r}', \omega) d\mathbf{r}'$ upon the replacement $\mathbf{D} \to \mathbf{D} + i\omega^{-1} \mathbf{J}_{\text{ind}}$. Notice that, with this new definition for the displacement vector, Maxwell's equations (2.3) can be rewritten as

$$\nabla \cdot \mathbf{D}(\mathbf{r}, \omega) = \rho_{\text{ext}}(\mathbf{r}, \omega) , \tag{2.8a}$$

$$\nabla \cdot \mathbf{H}(\mathbf{r}, \omega) = 0 , \tag{2.8b}$$

$$\nabla \times \mathbf{E}(\mathbf{r}, \omega) = i\omega \mathbf{B}(\mathbf{r}, \omega) , \tag{2.8c}$$

$$\nabla \times \mathbf{H}(\mathbf{r}, \omega) = -i\omega \mathbf{D}(\mathbf{r}, \omega) + \mathbf{J}_{\text{ext}}(\mathbf{r}, \omega) , \tag{2.8d}$$

that is, only external charges and currents enter in the scalar ρ_{ext} and the vector $\mathbf{J}_{\text{ext}}$, since all the conduction (i.e., free-carrier) charges and currents are now embodied in **D** alone. The benefit of this formulation becomes now apparent: in the absence of external charges and currents, one simply needs to consider a single quantity, **D**, and a single (total) response function, ϵ, which encapsulates the response of the induced background polarization as well as the induced current due to free-carriers.

2.1.2 Boundary Conditions at Interfaces

Consider an arbitrary interface between two different media, separating, say, medium 1 and medium 2. The presence of the interface thus makes the material properties of the compound system discontinuous across the boundary defined by the interface. However, Maxwell's equations and the constitutive relations outlined above naturally still hold within each domain. The relation between the fields on either side of the interface can be determined upon imposing the appropriate boundary conditions (derivable by considering the integral form of Maxwell's equations, followed by the application of the conventional pill-box arguments [1–3]); these can be written as [1]

$$\hat{\mathbf{n}} \times (\mathbf{E}_2 - \mathbf{E}_1) = 0\,, \tag{2.9a}$$

$$\hat{\mathbf{n}} \times (\mathbf{H}_2 - \mathbf{H}_1) = \mathbf{K}\,, \tag{2.9b}$$

for the tangential components of the fields, and

$$\hat{\mathbf{n}} \cdot (\mathbf{D}_2 - \mathbf{D}_1) = \varrho\,, \tag{2.10a}$$

$$\hat{\mathbf{n}} \cdot (\mathbf{B}_2 - \mathbf{B}_1) = 0\,, \tag{2.10b}$$

for the field components normal to the interface. Here, $\hat{\mathbf{n}}$ is a unit vector normal to the interface, pointing from medium 1 into medium 2, and the subscripts in the fields label the same in each medium. It should be noted that, in practice, the imposition of the boundary conditions for the tangential components of the fields, i.e., Eqs. (2.9a) and (2.9b), is sufficient; this is because, in doing so, the corresponding boundary conditions for the normal components, Eqs. (2.10a) and (2.10b), are automatically satisfied. Additionally, in Eqs. (2.9b) and (2.10a), the quantities $\mathbf{K}$ and ϱ anticipate, respectively, the hypothetical presence of a nonvanishing surface current density, and associated surface charge density (that together naturally satisfy the two-dimensional version of the continuity equation, $\nabla \cdot \mathbf{K} + \partial_t \varrho = 0$). In traditional nanophotonics such surface sources are absent in most situations [5, 6]. However, and as we will see explicitly in Chap. 4, it turns out that the aforementioned surface current density (and associated surface charge density) is a convenient way to take into account the presence of two-dimensional materials and their optical response [7].

2.1.3 Local-Response Approximation

As we have seen above, the constitutive relations linking the electric displacement or the current density to the (macroscopic) electric field are in general nontrivial in real space [cf. Eqs. (2.5b) and (2.5c)]. In particular, $\mathbf{D}$ can be cast, in the frequency domain, as (and similarly for $\mathbf{J}$)

$$\mathbf{D}(\mathbf{r},\omega) = \epsilon_0 \int \epsilon(\mathbf{r},\mathbf{r}';\omega)\mathbf{E}(\mathbf{r}',\omega)d\mathbf{r}', \tag{2.11}$$

implying that the displacement field at position $\mathbf{r}$ depends on the electric field at *all* positions $\mathbf{r}'$—hence the term *nonlocal*—, weighted by the system's response function $\epsilon(\mathbf{r},\mathbf{r}';\omega)$. This represents an obvious difficulty, which is aggravated by the fact that the full inhomogeneous $\epsilon(\mathbf{r},\mathbf{r}';\omega)$ is often unknown. This complication is typically circumvented by the application of the local-response approximation (LRA) in which nonlocal effects are neglected. It is worth noting that, albeit being somewhat drastic, this approximation has been able to describe and explain the vast majority of electromagnetic phenomena in nanophotonics, and in plasmonics in particular[4] (see, for instance, Refs. [5–7, 13–22] and references therein). Notwithstanding, when the characteristic dimension of the system decreases towards the inherent electronic length scales, the electrodynamics governing the system becomes significantly influenced by nonlocal effects, which then exhibit clear experimental signatures. We shall return to this point in Part II, where number of nonclassical effects influencing the system's optical response will be investigated in detail.

Mathematically, the LRA can be expressed in real space via $\epsilon(\mathbf{r},\mathbf{r}';\omega) = \epsilon_{\text{LRA}}(\omega)\delta(\mathbf{r}-\mathbf{r}')$, and therefore Eq. (2.11) reduces to the familiar multiplicative relation

$$\mathbf{D} = \epsilon_0\epsilon_{\text{LRA}}(\omega)\mathbf{E} \equiv \epsilon_0\epsilon(\omega)\mathbf{E}, \tag{2.12}$$

where $\epsilon_{\text{LRA}}(\omega)$ is the local, bulk dielectric function of the homogeneous system.[5] In the same fashion, the current density and the conductivity in the LRA are interrelated through

$$\mathbf{J} = \sigma_{\text{LRA}}(\omega)\mathbf{E} \equiv \sigma(\omega)\mathbf{E}. \tag{2.13}$$

Alternatively, the same result could be obtained by starting from the momentum space version of Eq. (2.11)—that is, Eq. (2.6a)—and then taking the $\mathbf{k}\to 0$ limit, leading to $\epsilon(\mathbf{k}\to 0,\omega)\equiv\epsilon(\omega)$. Furthermore, this also shows that the maintenance of the $\mathbf{k}$-dependence (or *spatial* dispersion) of the response functions is formally equivalent to treating the electrodynamics in a nonlocal manner in real space.

---

[4]The reason why this unarguably significant approximation works so well in many situations is due to the large mismatch between the wavelength associated with the optical excitations and the Fermi wavelength (i.e., $\lambda_{\text{exc}} \gg \lambda_F$), resulting in a small nonlocal parameter ($\xi_{\text{NL}} \sim v_F/\omega$) [10–12] and thus making nonlocality relatively short-ranged [10, 11].

[5]In a structure made from different materials, to each domain associated with a distinct medium it is assigned its corresponding local, bulk dielectric function [with the fields in each medium related via the boundary conditions (2.9)].

2.1.4 The Drude Model of the Free-Electron Gas

Although in principle the general concepts akin to the LRA outlined in Sect. 2.1.3 are valid irrespective of the specific model for the frequency-dependent dielectric function and/or conductivity, the terms 'LRA' and 'Drude model' are often used interchangeably in the literature,[6] as the latter is by far the most widely adopted model to describe the optical response of metals. Therefore, it is instructive to present here a succinct derivation of the AC (i.e., frequency-dependent) Drude response model for conductors.

Within the Drude model [23], metals are treated as a free-electron gas permeating a background formed by the immobile ions that make up the crystal's lattice. In this approach, the microscopic behavior of the (independent) conduction electrons subjected to an external electric field, $\mathbf{E}(t)$, is governed classically by Newton's second law, following the equation of motion [7, 8]

$$\frac{d}{dt}\mathbf{p}(t) = -e\mathbf{E}(t) - \frac{1}{\tau}\mathbf{p}(t)\,, \tag{2.14}$$

where $\mathbf{p}$ is the (average) momentum of the electron, $e > 0$ is the elementary charge, and τ denotes the momentum relaxation-time (i.e., the average time between instantaneous collisions of the electron with the heavier ions); thus, one may define a probability per unit time, or scattering rate, $\gamma = \tau^{-1}$. For a time-varying field $\mathbf{E}(t) = \mathbf{E}_0 e^{-i\omega t}$, the previous equation leads to (assuming that $\mathbf{p}$ follows the same time dependence as the driving field)

$$\mathbf{p}(t) = \frac{e}{i\omega - \gamma}\mathbf{E}(t)\,. \tag{2.15}$$

Since the current induced by the external field can also be written as $\mathbf{J} = -en_e\mathbf{v}$, with n_e and $\mathbf{v} = \mathbf{p}/m$ denoting the electron density and (average) velocity, then, using Eq. (2.15) one obtains

$$\mathbf{J} = \frac{e^2 n_e}{m}\frac{1}{\gamma - i\omega}\mathbf{E}\,. \tag{2.16}$$

Comparing this result with the constitutive relation (2.13), one readily identifies the frequency-dependent conductivity of the uniform electron gas as being given by [7, 8]

$$\boxed{\sigma(\omega) = \frac{e^2 n_e}{m}\frac{1}{\gamma - i\omega}}\,. \tag{2.17}$$

Moreover, and recalling the interrelation (2.7), the equivalent Drude dielectric function then reads [7, 8]

[6] Although, technically, the Drude model constitutes *one example* of the metal's response in the LRA.

$$\epsilon(\omega) = \epsilon_\infty(\omega) - \frac{\omega_p^2}{\omega^2 + i\gamma\omega}, \tag{2.18}$$

where $\omega_p^2 = \frac{e^2 n_e}{m\epsilon_0}$ is dubbed as the *plasma frequency* of the electron gas. In the so-called jellium approximation the contribution of the homogeneous ionic background simply amounts to $\epsilon_\infty = 1$. In practice, and when dealing with "real" metals, the frequency dependence of $\epsilon_\infty(\omega)$ is kept in order to incorporate, although ad hoc, (i) interband transitions; and, (ii) background screening arising from lower-lying orbitals (e.g., the *d*-band of noble metals). A typical way of phenomenologically taking into account these effects is to add to the free-carrier contribution a series of Lorentz oscillators[7] in such a way that the resulting dielectric function fits the experimental data in a given frequency window.

Despite its simplicity, the Drude model can reasonably explain the main features of metals (specially simple, alkali metals) in the infrared and visible parts of the electromagnetic spectrum, provided that the frequencies of interest are below the onset of interband transitions. This should not really constitute a surprise, since this model effectively considers only a single band. After Drude's seminal work in 1900 [23], Sommerfeld further developed the model based on a semiclassical description of the electron gas [28] (namely, by incorporating Fermi–Dirac statistics and introducing concepts from band theory). Another prominent semiclassical model for the conductivity is the linearized Boltzmann transport equation [9]; the latter, in the relaxation-time approximation, reduces to Drude's result in the zero-temperature limit.

Figure 2.1 shows the dielectric function of silver and gold. The corresponding Drude fits [cf. Eq. (2.18)] are presented as colored dashed lines (with the fitting parameters detailed in the respective insets). The fits have been performed based on Johnson and Christy's experimental data [29] (shown in the figure as colored circles). The outcome of the aforementioned fits echo what has been outlined in the previous paragraphs, namely: (i) the Drude model describes quite successfully the low frequency region where the metals' response is dominated by the intraband free-carrier contribution; (ii) its validity rapidly declines near and above the threshold of interband transitions [signaled by a step-like increase in $\text{Im}\,\epsilon(\omega)$].

As an example, we also show a Drude model combined with a Lorentzian-like critical points model (superimposed black lines) fitted to gold's experimental data. Clearly, the fit is greatly improved for higher frequencies, owing to the (albeit ad hoc) account of interband transitions.[8]

[7]It is worth noting that other variations of the Lorentz oscillator model are also common, such as models based on critical points and Gaussian integrals; cf. Refs. [24, 25] and Refs. [26, 27], respectively, for details.

[8]Caution should be taken, however, when attempting to draw physically meaningful conclusions from a direct interpretation of the fitting parameters that result from such fits, because they tend to misrepresent the *actual* plasma frequencies and the "resonant poles" do not necessarily correspond to the *proper* interband transitions associated with the material's bandstructure. Nevertheless, such fits are often valuable when modeling the optical properties of metals in specific frequency windows.

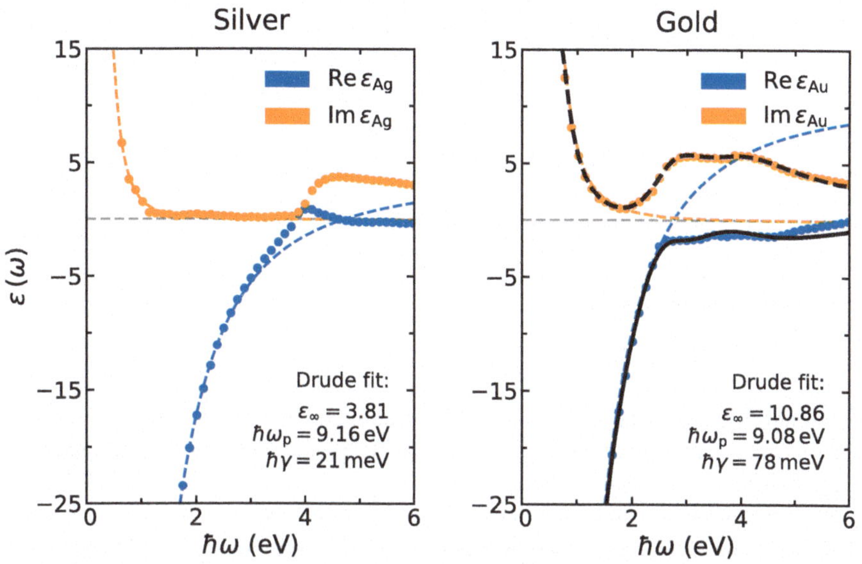

Fig. 2.1 Dielectric functions of selected noble metals typically used as plasmonic materials: silver (left panel) and gold (right panel). The colored circles correspond to experimentally determined values (from Johnson and Christy [29]). The associated Drude fits are shown as colored dashed lines, and where obtained using the fitting parameters indicated in the respective insets. The black lines (solid: real part; dashed: imaginary part) shown in the panel corresponding to gold's optical constants represent a composite fit containing the Drude model (for the free-carrier part) augmented with a critical points model [24] (in an attempt to capture the contribution of interband transitions)

Naturally, a proper, rigorous account of the optical properties of solids requires a quantum mechanical description of the governing physics [9, 30–32], and often necessitates the application of *ab initio* computational techniques [33–35]. Still, the simplicity and straightforward applicability of the Drude model—potentially augmented with other terms, as discussed above—remains an important tool to model the optical response of metals in nanophotonics.

2.2 Fundamentals of Plasmonics

After having established the theoretical foundations of classical electrodynamics in Sect. 2.1, we are now equipped with the basic knowledge that is needed in order to introduce the elementary concepts of classical plasmonics. Indeed, within the regime of validity of classical electrodynamics, the plasmonic excitations in a given system can be fully determined by applying Maxwell's equations—along with the associated boundary conditions—together with the specification of the materials' (local) response.

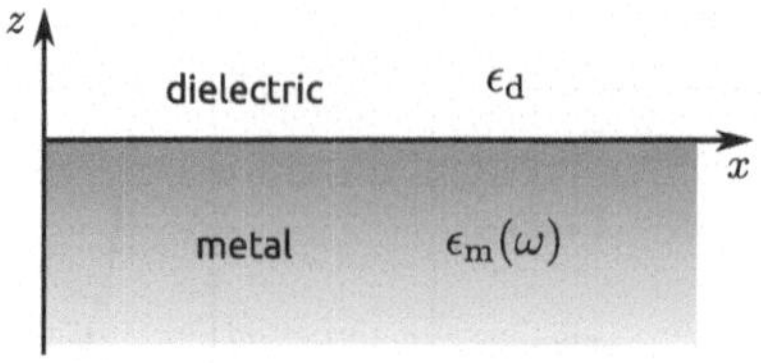

Fig. 2.2 Illustration of a planar dielectric–metal interface, defined by the $z = 0$ plane. The dielectric medium is characterized by a dielectric constant, ϵ_d (which, in principle, may also be frequency-dependent), whereas the electromagnetic properties of the metal are accounted for by its frequency-dependent dielectric function, $\epsilon_m(\omega)$

In this section, we provide an overview of the main features of propagating surface plasmon polaritons sustained at planar dielectric–metal interfaces, and localized surface plasmons supported by metal particles (with spherical plasmonic nanoparticles being the subject of special attention).

Furthermore, it should be emphasized that by summarizing here the core knowledge that defines the field of classical plasmonics we will not only set the stage for the topics discussed in the coming chapters, but also establish a reference point to which nonclassical effects in plasmonics can be compared to.

2.2.1 *Surface Plasmon Polaritons at Planar Interfaces*

In what follows we derive the dispersion relation of surface plasmon polaritons (SPP) propagating along flat dielectric–metal interfaces, and discuss their key properties. We begin by considering a single dielectric–metal interface, and, subsequently, move on to the double interface; in the latter, we introduce the concept of plasmons in thin-films and gap-plasmons. From here, one may straightforwardly generalize these results to arbitrary N-layer structures made from these building blocks.

2.2.1.1 Dielectric–Metal Interface

We consider a planar interface between two different media, defined by the boundary between a dielectric half-space ($z > 0$) and a metal half-space ($z < 0$), as schematically illustrated in Fig. 2.2. The system is assumed to be uniform along the y-direction. We seek p-polarized[9] [or transverse magnetic (TM)] solutions akin to surface electromagnetic waves propagating, say, along the x-direction, and decaying exponentially

[9]This is because SPPs sustained at traditional dielectric–metal interfaces only exist for p-polarization [7, 36]. Incidentally, in the case of two-dimensional (doped) graphene, both p- and s-polarized surface waves exist [7], albeit the latter exhibit weak confinement and thus are typically of little interest; the former, on the other end, can exhibit enormous field confinement as we shall see in Chap. 4.

along the vertical (z-) direction. Concretely, these can be written as (a harmonic time-dependence of form $e^{-i\omega t}$ is implicitly assumed herein)

$$\mathbf{E}_{\mathrm{d}}(\mathbf{r}) = \left[E_{x,\mathrm{d}}\,\hat{\mathbf{x}} + E_{z,\mathrm{d}}\,\hat{\mathbf{z}}\right] e^{iqx}\, e^{-\kappa_{\mathrm{d}} z}\,, \tag{2.19a}$$

$$\mathbf{H}_{\mathrm{d}}(\mathbf{r}) = H_{y,\mathrm{d}}\, e^{iqx}\, e^{-\kappa_{\mathrm{d}} z}\,\hat{\mathbf{y}}\,, \tag{2.19b}$$

in the dielectric medium, and

$$\mathbf{E}_{\mathrm{m}}(\mathbf{r}) = \left[E_{x,\mathrm{m}}\,\hat{\mathbf{x}} + E_{z,\mathrm{m}}\,\hat{\mathbf{z}}\right] e^{iqx}\, e^{\kappa_{\mathrm{m}} z}\,, \tag{2.19c}$$

$$\mathbf{H}_{\mathrm{m}}(\mathbf{r}) = H_{y,\mathrm{m}}\, e^{iqx}\, e^{\kappa_{\mathrm{m}} z}\,\hat{\mathbf{y}}\,, \tag{2.19d}$$

in the metal half-space. In the previous expressions, q denotes the SPP's wavevector (also known simply as propagation constant), whereas the quantities $\kappa_{\mathrm{d,m}}$ characterize the decay of the fields along the perpendicular direction (and thus control the vertical field confinement). The interrelations between the different components of the fields in each medium can be found by inserting Eqs. (2.19) into Maxwell's equations [cf., for instance, Eqs. (2.8)]; such a procedure leads to

$$H_{y,j} = -\frac{\omega}{q}\epsilon_0\epsilon_j E_{z,j} \quad \text{and} \quad H_{y,j} = -s_j\, i\frac{\omega}{\kappa_j}\epsilon_0\epsilon_j E_{x,j}\,, \tag{2.20a}$$

$$E_{z,j} = s_j\, i\frac{q}{\kappa_j} E_{x,j}\,, \tag{2.20b}$$

for $j \in \{\mathrm{d}, \mathrm{m}\}$, and where we have defined the auxiliary variable $s_j = \delta_{j\mathrm{d}} - \delta_{j\mathrm{m}}$ in terms of Kronecker deltas. Additionally, we also obtain the relation

$$\kappa_j^2 = q^2 - (\omega/c)^2\epsilon_j\,. \tag{2.21}$$

In the above expressions, ϵ_0 stands for the permittivity of vacuum, and $c = 1/\sqrt{\epsilon_0\mu_0}$ is the speed of light in free-space (with μ_0 denoting vacuum permeability).

At this point, all that is required is to match the fields at both sides of the interface in accordance with the boundary conditions specified by Eqs. (2.9a) and (2.9b)—expressing the continuity of the tangential components of the electromagnetic field across the interface—which yield

$$E_{x,\mathrm{d}} - E_{x,\mathrm{m}} = 0\,, \tag{2.22a}$$

$$H_{y,\mathrm{d}} - H_{y,\mathrm{m}} = 0\,. \tag{2.22b}$$

By making use of the relations (2.20) and substituting them into the system of equations (2.22), one readily finds the following implicit condition for the SPP dispersion relation [7, 36]:

$$\frac{\epsilon_{\mathrm{d}}}{\kappa_{\mathrm{d}}} + \frac{\epsilon_{\mathrm{m}}}{\kappa_{\mathrm{m}}} = 0\,. \tag{2.23}$$

This equation is an *implicit* condition because $\kappa_{\rm d,m}$ possess an explicit dependence of both frequency, ω, and (parallel) wavevector, q, that is, $\kappa_{\rm d,m} \equiv \kappa_{\rm d,m}(q, \omega)$; similarly, the (local) dielectric function of the metal[10] is frequency-dependent, $\epsilon_{\rm m} \equiv \epsilon_{\rm m}(\omega)$.

Furthermore, from Eq. (2.23) we can also determine a necessary condition for the existence of SPPs at the interface between two different media. In particular, in order to have an electromagnetic wave decaying along the perpendicular to the interface, it is required that $\text{Re}\, \kappa_{\rm d,m} > 0$ [cf. Eqs. (2.19)]. Hence, (the real parts of) $\epsilon_{\rm d}$ and $\epsilon_{\rm m}$ must have opposite signs; for insulators, one typically has $\epsilon_{\rm d} \equiv \text{Re}\, \epsilon_{\rm d} > 0$, and thus one must have a medium with $\text{Re}\, \epsilon_{\rm m} < 0$ in the opposite half-space [7, 36]. This is the case of most metals for frequencies in the visible/near-infrared and below.[11]

Alternatively, with the help of Eq. (2.21), one may express the SPP wavevector in closed-form, namely [6, 7, 18]

$$q_{\rm SPP} = \frac{\omega}{c} \sqrt{\frac{\epsilon_{\rm d} \epsilon_{\rm m}}{\epsilon_{\rm d} + \epsilon_{\rm m}}}, \tag{2.24}$$

which has the advantage of providing a simple link between the SPP's wavevector and its frequency.

Box 2.1—Dispersion relation of SPPs from the poles of r_p

An alternative route towards the derivation of the condition that yields the spectrum of SPPs is to identify the poles of the system's scattering coefficients; for the planar interface, these are known as Fresnel coefficients (for reflection and transmission) [1, 2, 5]. Since SPPs possess TM character, the Fresnel coefficients of interest are the ones corresponding to *p*-polarized waves. Specifically, the reflection coefficient associated with a planar interface is given by [5]

$$r_p = \frac{\epsilon_2 k_{z,1} - \epsilon_1 k_{z,2}}{\epsilon_2 k_{z,1} + \epsilon_1 k_{z,2}}, \tag{2.25}$$

for an excitation coming from medium 1. The resonances associated with excitations of the system are then determined by the poles of r_p; these are given by the condition $\epsilon_2 k_{z,1} + \epsilon_1 k_{z,2} = 0$. Moreover, for the dielectric–metal structure, one can make the following identifications: $\{1, 2\} \leftrightarrow \{\rm d, m\}$ and $k_{z,j} = i\kappa_j$. Hence, the previous condition becomes $\epsilon_{\rm m} \kappa_{\rm d} + \epsilon_{\rm d} \kappa_{\rm m} = 0$, or, equivalently,

[10] In principle, the permittivity of the dielectric medium may also exhibit frequency dependence. We emphasize that Eq. (2.23) remains valid nevertheless.

[11] For some metals (like Al), however, $\text{Re}\, \epsilon_{\rm m}$ remains negative up to the ultraviolet region [37]. On the other end, heavily-doped semiconductors can also support SPPs, albeit at significantly lower frequencies than metals [38, 39].

$$\frac{\epsilon_d}{\kappa_d} + \frac{\epsilon_m}{\kappa_m} = 0, \tag{2.26}$$

which is exactly the same SPP condition given previously by Eq. (2.23).

Figure 2.3 shows the dispersion relation of the polaritonic modes supported by a planar interface between a dielectric and a Drude metal. In the lossless case (cf. Fig. 2.3a), there are two distinct branches of the dispersion relation. One branch, located above the plasma frequency, corresponds to bulk plasmon polaritons (BPP); it represents light propagating inside the metal,[12] where the light field is coupled to the internal degrees of freedom of the plasma. The other branch—and this is the one of interest—corresponds to surface plasmon polaritons (SPP), which are *surface* electromagnetic waves coupled to collective oscillations of the free-carriers in the metal. Crucially, notice that the SPP dispersion curve lies to the right of the light line (dashed green line, defined by $\omega = qc/\sqrt{\epsilon_d}$), and, consequently, the SPPs' electromagnetic field is effectively confined near the interface. In addition, this also renders direct excitation of SPP by light impossible due to the mismatch between q_{SPP} and the photon's wavevector, and adequate coupling techniques are therefore required (e.g., using gratings, prism-coupling, or by exploiting surface roughness, etc) [6, 7, 36]. Moreover, the ratio between the wavelength of a photon in free-space and that of a SPP oscillating at the same frequency is given by $\frac{\lambda_0}{\lambda_{SPP}} = \sqrt{\frac{\epsilon_d \epsilon_m}{\epsilon_d + \epsilon_m}}$ [see Eq. (2.24)]. Hence, a particularly alluring feature of SPPs now becomes readily apparent: since $\lambda_{SPP}/\lambda_0 < 1$, these polaritons can be exploited in order to confine the electromagnetic field below the diffraction limit [6, 7, 13, 14]. In some systems, this effect can be quite substantial, i.e., with $\lambda_{SPP}/\lambda_0 \ll 1$ [6, 7, 40–43]. As we shall see further ahead, this property is pivotal to manipulate electromagnetic fields at the nanoscale and to enhance light–matter interactions. Lastly, Fig. 2.3a further shows that the SPP dispersion asymptotically approaches the surface plasmon frequency[13] $\omega_{SP} = \omega_p/\sqrt{\epsilon_\infty + \epsilon_d}$ at large wavevectors (high confinement region). This signals the realization of the quasi-static regime (or nonretarded limit); here, the mode becomes purely longitudinal and dispersionless. This contrasts with the behavior at low wavevectors, where $q \approx \sqrt{\epsilon_d}\omega/c$ and the mode is essentially photon-like and exhibits poor field localization.

Next, we relax the assumption of negligible loss and consider the case where the metal possesses finite damping ($\gamma \neq 0$). In this spirit, Fig. 2.3b depicts the SPP dispersion for a Drude metal with finite loss. While the basic features present in Fig. 2.3a remain somewhat discernible, there is, however, a striking difference: the

[12] The dispersion of BPPs follows from $q(\omega) = \sqrt{\epsilon_m(\omega)}\frac{\omega}{c}$. In the lossless case, this relation can be straightforwardly inverted, producing $\omega(q) = \sqrt{\omega_p^2 + (qc)^2}$.

[13] Corresponding to the solution of the SPP's condition in the nonretarded limit: $\epsilon_d + \epsilon_m = 0$.

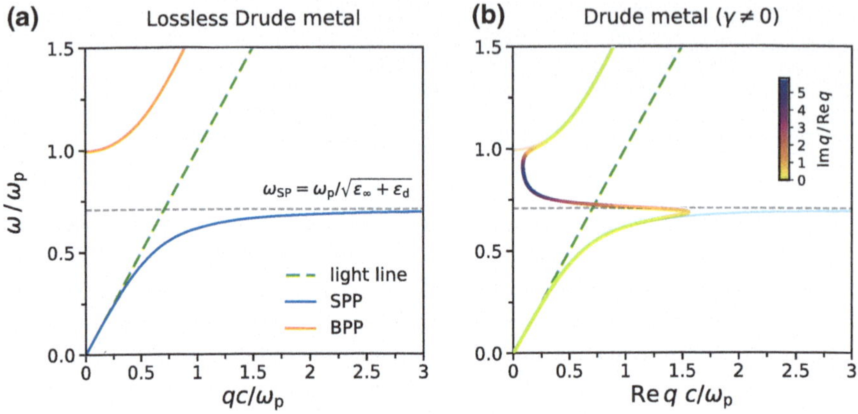

Fig. 2.3 Dispersion relation of plasmon polaritons at dielectric–metal interfaces. We assume that the dielectric medium is air ($\epsilon_d = 1$), and that the metal is well described by the Drude model [i.e., with dielectric function given by Eq. (2.18)]. We further assume that $\epsilon_\infty = 1$. **a** Lossless Drude metal ($\gamma = 0$). The dashed horizontal line indicates the nonretarded surface plasmon frequency, $\omega_{SP} = \omega_p/\sqrt{\epsilon_\infty + \epsilon_d}$. **b** Drude metal with finite damping ($\gamma = \omega_p/20$). The color scale indicates the magnitude of the loss experienced by the polariton, defined here by the figure of merit $Q^{-1} = \mathrm{Im}\, q/\mathrm{Re}\, q$ (akin to an inverse quality factor)

SPP dispersion curve reaches a maximum Re q and then bends backwards.[14] This is a manifestation of the fact that q is now a complex-valued quantity [6, 45]. Therefore, in the present case the SPP travels along the dielectric–metal interface with wavelength $\lambda_{SPP} = 2\pi/\mathrm{Re}\, q_{SPP}$, while being attenuated according to $e^{-x\mathrm{Im}\, q_{SPP}}$ as it propagates. In order to better quantify the impact of losses, we have plotted the SPP dispersion using a color code that reflects the figure of merit $Q^{-1} = \mathrm{Im}\, q/\mathrm{Re}\, q$ (inverse quality factor). Importantly, it is clear that albeit the field confinement ($\propto \mathrm{Re}\, q$) successively increases as the dispersion approaches ω_{SP}, so does the impact of damping ($\propto \mathrm{Im}\, q$); in the vicinity of the nonretarded surface plasmon frequency, the latter becomes indeed prohibitive. This trade-off between field confinement and losses is a well-known feature of plasmonics.

We finalize the discussion of the SPP's dispersion relation by plotting it for two prototypical plasmonic materials: gold and silver; cf. Fig. 2.4. Here, deviations from the Drude-like behavior reflect the influence of interband transitions, which cause damping to increase (recall Fig. 2.1). This effect is particularly significant for gold, in which SPPs are clearly beset by losses near its ω_{SP}. In opposition, silver exhibits fairly good plasmonic properties. Nonetheless, both materials remain widely used

[14] This back-bending is absent if, alternatively, one would solve Eq. (2.23) for a real-valued q, thereby obtaining a complex-valued frequency, $\omega = \omega' + i\omega''$. In this situation, the imaginary part of the SPP frequency dictates the SPP's life-time via $\tau = -1/(2\omega'')$. In principle, from a theoretical standpoint, it is equally legit to either solve for a real-valued frequency or for real-valued wavevector [44, 45]. However, when applying the theoretical modeling to experiments this choice is not arbitrary, and has to be made in conformity with the specific type of measurement being carried out [44, 46].

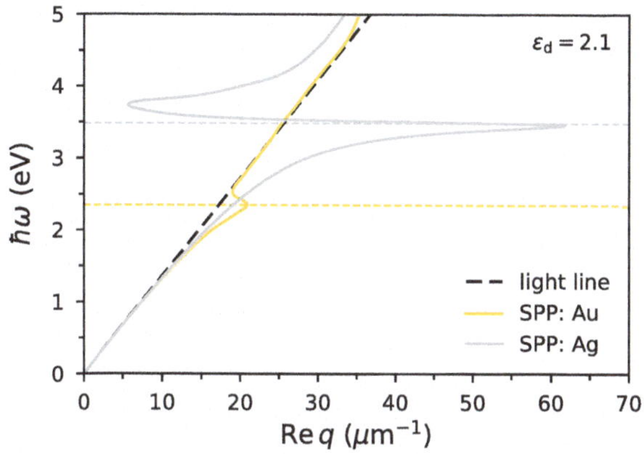

Fig. 2.4 Dispersion relation of SPPs in dielectric–gold (solid golden line) and dielectric–silver (solid gray line) planar interfaces. The dielectric medium is characterized by a relative permittivity of $\epsilon_d = 2.1$ (representative of SiO_2). The optical constants of gold and silver were obtained by interpolating Johnson and Christy's experimental data [29]. The dashed horizontal lines indicate the corresponding nonretarded surface plasmon frequencies (stemming from $\epsilon_m(\omega) + \epsilon_d = 0$)

in plasmonics, because although silver has superior plasmonic properties, it suffers from oxidation under ambient conditions, while gold is stable, despite being lossy. Plasmonic damping can be partly mitigated by engineering the SPP dispersion (effectively bringing the SPP resonances away from interband transitions) and/or by exploiting other geometries (e.g., SPPs in metal-gap structures [47, 48]).

2.2.1.2 Double-Interface: SPPs in DMD and MDM Structures

The derivation of the spectral properties of SPPs in planar double-interfaces, namely in the dielectric–metal–dielectric (DMD) and metal–dielectric–metal (MDM) configurations,[15] essentially follows the same steps previously outlined in the single-interface case. The only difference is that now, in the middle region, both growing and decaying exponentials are allowed (see, for instance, Ref. [7]). An equally valid approach to determine the resonances supported by a given system, is to look at the poles of the associated scattering coefficients (recall Box 2.1). In what follows, this is the avenue that we will pursue. We consider a three-layer system forming a double-interface as portrayed in Fig. 2.5. The outer media are taken as being semi-infinite, and encapsulate an inner slab with thickness d. As before, we assume that the whole system is uniform along the y-direction.

The Fresnel reflection coefficient for p-polarization associated with the 3-layer structure is given by[16]

[15]The dispersion of coupled SPP modes in planar DMD and MDM structures was first considered in the late 1960s by Kliewer and Fuchs [49] and by Economou [50].

[16]In writing the second equality, the relations $r_p^{jk} = -r_p^{kj}$ and $t_p^{jk} = 1 + r_p^{jk}$ have been exploited.

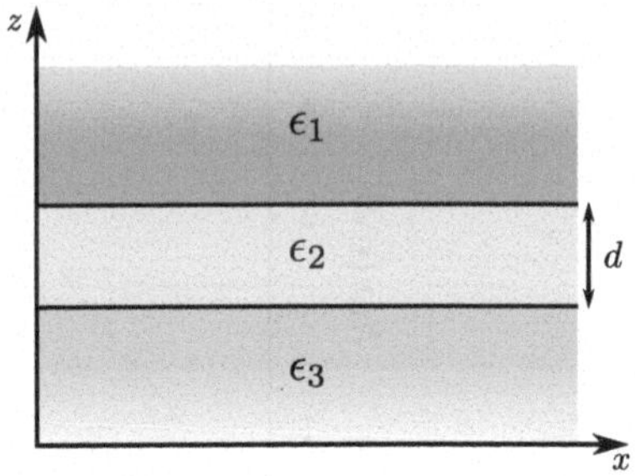

Fig. 2.5 Sketch of a planar double-interface separating three different media. Each medium is characterized by a dielectric function ϵ_j, where $j \in \{1, 2, 3\}$. The middle layer has a thickness given by d

$$r_p^{13} = r_p^{12} + \frac{t_p^{21} t_p^{12} r_p^{23} e^{i2k_{z,2}d}}{1 - r_p^{21} r_p^{23} e^{i2k_{z,2}d}} = \frac{r_p^{12} + r_p^{23} e^{i2k_{z,2}d}}{1 + r_p^{12} r_p^{23} e^{i2k_{z,2}d}}, \tag{2.27}$$

where $k_{z,j} = \sqrt{\epsilon_j k_0^2 - q^2} = i\kappa_j$, with $k_0 = \omega/c$. In the above, r_p^{jk} denotes the Fresnel coefficient for p-polarized waves incident from medium j and reflected by the interface separating the media j and k. Hence, the resonances supported by the three-layer system with a double-interface are given by the zeros of the denominator of r_p^{13}, that is, $1 + r_p^{12} r_p^{23} e^{i2k_{z,2}d} = 0$. Upon substituting the explicit formulas for the individual Fresnel reflection coefficients, we obtain (recalling that $k_{z,j} = i\kappa_j$) [7]

$$\boxed{\left(\frac{\epsilon_1}{\kappa_1} + \frac{\epsilon_2}{\kappa_2}\right)\left(\frac{\epsilon_2}{\kappa_2} + \frac{\epsilon_3}{\kappa_3}\right) + \left(\frac{\epsilon_2}{\kappa_2} - \frac{\epsilon_1}{\kappa_1}\right)\left(\frac{\epsilon_3}{\kappa_3} - \frac{\epsilon_2}{\kappa_2}\right) e^{-2\kappa_2 d} = 0}\,. \tag{2.28}$$

Clearly, the coupling between opposite interfaces—via the Coulomb interaction—is controlled by the term containing the exponential. Thus, for $\kappa_2 d \ll 1$ new hybrid modes appear, arising from the coupling between the bare modes akin to each interface. Contrarily, in the large thickness limit, $d \to \infty$, the uncoupled SPP dispersion for each interface is recovered; cf. Eq. (2.23).

For symmetric environments, that is, for $\epsilon_1 = \epsilon_3$ (and thus $\kappa_1 = \kappa_3$), Eq. (2.28) simplifies considerably, yielding two distinct solutions [6, 7, 36, 49, 50]:

$$\tanh\left(\frac{\kappa_2 d}{2}\right) = -\frac{\epsilon_2 \kappa_1}{\epsilon_1 \kappa_2}, \tag{2.29a}$$

$$\coth\left(\frac{\kappa_2 d}{2}\right) = -\frac{\epsilon_2 \kappa_1}{\epsilon_1 \kappa_2}. \tag{2.29b}$$

These two modes exhibit opposite symmetries: one is even and the other is odd. Which is which in relation to Eqs. (2.29a) and (2.29b), depends on the choice of the field component to which the symmetry is labeled after. For instance, if ones chooses

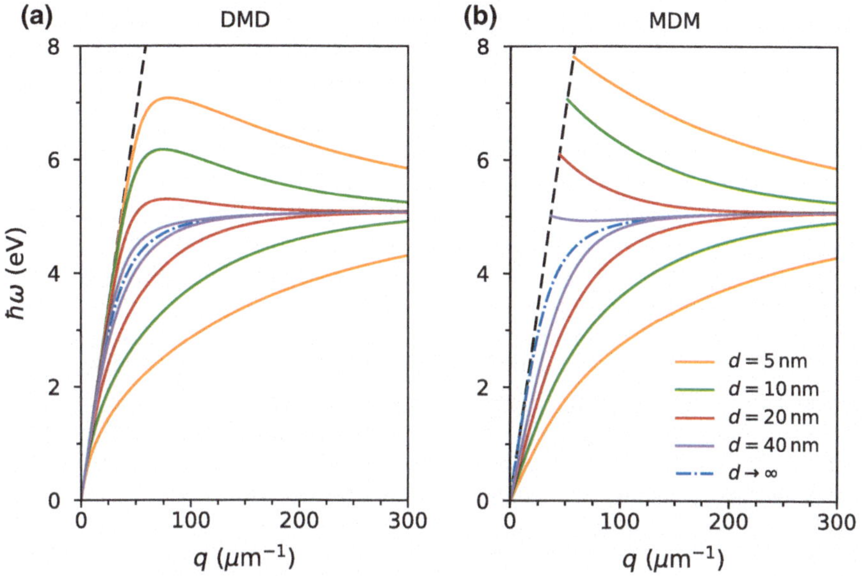

Fig. 2.6 Dispersion relation of hybrid SPP modes in planar DMD and MDM structures. Metallic media are assumed to have a dielectric function of the Drude form with negligible damping and $\hbar\omega_\mathrm{p} = 9\,\mathrm{eV}$ and $\epsilon_\infty = 1$. Dielectric media has $\epsilon_\mathrm{d} = 2.1$. The dashed black line represents the light line in the dielectric media, given by $\omega = qc/\sqrt{\epsilon_\mathrm{d}}$. In (**b**) the upper modes enter the light cone region (and similarly in (**a**)), that is, to the left of the light line; we have chosen not to plot them here because they are leaky modes

to define the symmetry with respect to $E_x(z)$, then Eqs. (2.29a) and (2.29b) are odd and even, respectively [and vice-versa if one would choose $E_z(z)$].

The dispersion relation of hybrid SPPs sustained at planar DMD and MDM structures is shown in Fig. 2.6. For the sake of clarity, we have chosen a symmetric dielectric environment (i.e., $\epsilon_1 = \epsilon_3$). In both configurations, for large d the modes do not hybridize and therefore are indistinguishable from that of a single dielectric–metal interface (dot-dashed line). However, as the thickness of the inner medium is decreased, the modes in opposite interface interact and hybridize, giving rise to a new set of coupled modes: one shifting towards higher energies, while the other shifts towards lower energies (relative to the uncoupled SPP dispersion). This mode splitting is naturally stronger for smaller d, since the coupling strength increases with decreasing separations. Lastly, for $\kappa_2 d \approx qd \gg 1$, the modes decouple again because the SPP's field bounded to one of the interfaces decays steeply along the vertical and therefore does not overlap with that of the SPP on the opposite interface.

Furthermore, notice that while the dispersion curves akin to DMD and MDM modes are qualitatively different in the retarded region—i.e., for small wavevectors, near the light line—they successively become more alike as we approach larger wavevectors. Indeed, at sufficiently large wavevectors, the dispersion relation of hybrid SPPs modes in DMD and MDM structures becomes identical. This is consequence of the realization of the nonretarded limit, for which Eqs. (2.29) reduce to[17]

$$\omega(\omega + i\gamma) = \omega_{\mathrm{p}}^2 \left[\epsilon_\infty + \epsilon_{\mathrm{d}} F_{\mathrm{h}}(qd/2)\right]^{-1} , \qquad (2.30)$$

with $F_{\mathrm{h}}(x) = \{\tanh(x), \coth(x)\}$, and where we have assumed that the metallic media are described by a Drude dielectric function.

In passing, we bring to the reader's attention that each mode of the pair of hybrid SPPs—both in the DMD and MDM configurations—are affected differently by losses upon decreasing d [6, 7, 13, 54]. Focusing, as an example, in the DMD geometry, one finds that the propagation length, $L_{\mathrm{p}} = (2\mathrm{Im}\, q)^{-1}$, associated with the upper SPP mode [odd in $E_x(z)$] increases with decreasing film thickness d, whereas the propagation length associated with the lower SPP mode [even in $E_x(z)$] decreases concomitantly with decreasing film thickness. Due to this behavior, these hybrid SPP modes are often referred to as long-range SPP and short-range SPP, respectively. Evidently, these considerations are of paramount importance when using such coupled SPP modes for plasmonic waveguiding and circuitry [55, 56].

2.2.2 *Localized Surface Plasmons in Metal Spheres*

Although the planar dielectric–metal interface can be regarded as the simplest SPP-supporting system, other prototypical geometries are also particularly elucidative. The most prominent member of the latter is unarguably the sphere. Naturally, the plasmonic modes supported by metal spheres share many of the features of SPPs in planar structures. However, plasmonic resonances in finite-sized structures—such as the sphere—are typically known as localized surface plasmons[18] (LSPs) [6, 36]. This terminology simply highlights that, unlike SPPs in extended planar systems, these surface plasmons do not propagate due to the obvious geometric constraint imparted by the systems' finite-size. Instead, when excited by an adequate source (e.g., a plane-wave or a fast electron), the displacement of free charge carriers in response to that stimulus creates a restoring force capable of maintaining a self-sustained collective oscillation of the free carriers. Consequently, LSP resonances in

[17] In the special case of a homogeneous 3DEG in vacuum ($\epsilon_\infty = \epsilon_{\mathrm{d}} = 1$), the SPP dispersion in DMD and MDM structures simplify further to $\omega_\pm(q) = \frac{\omega_{\mathrm{p}}}{\sqrt{2}}\sqrt{1 \pm e^{-qd}}$ (assuming negligible damping, for simplicity). We further note that, in the nonretarded limit, inverse structures exhibit complementary resonances (for each mode profile), in the sense that they obey the sum-rule $\omega_{\mathrm{p}}^2 = \omega_{\mathrm{o}}^2 + \omega_{\mathrm{c}}^2$, where the subscripts denote the "ordinary" geometry (defined arbitrarily) and its "complementary" (or inverse) [51]. Here, the DMD and MDM structures are complementary (one is obtained from the other by interchanging the material regions). Interestingly, this sum-rule can also be viewed as a statement of Babinet's principle [2, 52, 53].

[18] We note that localized surface plasmons, like surface plasmon polaritons, are also polaritons. However, the epithet 'polariton' is often absent in literature when referring to the former. For this reason, we adopt this designation here in order to conform with the majority of the literature.

plasmonic particles can couple to light directly without the need of special coupling techniques [36].

In the following, we outline the optical response of plasmonic spheres (Fig. 2.7) to external stimuli, focusing on the main features of LSPs (a complementary account will be also presented in Sect. 8.2, when discussing nonclassical effects in metal spheres).

Mie theory. The appropriate theoretical description of the scattering of electromagnetic waves by a spherical object is known as Mie theory[19] [58–61]. The mathematical details of the theory and the derivation of the so-called Mie coefficients are provided in the Appendix C (after taking the classical limit, since in the appendix we provide a generalization of the traditional Mie coefficients augmented with quantum corrections). Other good references are also Refs. [1, 3, 59, 60]. Hence, here we focus instead on their practical role in the calculation of the optical response of spherical metal particles, and, in particular, discuss the basic properties of the LSP resonances supported by them.

The Mie scattering coefficients associated with light scattered by a metallic sphere are given by[20] [1, 3, 59, 60]

$$a_l^{\mathrm{TM}} = \frac{\epsilon_{\mathrm{m}} j_l(x_{\mathrm{m}})\Psi_l'(x_{\mathrm{d}}) - \epsilon_{\mathrm{d}} j_l(x_{\mathrm{d}})\Psi_l'(x_{\mathrm{m}})}{\epsilon_{\mathrm{m}} j_l(x_{\mathrm{m}})\xi_l'(x_{\mathrm{d}}) - \epsilon_{\mathrm{d}} h_l^{(1)}(x_{\mathrm{d}})\Psi_l'(x_{\mathrm{m}})}, \tag{2.31a}$$

$$b_l^{\mathrm{TE}} = \frac{j_l(x_{\mathrm{m}})\Psi_l'(x_{\mathrm{d}}) - j_l(x_{\mathrm{d}})\Psi_l'(x_{\mathrm{m}})}{j_l(x_{\mathrm{m}})\xi_l'(x_{\mathrm{d}}) - h_l^{(1)}(x_{\mathrm{d}})\Psi_l'(x_{\mathrm{m}})}, \tag{2.31b}$$

where the dimensionless wavevectors $x_j = k_j R$ with $k_j = \sqrt{\epsilon_j}\omega/c$ ($j \in \{\mathrm{d}, \mathrm{m}\}$), have been introduced, along with the spherical Bessel and Hankel functions of the first kind $j_l(x)$ and $h_l^{(1)}(x)$, and the Riccati–Bessel functions $\Psi_l(x) \equiv x j_l(x)$ and $\xi_l(x) \equiv x h_l^{(1)}(x)$; primed functions denote their derivatives with respect to their arguments. We stress that, in essence, Eqs. (2.31) are nothing but the spherically symmetric equivalents of the Fresnel reflections coefficients—r_p and r_s, respectively—for the planar interface. Equipped with the Mie coefficients, the electromagnetic response of the sphere can be readily determined.

---

[19]Despite being most commonly known as 'Mie theory', there is, in fact, an earlier account of the theory given by Lorenz in 1890 [57], that is, almost two decades before Mie's seminal paper of 1908 [58].

[20]For the sake of completeness, the Mie coefficients associated with the fields inside the particle are [1, 3, 59, 60]:

$$c_l^{\mathrm{TE}} = \frac{j_l(x_{\mathrm{d}})\xi_l'(x_{\mathrm{d}}) - h_l^{(1)}(x_{\mathrm{d}})\Psi_l'(x_{\mathrm{d}})}{j_l(x_{\mathrm{m}})\xi_l'(x_{\mathrm{d}}) - h_l^{(1)}(x_{\mathrm{d}})\Psi_l'(x_{\mathrm{m}})},$$

$$d_l^{\mathrm{TM}} = \sqrt{\epsilon_{\mathrm{m}}\epsilon_{\mathrm{d}}}\frac{j_l(x_{\mathrm{d}})\xi_l'(x_{\mathrm{d}}) - h_l^{(1)}(x_{\mathrm{d}})\Psi_l'(x_{\mathrm{d}})}{\epsilon_{\mathrm{m}} j_l(x_{\mathrm{m}})\xi_l'(x_{\mathrm{d}}) - \epsilon_{\mathrm{d}} h_l^{(1)}(x_{\mathrm{d}})\Psi_l'(x_{\mathrm{m}})},$$

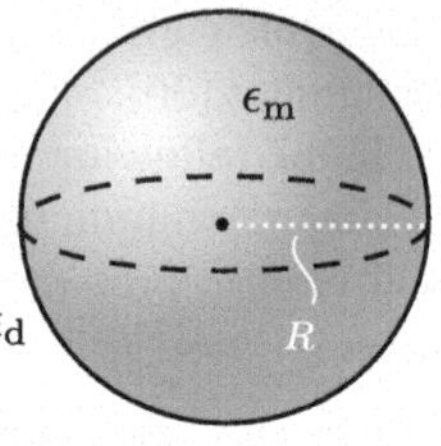

Fig. 2.7 Sketch of a spherical particle with radius R and dielectric function ϵ_m. The sphere is embedded in a homogeneous medium with relative permittivity ϵ_d

In particular, assuming that the sphere is illuminated by a monochromatic plane-wave, the corresponding cross-sections stem from [59]

$$\sigma_{ext} = \frac{2\pi}{k_d^2} \sum_{l=1}^{\infty} (2l+1) \mathrm{Re} \left(a_l^{TM} + b_l^{TE} \right) , \tag{2.32a}$$

$$\sigma_{scat} = \frac{2\pi}{k_d^2} \sum_{l=1}^{\infty} (2l+1) \left(\left| a_l^{TM} \right|^2 + \left| b_l^{TE} \right|^2 \right) , \tag{2.32b}$$

$$\sigma_{abs} = \sigma_{ext} - \sigma_{scat} , \tag{2.32c}$$

expressing, respectively, the extinction, scattering, and absorption cross-sections. In the above formulas, l stands for the angular momentum; hence, for metallic nanospheres, a_l^{TM} contains a series of poles associated with LSP resonances of electric dipole ($l = 1$), electric quadrupole ($l = 2$), etc, character [59, 62].

At this point, before evaluating and analyzing Eqs. (2.32) further, it is instructive to study the nonretarded version of the present problem, that is, of light scattering by a sphere in the quasi-static limit.

Polarizability of nanospheres and LSP resonances: nonretarded limit. The quasi-static limit is applicable whenever the characteristic dimension of the system—here a spherical particle—is significantly smaller than the excitation's wavelength. Mathematically, for a spherical object, this can be expressed by the condition $R \ll \lambda$, where R is the particle's radius and λ the wavelength of the impinging electromagnetic radiation. Within this regime, retardation effects (i.e., effects that reflect the finiteness of the speed of light) are rendered negligible and therefore a nonretarded formulation of the problem suffices. In practice, what this means is that, rather than solving the scattering problem in the framework of Maxwell's equations, one may simply work out the solution using electrostatics by solving Poisson's (or Laplace's) equation instead [1].

That can be carried out by expanding the potential inside and outside the sphere in terms of a series of multipoles and then match the corresponding boundary conditions at $r = R$. Such procedure allows the determination of the sphere's multipolar polarizability, α_l, which encodes the system's response to a potential of arbitrary multipolar order l; specifically, it is given by [62, 63]

$$\alpha_l = 4\pi R^{2l+1} \frac{\epsilon_\text{m} - \epsilon_\text{d}}{\epsilon_\text{m} + \frac{l+1}{l}\epsilon_\text{d}} . \tag{2.33}$$

Hence, the l-th LSP resonance simply follows from the solution of

$$l\epsilon_\text{m} + (l+1)\epsilon_\text{d} = 0 , \tag{2.34}$$

which, for a Drude metal sphere, may be written as

$$\omega_l(\omega_l + i\gamma) = \frac{\omega_\text{p}^2}{\epsilon_\infty + \frac{l+1}{l}\epsilon_\text{d}} \quad \overset{(\gamma=0)}{\longrightarrow} \quad \omega_l = \frac{\omega_\text{p}}{\sqrt{\epsilon_\infty + \frac{l+1}{l}\epsilon_\text{d}}} . \tag{2.35}$$

In the dipole limit, Eq. (2.33) reduces to the familiar dipolar polarizability [1, 59, 60] ($\alpha \equiv \alpha_1$)

$$\alpha = 4\pi R^3 \frac{\epsilon_\text{m} - \epsilon_\text{d}}{\epsilon_\text{m} + 2\epsilon_\text{d}} , \tag{2.36}$$

that describes the dipole moment $\mathbf{p} = \epsilon_0\epsilon_\text{d}\alpha\mathbf{E}_0$ set up by the sphere in response to a constant (or slowly varying across the sphere's size) electric field $\mathbf{E}_0$. From here, the so-called Fröhlich condition for the dipolar LSP becomes readily apparent: $\epsilon_\text{m} + 2\epsilon_\text{d} = 0$. Therefore, for a lossless homogeneous three-dimensional electron gas (3DEG) in vacuum, the dipolar LSP resonance occurs at $\omega = \omega_\text{p}/\sqrt{3}$. Finally, the corresponding cross-sections in the nonretarded limit take the form[21] [59]

$$\sigma_\text{abs} = k_\text{d}\,\text{Im}\,\alpha \quad \text{and} \quad \sigma_\text{scat} = \frac{k_\text{d}^4}{6\pi}|\alpha|^2 , \tag{2.37}$$

and where $\sigma_\text{ext} = \sigma_\text{abs} + \sigma_\text{scat}$.

Results. Figure 2.8 summarizes the key spectral features of LSP resonances in spherical metal particles. As discussed above, the optical response of plasmonic spheres is dominated by set of LSP resonances with angular momentum l, which are determined by the poles of the Mie coefficient a_l^TM (Fig. 2.8a). Clearly, as the sphere's radius is reduced, the LSP resonances become dispersionless and approach their nonretarded values [cf. Eq. (2.35)]. Incidentally, the corresponding resonance widths become independent of l in the nonretarded regime; in that limit, they equal $\text{Im}\,\omega_l = -\gamma/2$ (see inset). Contrasting this, in the retarded regime the widths are larger for larger spheres due to radiation damping. Furthermore, notice that the relative contribution of each multipole for the extinction cross-section is strongly dependent on the particle's radius (Fig. 2.8b, c). Crucially, the contribution of higher-order LSPs to the overall cross-section vanishes steeply upon reducing the sphere's radius (Fig. 2.8c). For small nanospheres, the entire cross-section is eventually dominated by the dipole LSP ($l = 1$), which is simply a consequence of the dipole limit.

[21] Neglecting the effect of radiation reaction [5, 64].

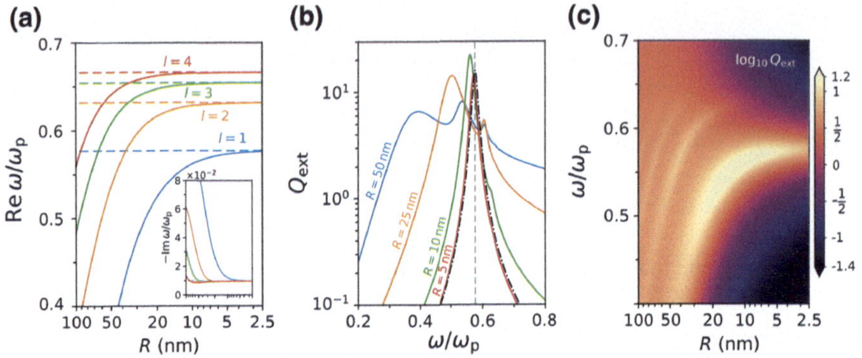

Fig. 2.8 Spectral features of LSP resonances in metal spheres. **a** Localized surface plasmon resonances for $l = \{1, 2, 3, 4\}$ as a function of the particle's radius (solid: Mie theory; dashed: quasi-statics). **b** Extinction cross-section normalized to the geometric one, $Q_{\text{ext}} \equiv \sigma_{\text{ext}}/(\pi R^2)$, for plasmonics spheres with different radii. The black dot-dashed line corresponds to a quasi-static calculation (for $R = 5$ nm only). The vertical gray dashed line indicates the dipole LSP resonance. **c** Intensity plot of the (normalized) extinction cross-section for metal spheres of varying radii under plane-wave illumination. Material setup: Drude metal sphere ($\hbar\omega_{\text{p}} = 9\,\text{eV}$, $\gamma = \omega_{\text{p}}/50$, and $\epsilon_\infty = 1$) in air ($\epsilon_{\text{d}} = 1$)

Here, we have presented a brief review of the salient features of LSPs in metal spheres[22] based on the workhorse that is the theory of classical electrodynamics. We shall return to the discussion of the fundamental properties of LSPs in metal spheres in Sect. 8.2, where we will describe the governing physics in the mesoscopic regime, i.e., in a framework that accommodates quantum mechanical effects such as nonlocality, electronic spill-out, and surface-enabled Landau damping.

2.2.3 *Other Geometries*

The planar dielectric–metal interface and the spherical metal particle constitute two archetypal plasmon-supporting structures. Their basic, highly symmetric geometry makes them naturally well-suited for investigating the elementary properties of plasmonic excitations, both from theoretical and experimental viewpoints. In spite of this, the zoo of plasmonic architectures is considerably wealthier; in fact, one may argue that there are infinitely many plasmonic systems, and, as such, enumerating all of them is as unforgiving as it is futile. Therefore, here we shall only enumerate a modest selection of the most popular plasmonic architectures considered in the scientific literature. Our aim is not to be exhaustive or utterly detailed, but rather (i) to highlight the plethora of plasmonic systems; and (ii) to provide a listing of the

[22]Though we have focused on plasmonic metal spheres, the theory outlined above is naturally also applicable to spheres made out of any (local and uniform) material.

most studied plasmonic structures accompanied by references where the interested reader may find more details about each particular geometry.

Starting from planar-like configurations, there has been a growing interest in layered insulating–conducting structures for producing novel plasmonic crystals [65] and plasmonic metamaterials [66–69], including uniaxial metamaterials capable of supporting hyperbolic SPPs [70–74]. Other prominent examples are Tamm plasmons [75–77], plasmonic gratings [18, 78–81] (including the ones realizing spoof plasmons [82, 83]), and also plasmons in metal gap structures [47, 84–87] and channel plasmons [47, 48, 88].

On the finite-sized end, there is a copious amount of different nanoparticles capable of supporting LSP resonances [6, 20]. Other than the metal sphere studied in Sect. 2.2.2, there have been many investigations of LSPs in plasmonic disks [89–92], triangles [93–95], cubes [16, 96, 97], nanoplatelets [98], and nanorods [99], among others [100–102]. Furthermore, the pairing of the said plasmonic particles can be used to build new coupled structures which may be exploited for extending the library of plasmon resonances further, e.g., using plasmonic dimers [103–108] (including bowties [109–112]) and oligomers [113–116], or multilayered nanoparticles [117–119] such as core-shell spheres [59, 120–124]. Alternatively, the individual plasmonic particles can be assembled periodically into plasmonic lattices [125–129], which are then also capable of sustaining surface plasmon lattice resonances (arising from diffractively coupled LSP resonances).

In passing, it is worthwhile to mention that analytical solutions can only be obtained for a small subset of the above-listed plasmonic systems, primarily the ones possessing sufficient symmetry (e.g., the planar interface, the sphere, or the cylinder). In the quasi-static limit, an ellipsoidal particle can also be straightforwardly solved [59]. Notably, it has recently been shown in Ref. [130] that even plasmonic particles with relatively complicated morphologies can be parameterized quasi-analytically—though utilizing previous knowledge obtained from numerical calculations—provided that the system is deeply subwavelength (so that retardation effects are negligible). Notwithstanding, in general the optical response of less symmetric systems, which arguably constitute the majority of plasmon-supporting geometries, can only be calculated using numerical techniques. Fortunately, in the past couple of decades, the growth in number, quality, performance, and availability of numerical methods for nanophotonics [131] has been as agile as the advances seen in nanofabrication and experimental techniques. Examples of numerical treatments frequently employed in plasmonics and nanophotonics are: the finite-difference time-domain (FDTD) method [132, 133], modal and Fourier methods [132], the finite element method (FEM) [134], the Green's function integral equation method (GFIEM) [135], the boundary element method (BEM) [136–139], the discrete-dipole approximation (DDA) [140], etc. Each of the aforementioned techniques is then rich in specialized variations depending on how the problem is considered from a numerical analysis standpoint.

2.3 Dipole Emission: Spontaneous Decay Rate and Optical Local Density of States

The emission of radiation by a classical oscillating electric dipole is a textbook example for illustrating the theory of classical electrodynamics. A rigorous treatment of light–matter interactions requires, however, a formulation in terms of quantum electrodynamics (QED). We will return to the quantum mechanical formalism in Sect. 8.3, and therefore in the following we concern ourselves with the classical theory alone. Strikingly, the expression for the power emitted by a classical dipole has the same mathematical structure[23] as the spontaneous emission rate of a (two-level) quantum emitter derived within the proper QED framework [5, 141–143]. Nevertheless, even a purely classical description of the optical interaction between a dipole and its surroundings is useful in order to become acquainted with concepts such as the optical local density of states (LDOS), the Purcell effect, and the dyadic Green's functions' formalism [5, 7, 64]. Moreover, as we shall see in Sect. 8.3 (and Appendix D), the Green's functions framework provides a practical link between classical electrodynamics and macroscopic quantum electrodynamics.

2.3.1 *Electric Dipole in a Homogeneous Medium*

Power radiated by a classical dipole. Considering a current distribution with a harmonic time-dependence, in a linear and homogeneous medium, it follows from the Poynting theorem that the time-averaged dissipated power due to a current density $\mathbf{j}(\mathbf{r})$ can be calculated via [5, 142]

$$P = -\frac{1}{2}\int \mathrm{Re}\{\mathbf{j}^*(\mathbf{r})\cdot\mathbf{E}(\mathbf{r})\}d\mathbf{r}\,. \tag{2.38}$$

For an ideal, point-like electric dipole source located at $\mathbf{r}_0$, the corresponding current density is given by $\mathbf{j}(\mathbf{r}) = -i\omega\boldsymbol{\mu}\delta(\mathbf{r}-\mathbf{r}_0)$, where $\boldsymbol{\mu}$ designates the electric dipole moment. Hence, Eq. (2.38) can be cast as

$$P = \frac{\omega}{2}\mathrm{Im}\{\boldsymbol{\mu}^*\cdot\mathbf{E}(\mathbf{r}_0)\}\,, \tag{2.39}$$

where $\mathbf{E}(\mathbf{r}_0)$ is the electric field at the dipole's position. We now invoke the electromagnetic dyadic Green's function formalism to write the electric field in terms of the dyadic Green's function [1, 5, 142], i.e.,

[23]Provided that the (classical) dipole moment $\boldsymbol{\mu}$ is replaced by (twice) the quantum mechanical matrix element, $2\langle g\hat{\boldsymbol{\mu}}|e\rangle$, where $|g\rangle$ ($|e\rangle$) refers to the ground (excited) state of the emitter and $\hat{\boldsymbol{\mu}} = -e\hat{\mathbf{r}}$ denotes the quantum mechanical dipole moment operator.

$$\mathbf{E}(\mathbf{r}) = \omega^2 \mu_0 \overleftrightarrow{\mathbf{G}}(\mathbf{r}, \mathbf{r}_0; \omega) \cdot \boldsymbol{\mu}\,, \tag{2.40}$$

where the Green's dyadic satisfies $\nabla \times \nabla \times \overleftrightarrow{\mathbf{G}}(\mathbf{r}, \mathbf{r}'; \omega) - k_0^2 \epsilon(\mathbf{r}, \omega) \overleftrightarrow{\mathbf{G}}(\mathbf{r}, \mathbf{r}'; \omega) = \overleftrightarrow{\mathbf{I}} \delta(\mathbf{r} - \mathbf{r}')$. Equation (2.40) describes the electric field at position $\mathbf{r}$ due to an electric dipole source placed at position $\mathbf{r}_0$. Inserting the previous equation into Eq. (2.39) leads to

$$P = \frac{\omega^3 |\boldsymbol{\mu}|^2}{2c^2 \epsilon_0} \left[\hat{\mathbf{n}}_\mu \cdot \mathrm{Im} \left\{ \overleftrightarrow{\mathbf{G}}(\mathbf{r}_0, \mathbf{r}_0; \omega) \right\} \cdot \hat{\mathbf{n}}_\mu \right]. \tag{2.41}$$

where $\hat{\mathbf{n}}_\mu$ is a unit vector describing the orientation of the dipole. In a homogeneous medium, we have $\overleftrightarrow{\mathbf{G}} \equiv \overleftrightarrow{\mathbf{G}}_0$, where the free-space Green's function takes the form [5]

$$\overleftrightarrow{\mathbf{G}}_0(\mathbf{r}, \mathbf{r}_0; \omega) = \left[\overleftrightarrow{\mathbf{I}} + \frac{1}{k^2} \nabla \otimes \nabla \right] G_0(\mathbf{r}, \mathbf{r}_0; \omega) \qquad \text{with} \qquad G_0(\mathbf{r}, \mathbf{r}_0; \omega) = \frac{e^{ik|\mathbf{r} - \mathbf{r}_0|}}{4\pi |\mathbf{r} - \mathbf{r}_0|}. \tag{2.42}$$

Here, $k = \sqrt{\epsilon} k_0$ (where ϵ denotes the medium's permittivity) and $\overleftrightarrow{\mathbf{I}}$ represents the unit dyad. Since the imaginary part of the free-space Green's function is $\mathrm{Im}\overleftrightarrow{\mathbf{G}}_0(\mathbf{r}_0, \mathbf{r}_0; \omega) = \frac{k}{6\pi} \overleftrightarrow{\mathbf{I}}$, we thus obtain (after orientation averaging[24])

$$P_0 = \frac{|\boldsymbol{\mu}|^2}{12\pi \epsilon_0} \frac{\omega^3 k}{c^2}\,, \tag{2.43}$$

which is the well-known Larmor formula for the power radiated by a classical oscillating electric dipole [1, 5, 144].

2.3.2 *Electric Dipole in an Inhomogeneous Medium: LDOS and Purcell Enhancement*

The power of the dyadic Green's functions formalism becomes readily apparent when dealing with inhomogeneous media. In particular, we stress that Eq. (2.41) remains valid even in such a scenario. Hence, provided that the system's Green's dyadic is known—or obtainable in some way, e.g., using normal (or quasinormal) modes—then the calculation of the governing electrodynamics becomes relatively straightforward.

When considering a inhomogeneous environment, for reasons that will become clear ahead, it is useful to decompose the total electric field at the dipole's origin—i.e., the field that enters in Eq. (2.39)—into a primary field, $\mathbf{E}_0$, plus a scattered part, namely

$$\mathbf{E}(\mathbf{r}_0) = \mathbf{E}_0(\mathbf{r}_0) + \mathbf{E}_{\mathrm{scat}}(\mathbf{r}_0)\,, \tag{2.44}$$

---

[24] Note that $\left\langle \hat{\mathbf{n}}_\mu \cdot \mathrm{Im} \left\{ \overleftrightarrow{\mathbf{G}}_0(\mathbf{r}_0, \mathbf{r}_0; \omega) \right\} \cdot \hat{\mathbf{n}}_\mu \right\rangle = \frac{1}{3} \mathrm{Im} \left\{ \mathrm{Tr} \overleftrightarrow{\mathbf{G}}_0(\mathbf{r}_0, \mathbf{r}_0; \omega) \right\} = \frac{k}{6\pi}$.

so that

$$P = P_0 + \frac{\omega}{2}\text{Im}\{\boldsymbol{\mu}^* \cdot \mathbf{E}_{\text{scat}}(\mathbf{r}_0)\}\,,$$
$$\Rightarrow \frac{P}{P_0} = 1 + \frac{6\pi\epsilon_0\epsilon_1}{|\boldsymbol{\mu}|^2 k_1^3}\text{Im}\{\boldsymbol{\mu}^* \cdot \mathbf{E}_{\text{scat}}(\mathbf{r}_0)\}\,, \tag{2.45}$$

where $k_1 = \sqrt{\epsilon_1}k_0$. Notice that the presence of an inhomogeneity in the medium changes the rate of energy's dissipation. This is because the (total) field at $\mathbf{r}_0$ is the sum of the dipole's primary field together with the dipole's field (produced at an *earlier* time) after it has been scattered in the inhomogeneous environment.

Spontaneous emission rate and LDOS

The spontaneous decay rate of a quantum emitter normalized to its value in vacuum is equal to the power radiated by a classical dipole normalized with respect to P_0, that is, $\gamma/\gamma_0 = P/P_0$ [5]. Thus, we may write [5]

$$\frac{\gamma}{\gamma_0} = 1 + \frac{6\pi\epsilon_0\epsilon_1}{k_1^3}\frac{1}{|\boldsymbol{\mu}|^2}\text{Im}\{\boldsymbol{\mu}^* \cdot \mathbf{E}_{\text{scat}}(\mathbf{r}_0)\}\,. \tag{2.46}$$

Furthermore, we note that a quantum mechanical calculation of the spontaneous emission rate using Fermi's golden rule establishes a connection between the former and the (electric[25]) LDOS, $\rho^{\text{E}}(\mathbf{r}_0, \omega)$. Explicitly, one finds [5]

$$\gamma = \frac{\pi\omega}{3\hbar\epsilon_0}|\boldsymbol{\mu}|^2\rho^{\text{E}}_{\mathbf{n}}(\mathbf{r}_0, \omega)\,, \tag{2.47}$$

where $\rho^{\text{E}}_{\mathbf{n}}(\mathbf{r}_0, \omega)$ stands for the **n**-projected (or partial) LDOS at position $\mathbf{r}_0$ and frequency ω. Importantly, $\rho^{\text{E}}_{\mathbf{n}}(\mathbf{r}_0, \omega)$ is linked with the imaginary part of the Green's dyadic via [5]

$$\rho^{\text{E}}_{\mathbf{n}}(\mathbf{r}_0, \omega) = \frac{6\omega}{\pi c^2}\left[\hat{\mathbf{n}}_\mu \cdot \text{Im}\left\{\overleftrightarrow{\mathbf{G}}(\mathbf{r}_0, \mathbf{r}_0; \omega)\right\} \cdot \hat{\mathbf{n}}_\mu\right]. \tag{2.48}$$

In vacuum, the previous two equations lead to $\rho^{\text{E}}_0 = \frac{\omega^2}{\pi^2 c^3}$, for the free-space LDOS, and $\gamma_0 = \frac{\omega^3|\boldsymbol{\mu}|^2}{3\pi\epsilon_0\hbar c^3}$, for the spontaneous decay rate. In a inhomogeneous medium, the normalized electric LDOS can then be written as

$$\frac{\rho^{\text{E}}_{\mathbf{n}}(\mathbf{r}_0, \omega)}{\rho^{\text{E}}_0(\omega)} = 1 + \frac{6\pi}{k_1}\left[\hat{\mathbf{n}}_\mu \cdot \text{Im}\left\{\overleftrightarrow{\mathbf{G}}_{\text{scat}}(\mathbf{r}_0, \mathbf{r}_0; \omega)\right\} \cdot \hat{\mathbf{n}}_\mu\right]. \tag{2.49}$$

[25] In passing, for the sake of correctness, we note that the optical (or electromagnetic) LDOS actually has both electric and magnetic contributions, i.e., $\rho(\mathbf{r}, \omega) = \rho^{\text{E}}(\mathbf{r}, \omega) + \rho^{\text{H}}(\mathbf{r}, \omega)$ [145]. Nevertheless, in a nanophotonics setting, and in plasmonics in particular, the former is typically larger than the latter by orders of magnitude, so that, effectively, $\rho(\mathbf{r}, \omega) \approx \rho^{\text{E}}(\mathbf{r}, \omega)$; for this reason, it is common to refer to $\rho^{\text{E}}(\mathbf{r}, \omega)$ as the entire LDOS. Additionally, in general emitters possess essentially an electric dipole character and hence $\rho^{\text{E}}(\mathbf{r}, \omega)$ is the relevant quantity.

where the system's Green's function $\overleftrightarrow{\mathbf{G}} = \overleftrightarrow{\mathbf{G}}_0 + \overleftrightarrow{\mathbf{G}}_{\text{scat}}$ has been decomposed into a homogeneous part, $\overleftrightarrow{\mathbf{G}}_0$, and a scattered part (associated with the change in the Green's function due to the structured environment), $\overleftrightarrow{\mathbf{G}}_{\text{scat}}$.

Crucially, Eq. (2.49) represents the Purcell factor, $F_{\text{p}} \equiv \rho^{\text{E}}_{\mathbf{n}}(\mathbf{r}_0, \omega)/\rho_0(\omega)$; it indicates the enhancement of the spontaneous decay rate of an emitter by the environment. Indeed, as it has been known since Purcell's work [146], the rate of spontaneous emission is not an intrinsic property of the emitter, and that it can be modified by the environment [5, 142, 147–149]. For instance, the decay rate dynamics of emitters can be substantially enhanced near material interfaces. In the case of plasmonics, the LDOS experienced by an emitter can be extremely large (by orders of magnitude) due to the existence of plasmon resonances [16, 17, 150]; such plasmon-assisted Purcell enhancement is an emblematic feature of plasmonics.

2.3.3 Electric Dipole Above a Planar Interface

We now consider an electric dipole embedded in a medium with relative permittivity ϵ_1, and located at a distance h above the surface of a semi-infinite substrate with relative permittivity ϵ_2 (which can be either a dielectric or a metal). The interface between the two media is defined by the $z = 0$ plane. Here it is where the work that we have done in the previous subsections pays off, since the LDOS enhancement can be computed using Eq. (2.49). Specifically, for an emitter above a planar interface, that equation can now be expressed as

$$\frac{\rho^{\text{E}}_{\mathbf{n}}(\mathbf{r}_0, \omega)}{\rho^{\text{E}}_0(\omega)} = 1 + \frac{6\pi}{k_1}\left[\hat{\mathbf{n}}_\mu \cdot \text{Im}\left\{\overleftrightarrow{\mathbf{G}}_{\text{ref}}(\mathbf{r}_0, \mathbf{r}_0; \omega)\right\} \cdot \hat{\mathbf{n}}_\mu\right], \tag{2.50}$$

where $\overleftrightarrow{\mathbf{G}}_{\text{ref}}$ refers to the Green's function associated with the reflected electric field due to the presence of the substrate. A suitable representation for $\overleftrightarrow{\mathbf{G}}_{\text{ref}}(\mathbf{r}_0, \mathbf{r}_0; \omega)$ is [5]

$$\overleftrightarrow{\mathbf{G}}_{\text{ref}}(\mathbf{r}_0, \mathbf{r}_0; \omega) = \frac{i}{8\pi k_1^2}\int_0^\infty dq\, \frac{q}{k_{z,1}} e^{i2k_{z,1}h} \begin{bmatrix} k_1^2 r_s - k_{z,1}^2 r_p & 0 & 0 \\ 0 & k_1^2 r_s - k_{z,1}^2 r_p & 0 \\ 0 & 0 & 2q^2 r_p \end{bmatrix}, \tag{2.51}$$

where r_s and r_p denote Fresnel's reflection coefficients for s- and p-polarization, while q and $k_{z,1} = \sqrt{\epsilon_1 k_0^2 - q^2}$ refer to the in-plane wavevector and to the projection of the wavevector onto the z-axis, respectively.

Spontaneous emission enhancement near a metal surface

In view of the above-noted considerations, the spectral dependence of the LDOS experienced by an emitter with dipole moment $\boldsymbol{\mu}$ and located at a height h above a planar dielectric–metal interface can be expressed as

$$\frac{\rho_{\mathbf{n}}^{\mathrm{E}}(h,\omega)}{\rho_0^{\mathrm{E}}(\omega)} = 1 + \frac{3}{4}\frac{\mu_x^2+\mu_y^2}{|\boldsymbol{\mu}|^2}\mathrm{Re}\int_0^\infty du\, \frac{u}{\sqrt{1-u^2}}\left[r_s - (1-u^2)r_p\right]e^{2ik_1h\sqrt{1-u^2}}$$
$$+ \frac{3}{2}\frac{\mu_z^2}{|\boldsymbol{\mu}|^2}\mathrm{Re}\int_0^\infty du\, \frac{u^3}{\sqrt{1-u^2}}\, r_p\, e^{2ik_1h\sqrt{1-u^2}}, \quad (2.52)$$

where we have normalized the LDOS to its value in a homogeneous dielectric medium with relative permittivity ϵ_1 (i.e., in the absence of the metal half-space). Note that, in the above, we have introduced the dimensionless wavevector $u \equiv q/k_1$ (while recalling that $k_1 = \sqrt{\epsilon_1}k_0$), and $r_p \equiv r_p(u,\omega)$ and $r_s \equiv r_s(u,\omega)$ are the reflection coefficients of the planar dielectric–metal interface. Considering separately the cases of a dipole emitter with perpendicular and parallel dipole moments (with the respect to the interface), we have

$$\frac{\rho_\perp^{\mathrm{E}}(h,\omega)}{\rho_0^{\mathrm{E}}(\omega)} = 1 + \frac{3}{2}\mathrm{Re}\int_0^\infty du\, \frac{u^3}{\sqrt{1-u^2}}\, r_p\, e^{2ik_1h\sqrt{1-u^2}}, \quad (2.53a)$$

and

$$\frac{\rho_\parallel^{\mathrm{E}}(h,\omega)}{\rho_0^{\mathrm{E}}(\omega)} = 1 + \frac{3}{4}\mathrm{Re}\int_0^\infty du\, \frac{u}{\sqrt{1-u^2}}\left[r_s - (1-u^2)r_p\right]e^{2ik_1h\sqrt{1-u^2}}, \quad (2.53b)$$

respectively. The corresponding orientation-averaged LDOS is then $\langle\rho^{\mathrm{E}}\rangle = \frac{1}{3}\rho_\perp^{\mathrm{E}} + \frac{2}{3}\rho_\parallel^{\mathrm{E}}$, which can be particularly useful when dealing with an ensemble of randomly-oriented dipoles (e.g., in dyes).

These equations can be used to calculate the spectral LDOS for an emitter above an arbitrary planar interface (in fact, they even hold in the case of a multi-layered

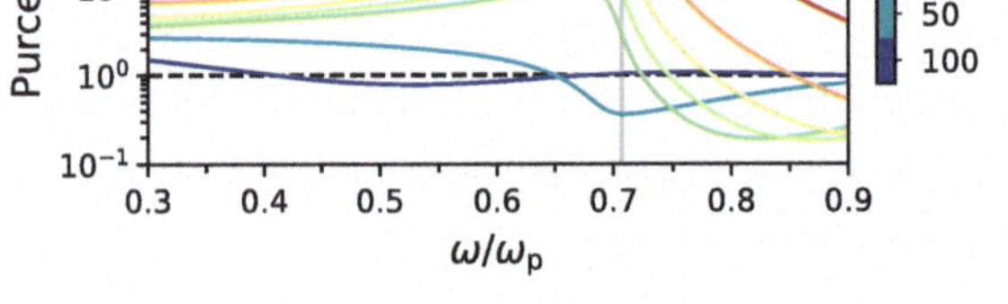

Fig. 2.9 Purcell enhancement, or normalized LDOS $\rho_\perp^{\mathrm{E}}/\rho_0^{\mathrm{E}}$, associated with the spontaneous decay rate of a dipole emitter with its dipole moment orientated perpendicularly to the planar metal surface; different emitter–surface separations are considered. We assume a Drude metal ($\hbar\omega_{\mathrm{p}} = 9\,\mathrm{eV}$, $\gamma = \omega_{\mathrm{p}}/50$, and $\epsilon_\infty = 1$) filling the lower half-space, while the upper half-space is air ($\epsilon_1 = 1$)

substrate, provided that the appropriate reflection coefficients are employed). In the specific case of a dielectric–metal interface, the LDOS is enhanced in the vicinity of the SPP resonance frequency—cf. Fig. 2.9. In the previous equations, this mechanism is embodied in the poles of r_p. In what follows, we shall discriminate the role of the different processes that contribute to the modification of the decay dynamics of an emitter near a metal surface.

Box 2.2—LDOS in the nonretarded limit for an emitter near a metal surface

In the nonretarded limit, and for a point-like emitter close to the metal's surface (i.e., $h \ll k_1^{-1}$), the Purcell enhancement (2.52) takes the simpler form:

$$F_p = \frac{3}{2}\left(\mu_\perp^2 + \frac{1}{2}\mu_\parallel^2\right)\int_0^\infty du\, u^2 e^{-2uk_1 h}\operatorname{Im} r_p\,. \tag{2.54}$$

Notice that, within the local-response approximation, the *nonretarded* reflection coefficient for p-polarization is simply $r_p = \frac{\epsilon_2-\epsilon_1}{\epsilon_2+\epsilon_1}$, and can thus be taken outside the integral. In turn, in this regime, the integral admits an analytical solution: $\int_0^\infty du\, u^2 e^{-2uk_1 h} = 2/(2k_1 h)^3$ (provided that $\operatorname{Re} k_1 h > 0$). These considerations therefore allow us to write ($\epsilon_1 \equiv \epsilon_\mathrm{d}$ and $\epsilon_2 \equiv \epsilon_\mathrm{m}$)

$$\begin{aligned} F_p &= \left(\mu_\perp^2 + \frac{1}{2}\mu_\parallel^2\right)\frac{3}{8k_\mathrm{d}^3 h^3}\operatorname{Im}\left\{\frac{\epsilon_\mathrm{m}-\epsilon_\mathrm{d}}{\epsilon_\mathrm{m}+\epsilon_\mathrm{d}}\right\}, \\ &= \left(\mu_\perp^2 + \frac{1}{2}\mu_\parallel^2\right)\frac{3\epsilon_\mathrm{d}}{4k_\mathrm{d}^3 h^3}\frac{\operatorname{Im}\epsilon_\mathrm{m}}{|\epsilon_\mathrm{m}+\epsilon_\mathrm{d}|^2}, \end{aligned} \tag{2.55}$$

which exhibits the well-known $F_p \propto h^{-3}$ dependence of the Purcell enhancement experienced by an emitter in the near vicinity of a metal surface. In the last step, we have implicitly assumed, for the sake of clarity, that the dielectric medium is lossless (which is often the case). Finally, notice that the LDOS diverges for $h \to 0$; such behavior is evidently unphysical, and it is consequence of the local-response approximation. In order to remedy this deficiency, a full quantum nonlocal theory must be employed.

Excitation of surface plasmon polaritons by an electric dipole

In free-space, an excited emitter can only spontaneously decay by emitting freely propagating photons. However, other decay channels may become available when the emitter is placed in the neighborhood of a structured electromagnetic environment. A prototypical example of the latter is the planar dielectric–metal interface considered above ($\epsilon_1 \equiv \epsilon_\mathrm{d}$ and $\epsilon_2 \equiv \epsilon_\mathrm{m}$). Having established that the spontaneous decay rate of emitters can be enhanced near metal interfaces (Fig. 2.9), it is now instructive to investigate the role of the different decay channels in the said enhancement. To that end, it is useful to analyze the differential dissipated power (or momentum-space power spectrum), dP/dq, hereupon defined through $P = \int_0^\infty \frac{dP}{dq} dq$, and where [144]

$$P = \frac{\omega}{8\pi\epsilon_0\epsilon_{\rm d}} \mathrm{Re} \int_0^\infty dq \frac{q}{k_{z,{\rm d}}} \left\{ q^2 \left[1 + r_p\, e^{2ik_{z,{\rm d}}h}\right] \mu_\perp^2 + k_{\rm d}^2 \left[1 + r_s\, e^{2ik_{z,{\rm d}}h}\right] \frac{1}{2}\mu_\parallel^2 \right.$$
$$\left. + k_{z,{\rm d}}^2 \left[1 - r_s\, e^{2ik_{z,{\rm d}}h}\right] \frac{1}{2}\mu_\parallel^2 \right\} . \qquad (2.56)$$

Focusing on the case of a perpendicular electric dipole—since this corresponds to the configuration that maximizes the coupling to SPPs—, we have plotted in Fig. 2.10 the (normalized) differential dissipated power. As indicated in the figure, there are three distinct channels contributing to the total dissipated power (or, equivalently, to the total spontaneous decay rate): (i) a contribution associated with emission into free-space photons (shaded region), (ii) a resonant channel due the emitter's decay into SPPs, manifesting itself as sharp peak at $u_{\rm SPP} \equiv \mathrm{Re}\, q_{\rm SPP}/k_{\rm d}$ in the q-space power spectrum, and (iii) a broad, non-resonant contribution associated with the metal's damping pathways (e.g., phonons, impurities, etc). Clearly, the relative contribution

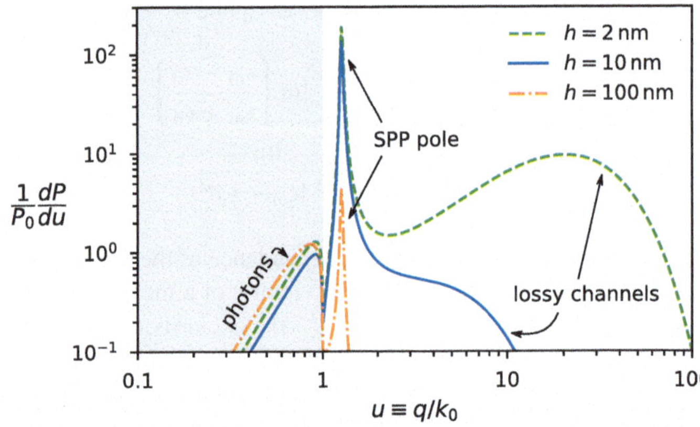

Fig. 2.10 Normalized differential dissipated power, $P_0^{-1}\,(dP/du)$ [also, notice that this is equivalent to $(\rho_0^{\rm E})^{-1}\,(d\rho^{\rm E}/du)$], for a vertically-oriented dipole placed in air above a flat metal surface, for different emitter–metal separations h. The dipole emission is assumed to occur at $\omega = 0.75\omega_{\rm SP}$. The material parameters and setup are the same as in Fig. 2.9. The shaded area in gray indicates the region associated with propagating waves (photons) for which $k_{z,{\rm d}}$ is real. On the other hand, for $u > 1$ (i.e., $q > k_0$) the wavevector $k_{z,{\rm d}}$ becomes purely imaginary and therefore represents modes bounded to the interface; these can be either SPPs or nonresonant contributions due to Ohmic losses in the metal

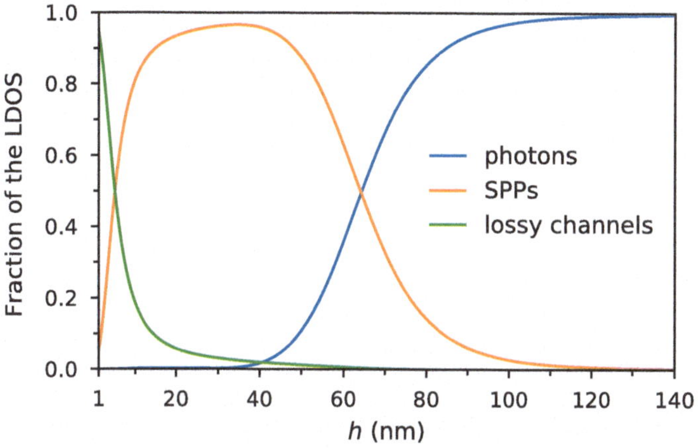

Fig. 2.11 Decomposition of the LDOS (or, equivalently, of the dissipated power or of the spontaneous decay rate) in terms of the three contributions described in the text—i.e., due to photons, SPPs, and lossy pathways—as a function of the emitter–surface separation h. A vertically-orientated dipole with transition frequency $\omega = 0.9\omega_{SP}$ is assumed. The setup and material parameters are the same as in Fig. 2.9

of each of the aforementioned channels depends strongly on the emitter–surface separation.[26]

Generically speaking, the emission of radiation via photons is the dominant channel for emitters lying far away from the metal surface. The decay into SPP is only substantial when the emitter–surface separation is enough so that the near-field of the dipole (that contains a distribution of large in-plane wavevectors capable of exciting SPPs) effectively reaches the metal's surface. If the emitter–surface separation is reduced further, one eventually enters the regime were the decay enhancement by lossy channels becomes dominant.

Figure 2.11 supports the interpretation described in the previous paragraph in a more objective and quantitative fashion: it shows the fraction of the LDOS (or of the dissipated power, or of the spontaneous decay rate) that corresponds to each one of the three distinct channels mentioned above.[27] In particular, the three regimes outlined in the preceding paragraph are well separated. Finally, the suppression of the

[26]Other than that, it also depends on the ratio between the emitter's resonant frequency and the surface plasmon frequency, and on the amount of losses.

[27]Here, the decomposition into the three different relaxation mechanisms is calculated using Eq. (2.56) as follows: the integral $\int_0^{k_d} dq[...]$ is associated with the emission into photons; the emission into SPPs can be obtained by evaluating the SPP pole contribution by computing the integral $\int_{k_d}^{\infty} dq[...]$ in which the reflection coefficient for p-polarization has been replaced by $r_p \approx \frac{2\epsilon_d\epsilon_m}{\epsilon_d^2 - \epsilon_m^2}\frac{q_{SPP}}{q - q_{SPP}}$ (this has to be done judiciously in some cases though); lastly, the nonresonant lossy contribution is whatever remains from the integral $\int_{k_d}^{\infty} dq[...]$ after the resonant SPP pole contribution has been subtracted. Each part is then divided by the total dissipated power in order to obtain the corresponding fractions.

radiative free-space photons channel and concomitant increase of the contribution due to SPPs, and, at extremely small separations, also due to lossy channels, is known as *quenching* [151–153]. This regime is marked by the strong reduction of the fluorescence observed in the far-field despite the the overall enhancement of spontaneous emission rate.

References

1. Jackson JD (1998) Classical electrodynamics, 3rd edn. Wiley, New York
2. Born M, Wolf E (1999) Principles of optics, 7th edn. Cambridge University Press
3. Stratton JA (1941) Electromagnetic theory. McGraw-Hill, New York
4. Landau LD, Lifshitz EM (1960) Electrodynamics of continuous media. Pergamon Press, New York
5. Novotny L, Hecht B (2012) Principles of nano-optics, 2nd edn. Cambridge University Press
6. Maradudin AA, Barnes WL, Sambles JR (eds) (2014) Modern plasmonics, 1st edn. Elsevier
7. Gonçalves PAD, Peres NMR (2016) An introduction to graphene plasmonics, 1st edn. World Scientific, Singapore. https://doi.org/10.1142/9948
8. Ashcroft NW, Mermin ND (1976) Solid state physics. Harcourt College Publishers, New York
9. Grosso G, Parravicini GP (2013) Solid state physics, 2nd edn. Academic Press
10. Ginzburg P, Zayats AV (2013) ACS Nano 7(5):4334. https://doi.org/10.1021/nn400842m
11. Mortensen NA (2013) Photonics nanostructures. Fundam Appl 11(4):303. https://doi.org/10.1016/j.photonics.2013.06.002
12. Raza S, Bozhevolnyi SI, Wubs M, Mortensen NA (2015) J Phys Condens Matter 27(18):183–204. https://doi.org/10.1088/0953-8984/27/18/183204
13. Gramotnev DK, Bozhevolnyi SI (2010) Nat Photonics 4(2):83. https://doi.org/10.1038/nphoton.2010.282
14. Barnes WL, Dereux A, Ebbesen TW (2003) Nature 424:824. https://doi.org/10.1038/nature01937
15. Stockman MI (2011) Opt Express 19(22):22029. https://doi.org/10.1364/OE.19.022029
16. Akselrod GM, Argyropoulos C, Hoang TB, Ciracì C, Fang C, Huang J, Smith DR, Mikkelsen MH (2014) Nat Photonics 8:835. https://doi.org/10.1038/nphoton.2014.228
17. Russell KJ, Liu TL, Cui S, Hu EL (2012) Nat Photonics 6:459. https://doi.org/10.1038/nphoton.2012.112
18. Raether H (1988) Surface plasmons on smooth and rough surfaces and on gratings. Springer
19. Fox M (2010) Optical properties of solids. Oxford master series in physics, 2nd edn. Oxford University Press
20. Pelton M, Bryant GW (2013) Introduction to metal-nanoparticle plasmonics. Wiley, New York
21. Gonçalves MR (2014) J Phys D Appl Phys 47(21):213001. https://doi.org/10.1088/0022-3727/47/21/213001
22. Raza S, Stenger N, Pors A, Holmgaard T, Kadkhodazadeh S, Wagner JB, Pedersen K, Wubs M, Bozhevolnyi SI, Mortensen NA (2014) Nat Commun 5:4125. https://doi.org/10.1038/ncomms5125
23. Drude P (1900) Ann Phys 306(3):566. https://doi.org/10.1002/andp.19003060312
24. Etchegoin PG, Le Ru EC, Meyer M (2006) J Chem Phys 125(16):164705 [Erratum: see https://doi.org/10.1063/1.2802403]
25. Sehmi HS, Langbein W, Muljarov EA (2017) Phys Rev B 95:115444. https://doi.org/10.1103/PhysRevB.95.115444
26. Brendel R, Bormann D (1992) J Appl Phys 71(1):1. https://doi.org/10.1063/1.350737

27. Rakić AD, Djurišić AB, Elazar JM, Majewski ML (1998) Appl Opt 37(22):5271. https://doi.org/10.1364/AO.37.005271
28. Sommerfeld A (1928) Z Phys 47(1):1. https://doi.org/10.1007/BF01391052
29. Johnson PB, Christy RW (1972) Phys Rev B 6:4370. https://doi.org/10.1103/PhysRevB.6.4370
30. Giuliani G, Vignale G (2005) Quantum theory of the electron liquid. Cambridge University Press. https://doi.org/10.1017/CBO9780511619915
31. Bruus H, Flensberg K (2004) Many-body quantum theory in condensed matter physics: an introduction. Oxford graduate texts. Oxford University Press
32. Mahan GD (2000) Many-particle physics, 3rd edn. Springer, New York
33. Fiolhais C, Marques MA, Ullrich CA, Nogueira F (eds) (2003) A primer in density functional theory. Lecture notes in physics. Springer, New York
34. Burke K, Werschnik J, Gross EKU (2005) J Chem Phys 123(6):062206. https://doi.org/10.1063/1.1904586
35. Kohanoff J (2006) Electronic structure calculations for solids and molecules: theory and computational methods. Cambridge University Press. https://doi.org/10.1017/CBO9780511755613
36. Maier SA (2007) Plasmonics: fundamentals and applications. Springer
37. Rakić AD (1995) Appl Opt 34(22):4755. https://doi.org/10.1364/AO.34.004755
38. Ginn JC, Jarecki RL, Shaner EA, Davids PS (2011) J Appl Phys 110(4):043110. https://doi.org/10.1063/1.3626050
39. N'Tsame Guilengui V, Cerutti L, Rodriguez JB, Tournié E, Taliercio T (2012) Appl Phys Lett 101(16):161113. https://doi.org/10.1063/1.4760281
40. Ozbay E (2006) Science 311(5758):189. https://doi.org/10.1126/science.1114849
41. Low T, Chaves A, Caldwell JD, Kumar A, Fang NX, Avouris P, Heinz TF, Guinea F, Martin-Moreno L, Koppens F (2017) Nat Mater 16:182. https://doi.org/10.1038/nmat4792
42. García de Abajo FJ (2014) ACS Photonics 1(3):135. https://doi.org/10.1021/ph400147y
43. Lundeberg MB, Gao Y, Asgari R, Tan C, Van Duppen B, Autore M, Alonso-González P, Woessner A, Watanabe K, Taniguchi T, Hillenbrand R, Hone J, Polini M, Koppens FHL (2017) Science 357(6347):187. https://doi.org/10.1126/science.aan2735
44. Archambault A, Teperik TV, Marquier F, Greffet JJ (2009) Phys Rev B 79:195414. https://doi.org/10.1103/PhysRevB.79.195414
45. Wolff C, Busch K, Mortensen NA (2018) Phys Rev B 97:104203. https://doi.org/10.1103/PhysRevB.97.104203
46. Archambault A, Besbes M, Greffet JJ (2012) Phys Rev Lett 109:097405. https://doi.org/10.1103/PhysRevLett.109.097405
47. Smith CLC, Stenger N, Kristensen A, Mortensen NA, Bozhevolnyi SI (2015) Nanoscale 7:9355. https://doi.org/10.1039/C5NR01282A
48. Bozhevolnyi SI, Volkov VS, Devaux E, Ebbesen TW (2005) Phys Rev Lett 95:046802. https://doi.org/10.1103/PhysRevLett.95.046802
49. Kliewer KL, Fuchs R (1967) Phys Rev 153:498. https://doi.org/10.1103/PhysRev.153.498
50. Economou EN (1969) Phys Rev 182:539. https://doi.org/10.1103/PhysRev.182.539
51. Apell SP, Echenique PM, Ritchie RH (1996) Ultramicroscopy 65(1):53. https://doi.org/10.1016/S0304-3991(96)00055-1
52. Ögüt B, Vogelgesang R, Sigle W, Talebi N, Koch CT, van Aken PA (2011) ACS Nano 5(8):6701. https://doi.org/10.1021/nn2022414
53. Rossouw D, Botton GA (2012) Opt Express 20(7):6968. https://doi.org/10.1364/OE.20.006968
54. Dionne JA, Sweatlock LA, Atwater HA, Polman A (2005) Phys Rev B 72:075405. https://doi.org/10.1103/PhysRevB.72.075405
55. Bozhevolnyi S (2008) Plasmonic nanoguides and circuits, 1st edn. Pan Stanford
56. Fang Y, Sun M (2015) Light Sci Appl 4(6):e294. https://doi.org/10.1038/lsa.2015.67
57. Lorenz L, Dan K (1890) Vidensk Selsk Skrift 6(6). http://www.imm.dtu.dk/~jerf/Lorenz_Danish.pdf

58. Mie G (1908) Ann Phys 330(3):377. https://doi.org/10.1002/andp.19083300302
59. Bohren CF, Huffman DR (1983) Absorption and scattering of light by small particles. Wiley, New York
60. van de Hulst HC (1981) Light scattering by small particles. Dover
61. Hergert W, Wriedt T (eds) (2012) The Mie theory: basics and applications. Springer series in optical sciences. Springer, New York
62. Christensen T, Yan W, Raza S, Jauho AP, Mortensen NA, Wubs M (2014) ACS Nano 8(2):1745. https://doi.org/10.1021/nn406153k
63. Myroshnychenko V, Rodríguez-Fernández J, Pastoriza-Santos I, Funston AM, Novo C, Mulvaney P, Liz-Marzán LM, García de Abajo FJ (2008) Chem Soc Rev 37:1792. https://doi.org/10.1039/B711486A
64. Amorim B, Gonçalves PAD, Vasilevskiy MI, Peres NMR (2017) Appl Sci 7(11). http://www.mdpi.com/2076-3417/7/11/1158
65. Pohl M, Belotelov VI, Akimov IA, Kasture S, Vengurlekar AS, Gopal AV, Zvezdin AK, Yakovlev DR, Bayer M (2012) Phys Rev B 85:081401. https://doi.org/10.1103/PhysRevB.85.081401
66. Boltasseva A, Atwater HA (2011) Science 331(6015):290. https://doi.org/10.1126/science.1198258
67. Engheta N, Ziolkowski RW (eds) (2006) Metamaterials: physics and engineering explorations. Wiley
68. Shvets G, Tsukerman I (eds) (2011) Plasmonics and plasmonic metamaterials. World Scientific
69. Takayama O, Shkondin E, Bodganov A, Aryaee Panah ME, Golenitskii K, Dmitriev P, Repän T, Malureanu R, Belov P, Jensen F, Lavrinenko AV (2017) ACS Photonics 4(11):2899. https://doi.org/10.1021/acsphotonics.7b00924
70. Poddubny A, Iorsh I, Belov P, Kivshar Y (2013) Nat Photonics 7(12):948. https://doi.org/10.1038/nphoton.2013.243
71. Gomez-Diaz JS, Tymchenko M, Alù A (2015) Phys Rev Lett 114:233901. https://doi.org/10.1103/PhysRevLett.114.233901
72. Sreekanth KV, Alapan Y, ElKabbash M, Ilker E, Hinczewski M, Gurkan UA, De Luca A, Strangi G (2016) Nat Mater 15(6):621. https://doi.org/10.1038/nmat4609
73. Peragut F, Cerutti L, Baranov A, Hugonin JP, Taliercio T, Wilde YD, Greffet JJ (2017) Optica 4(11):1409. https://doi.org/10.1364/OPTICA.4.001409
74. Lu L, Simpson RE, Valiyaveedu SK (2018) J Opt 20(10):103001. https://doi.org/10.1088/2040-8986/aade68
75. Sasin ME, Seisyan RP, Kalitteevski MA, Brand S, Abram RA, Chamberlain JM, Egorov AY, Vasil'ev AP, Mikhrin VS, Kavokin AV (2008) Appl Phy Lett 92(25):251112. https://doi.org/10.1063/1.2952486
76. Grossmann C, Coulson C, Christmann G, Farrer I, Beere HE, Ritchie DA, Baumberg JJ (2011) Appl Phys Lett 98(23):231105. https://doi.org/10.1063/1.3597304
77. Symonds C, Lheureux G, Hugonin JP, Greffet JJ, Laverdant J, Brucoli G, Lemaitre A, Senellart P, Bellessa J (2013) Nano Lett 13(7):3179. https://doi.org/10.1021/nl401210b
78. Barnes WL, Preist TW, Kitson SC, Sambles JR (1996) Phys Rev B 54:6227. https://doi.org/10.1103/PhysRevB.54.6227
79. Perney NMB, Baumberg JJ, Zoorob ME, Charlton MDB, Mahnkopf S, Netti CM (2006) Opt Express 14(2):847. https://doi.org/10.1364/OPEX.14.000847
80. Lucas BD, Kim JS, Chin C, Guo LJ (2008) Adv Mater 20(6):1129. https://doi.org/10.1002/adma.200700225
81. Adato R, Yanik AA, Amsden JJ, Kaplan DL, Omenetto FG, Hong MK, Erramilli S, Altug H (2009) Proc Natl Acad Sci USA 106(46):19227. https://doi.org/10.1073/pnas.0907459106
82. Pendry JB, Martín-Moreno L, García-Vidal FJ (2004) Science 305(5685):847. https://doi.org/10.1126/science.1098999
83. Rusina A, Durach M, Stockman MI (2010) Appl Phys A Mater Sci Process 100(2):375. https://doi.org/10.1007/s00339-010-5866-y

84. Jung J, Søndergaard T, Bozhevolnyi SI (2009) Phys Rev B 79:035401. https://doi.org/10.1103/PhysRevB.79.035401
85. Neutens P, Van Dorpe P, De Vlaminck I, Lagae L, Borghs G (2009) Nat Photonics 3(5):283. https://doi.org/10.1038/nphoton.2009.47
86. Im H, Bantz KC, Lindquist NC, Haynes CL, Oh SH (2010) Nano Lett 10(6):2231. https://doi.org/10.1021/nl1012085
87. Nielsen MG, Pors A, Albrektsen O, Bozhevolnyi SI (2012) Opt Express 20(12):13311. https://doi.org/10.1364/OE.20.013311
88. Novikov IV, Maradudin AA (2002) Phys Rev B 66:035403. https://doi.org/10.1103/PhysRevB.66.035403
89. Geisler M, Cui X, Wang J, Rindzevicius T, Gammelgaard L, Jessen BS, Goncalves PAD, Todisco F, Bøggild P, Boisen A, Wubs M, Mortensen NA, Xiao S, Stenger N (2019) ACS Photonics 6(4):994. https://doi.org/10.1021/acsphotonics.8b01766
90. Langhammer C, Kasemo B, Zorić I (2007) J Chem Phys 126(19):194702. https://doi.org/10.1063/1.2734550
91. Langhammer C, Schwind M, Kasemo B, Zorić I (2008) Nano Lett 8(5):1461. https://doi.org/10.1021/nl080453i
92. Schmidt FP, Ditlbacher H, Hohenester U, Hohenau A, Hofer F, Krenn JR (2012) Nano Lett 12(11):5780. https://doi.org/10.1021/nl3030938
93. Nelayah J, Kociak M, Stéphan O, de Abajo FJG, Tencé M, Henrard L, Taverna D, Pastoriza-Santos I, Liz-Marzán LM, Colliex C (2007) Nat Phys 3(5):348. https://doi.org/10.1038/nphys575
94. Schmidt FP, Ditlbacher H, Hofer F, Krenn JR, Hohenester U (2014) Nano Lett 14(8):4810. https://doi.org/10.1021/nl502027r
95. Campos A, Arbouet A, Martin J, Gérard D, Proust J, Plain J, Kociak M (2017) ACS Photonics 4(5):1257. https://doi.org/10.1021/acsphotonics.7b00204
96. Gao B, Arya G, Tao AR (2012) Nat Nanotechnol 7(7):433. https://doi.org/10.1038/nnano.2012.83
97. Zhang S, Bao K, Halas NJ, Xu H, Nordlander P (2011) Nano Lett 11(4):1657. https://doi.org/10.1021/nl200135r
98. Qin F, Zhao T, Jiang R, Jiang N, Ruan Q, Wang J, Sun LD, Yan CH, Lin HQ (2016) Adv Opt Mater 4(1):76. https://doi.org/10.1002/adom.201500496
99. Mahmoud MA, El-Sayed MA (2013) J Phys Chem Lett 4(9):1541. https://doi.org/10.1021/jz4005015
100. Grzelczak M, Pérez-Juste J, Mulvaney P, Liz-Marzán LM (2008) Chem Soc Rev 37:1783. https://doi.org/10.1039/B711490G
101. Tao AR, Habas S, Yang P (2008) Small 4(3):310. https://doi.org/10.1002/smll.200701295
102. Rycenga M, Cobley CM, Zeng J, Li W, Moran CH, Zhang Q, Qin D, Xia Y (2011) Chem Rev 111(6):3669. https://doi.org/10.1021/cr100275d
103. Nordlander P, Oubre C, Prodan E, Li K, Stockman MI (2004) Nano Lett 4(5):899. https://doi.org/10.1021/nl049681c
104. Hao E, Schatz GC (2004) J Chem Phys 120(1):357. https://doi.org/10.1063/1.1629280
105. Romero I, Aizpurua J, Bryant GW, García de Abajo FJ (2006) Opt Express 14(21):9988. https://doi.org/10.1364/OE.14.009988
106. Prodan E, Nordlander P (2004) J Chem Phys 120(11):5444. https://doi.org/10.1063/1.1647518
107. Sönnichsen C, Reinhard BM, Liphardt J, Alivisatos AP (2005) Nat Biotechnol 23(6):741. https://doi.org/10.1038/nbt1100
108. Zuloaga J, Prodan E, Nordlander P (2009) Nano Lett 9(2):887. https://doi.org/10.1021/nl803811g
109. Fischer H, Martin OJF (2008) Opt Express 16(12):9144. https://doi.org/10.1364/OE.16.009144
110. Kinkhabwala A, Yu Z, Fan S, Avlasevich Y, Müllen K, Moerner W (2009) Nat Photonics 3(11):654. https://doi.org/10.1038/nphoton.2009.187

111. Tabor C, Murali R, Mahmoud M, El-Sayed MA (2009) J Phys Chem A 113(10):1946. https://doi.org/10.1021/jp807904s
112. Roxworthy BJ, Ko KD, Kumar A, Fung KH, Chow EKC, Liu GL, Fang NX, Toussaint KC (2012) Nano Lett 12(2):796. https://doi.org/10.1021/nl203811q
113. Hentschel M, Saliba M, Vogelgesang R, Giessen H, Alivisatos AP, Liu N (2010) Nano Lett 10(7):2721. https://doi.org/10.1021/nl101938p
114. Hentschel M, Dregely D, Vogelgesang R, Giessen H, Liu N (2011) ACS Nano 5(3):2042. https://doi.org/10.1021/nn103172t
115. Gallinet B, Martin OJF (2011) ACS Nano 5(11):8999. https://doi.org/10.1021/nn203173r
116. Hentschel M, Schäferling M, Weiss T, Liu N, Giessen H (2012) Nano Lett 12(5):2542. https://doi.org/10.1021/nl300769x
117. Wang H, Brandl DW, Le F, Nordlander P, Halas NJ (2006) Nano Lett 6(4):827. https://doi.org/10.1021/nl060209w
118. Radloff C, Halas NJ (2004) Nano Lett 4(7):1323. https://doi.org/10.1021/nl049597x
119. Mukherjee S, Sobhani H, Lassiter JB, Bardhan R, Nordlander P, Halas NJ (2010) Nano Lett 10(7):2694. https://doi.org/10.1021/nl1016392
120. Oldenburg S, Averitt R, Westcott S, Halas N (1998) Chem Phys Lett 288(2):243. https://doi.org/10.1016/S0009-2614(98)00277-2
121. Wu Y, Nordlander P (2006) J Chem Phys 125(12):124708. https://doi.org/10.1063/1.2352750
122. Tam F, Chen AL, Kundu J, Wang H, Halas NJ (2007) J Chem Phys 127(20):204703. https://doi.org/10.1063/1.2796169
123. Gobin AM, Lee MH, Halas NJ, James WD, Drezek RA, West JL (2007) Nano Lett 7(7):1929. https://doi.org/10.1021/nl070610y
124. Knight MW, Halas NJ (2008) New J Phys 10(10):105006. https://doi.org/10.1088/1367-2630/10/10/105006
125. Kravets VG, Schedin F, Grigorenko AN (2008) Phys Rev Lett 101:087403. https://doi.org/10.1103/PhysRevLett.101.087403
126. Vecchi G, Giannini V, Gómez Rivas (2009) J Phys Rev Lett 102:146807. https://doi.org/10.1103/PhysRevLett.102.146807
127. Auguié B, Barnes WL (2008) Phys Rev Lett 101:143902. https://doi.org/10.1103/PhysRevLett.101.143902
128. García de Abajo FJ (2007) Rev Mod Phys 79:1267. https://doi.org/10.1103/RevModPhys.79.1267
129. Kravets VG, Kabashin AV, Barnes WL, Grigorenko AN (2018) Chem Rev 118(12):5912. https://doi.org/10.1021/acs.chemrev.8b00243
130. Yu R, Liz-Marzán LM, García de Abajo FJ (2017) Chem Soc Rev 46:6710. https://doi.org/10.1039/C6CS00919K
131. Gallinet B, Butet J, Martin OJF (2015) Laser Photonics Rev 9(6):577. https://doi.org/10.1002/lpor.201500122
132. Lavrinenko AV, Lagsgaard J, Gregersen N, Schmidt F, Søndergaard T (2014) Numerical methods in photonics, 1st edn. CRC Press
133. Inan US, Marshall RA (2011) Numerical electromagnetics: the FDTD method. Cambridge University Press, New York
134. Jin JM (2014) The finite element method in electromagnetics, 3rd edn. Wiley
135. Søndergaard TM (2019) Green's function integral equation methods in nano-optics, 1st edn. CRC Press
136. García de Abajo FJ, Howie A (2002) Phys Rev B 65:115418. https://doi.org/10.1103/PhysRevB.65.115418
137. Myroshnychenko V, Carbó-Argibay E, Pastoriza-Santos I, Pérez-Juste J, Liz-Marzán LM, García de Abajo FJ (2008) Adv Mater 20(22):4288. https://doi.org/10.1002/adma.200703214
138. Hohenester U, Trügler A (2012) Comput Phys Commun 183(2):370. https://doi.org/10.1016/j.cpc.2011.09.009
139. Waxenegger J, Trügler A, Hohenester U (2015) Comput Phys Commun 193:138. https://doi.org/10.1016/j.cpc.2015.03.023

140. Draine BT, Flatau PJ (1994) J Opt Soc Am A 11(4):1491. https://doi.org/10.1364/JOSAA.11.001491
141. Wylie JM, Sipe JE (1984) Phys Rev A 30:1185. https://doi.org/10.1103/PhysRevA.30.1185
142. Carminati R, Cazé A, Cao D, Peragut F, Krachmalnicoff V, Pierrat R, Wilde YD (2015) Surf Sci Rep 70(1):1. https://doi.org/10.1016/j.surfrep.2014.11.001
143. Griffiths DJ (2005) Introduction to quantum mechanics, 2nd edn. Prentice-Hall, New Jersey
144. Ford GW, Weber WH (1984) Phys Rep 113(4):195. https://doi.org/10.1016/0370-1573(84)90098-X
145. Joulain K, Carminati R, Mulet JP, Greffet JJ (2003) Phys Rev B 68:245405. https://doi.org/10.1103/PhysRevB.68.245405
146. Purcell EM (1946) Phys Rev 69:681
147. Pelton M (2015) Nat Photonics 9:427. https://doi.org/10.1038/nphoton.2015.103
148. Yablonovitch E (1987) Phys Rev Lett 58:2059. https://doi.org/10.1103/PhysRevLett.58.2059
149. Bouchet D, Carminati R (2019) J Opt Soc Am A 36(2):186. https://doi.org/10.1364/JOSAA.36.000186
150. Koenderink AF (2010) Opt Lett 35(24):4208. https://doi.org/10.1364/OL.35.004208
151. Lakowicz JR (2005) Anal Biochem 337(2):171. https://doi.org/10.1016/j.ab.2004.11.026
152. Anger P, Bharadwaj P, Novotny L (2006) Phys Rev Lett 96:113002. https://doi.org/10.1103/PhysRevLett.96.113002
153. Sorger VJ, Pholchai N, Cubukcu E, Oulton RF, Kolchin P, Borschel C, Gnauck M, Ronning C, Zhang X (2011) Nano Lett 11(11):4907. https://doi.org/10.1021/nl202825s

Chapter 3
Electronic and Optical Properties of Graphene

3.1 Electronic Structure of Graphene

Graphene has been hailed as a "wonder material" primarily due to its remarkable, singular optoelectronic properties. Notably, most of the awe-inspiring electronic properties of monolayer graphene can be understood from elementary considerations about its crystal structure and through the application of standard models of bandstructure theory (e.g., the tight-binding model or the $\mathbf{k} \cdot \mathbf{p}$ method) in two-dimensions. In this spirit, below we provide a cursory review of graphene's bandstructure and its basic electronic properties [1]. These considerations will then supply us with the necessary ingredients for determining the optical response of graphene [2].

3.1.1 Crystal Structure of Monolayer Graphene

Graphene is a two-dimensional carbon allotrope in which the carbon atoms are arranged in a honeycomb lattice. Being a monoelemental material, we start by recalling the basic chemical properties of carbon. Carbon is the sixth element in the periodic table, with a ground state atomic configuration $[\mathrm{C}] = 1\mathrm{s}^2 2\mathrm{s}^2 2\mathrm{p}^2$, and it is therefore tetravalent, that is, it has four valence electrons that can participate in the formation of chemical bonds. Determinantely, in the presence of other atoms (e.g., like C itself, or else, like H) it is nevertheless energetically favorable to promote a 2s electron to a 2p orbital due sp-hybridization (arising from the orbital mixing between an s orbital and a p orbital) [3–6]. In graphene, the material's crystalline structure is consequence of the planar sp^2-hybridization[1] and ensuing formation of three strong

[1] Specifically, the sp^2-hybridization—occurring in both graphene and graphite—involves the superposition of the 2s orbital with two 2p orbitals, say, the $2\mathrm{p}_x$ and the $2\mathrm{p}_y$ states. On the other hand, the chemical bonding in diamond is due sp^3-hybridization where the four chemical bonds are equivalent

P. A. D. Gonçalves, *Plasmonics and Light–Matter Interactions in Two-Dimensional Materials and in Metal Nanostructures*, Springer Theses,
https://doi.org/10.1007/978-3-030-38291-9_3

covalent σ-bonds (mutually separated by 120°) between the carbon atoms. The last unpaired valence electron in the unhybridized 2p orbital (say, the $2p_z$ orbital), lying perpendicular to the σ-bonds, mixes with the other $2p_z$ orbitals of adjacent carbon atoms thereby forming the so-called π-bonds, ultimately leading to the formation of half-filled π-bands [1, 11]. The latter are responsible for graphene's emblematic low-energy electronic properties [1, 3].

The single-atom-thick honeycomb structure of graphene is depicted in Fig. 3.1a. The graphene lattice can be viewed as the result of two offsetted triangular (sub)lattices, or, alternatively, as a triangular lattice with a basis of two atoms per unit cell. The primitive lattice vectors, $\mathbf{a}_1$ and $\mathbf{a}_2$, are given by

$$\mathbf{a}_1 = \frac{a_{\mathrm{CC}}}{2}\left(3, \sqrt{3}\right) \quad \text{and} \quad \mathbf{a}_2 = \frac{a_{\mathrm{CC}}}{2}\left(3, -\sqrt{3}\right), \tag{3.1}$$

where $a_{\mathrm{CC}} \simeq 1.42$ Å is the distance between neighboring carbon atoms. The lattice parameter a, hereby defined by the length of the primitive lattice vectors, is thus $a = \left|\mathbf{a}_{1,2}\right| = a_{\mathrm{CC}}\sqrt{3} \simeq 2.46$ Å. On the other hand, the nearest-neighbor vectors are given by

$$\boldsymbol{\delta}_1 = a_{\mathrm{CC}}\,(1, 0)\,, \quad \boldsymbol{\delta}_2 = \frac{a_{\mathrm{CC}}}{2}\left(-1, \sqrt{3}\right) \quad \text{and} \quad \boldsymbol{\delta}_3 = \frac{a_{\mathrm{CC}}}{2}\left(-1, -\sqrt{3}\right), \tag{3.2}$$

which will reveal useful when applying the tight-binding model to graphene (in Sect. 3.1.2).

The corresponding graphene's reciprocal lattice is portrayed in Fig. 3.1b, and it is spanned by the reciprocal lattice vectors[2]

$$\mathbf{b}_1 = \frac{2\pi}{3a_{\mathrm{CC}}}\left(1, \sqrt{3}\right) \quad \text{and} \quad \mathbf{b}_2 = \frac{2\pi}{3a_{\mathrm{CC}}}\left(1, -\sqrt{3}\right). \tag{3.3}$$

Furthermore, and as it will become apparent shortly, the points at the vertices of the first Brillouin zone—shaded hexagonal region in Fig. 3.1b—are of particular importance. Specifically, there are two inequivalent points (each with three-fold symmetry) located at the corners of the first Brillouin zone, denoted by $\boldsymbol{K}$ and $\boldsymbol{K}'$:

$$\boldsymbol{K} = \frac{2\pi}{3a_{\mathrm{CC}}}\left(1, \sqrt{3}\right) \quad \text{and} \quad \boldsymbol{K}' = \frac{2\pi}{3a_{\mathrm{CC}}}\left(1, -\sqrt{3}\right). \tag{3.4}$$

The low-energy physics of graphene can be described by in terms of the electronic spectrum around these points.

(this is also the case, for instance, in methane (CH_4), and in graphane (hydrogenated graphene) [7, 8] or fluorographene (fluorinated graphene) [9, 10]).

[2]These can be determined by exploiting the relationship $\mathbf{a}_i \cdot \mathbf{b}_j = 2\pi\delta_{ij}$ [12].

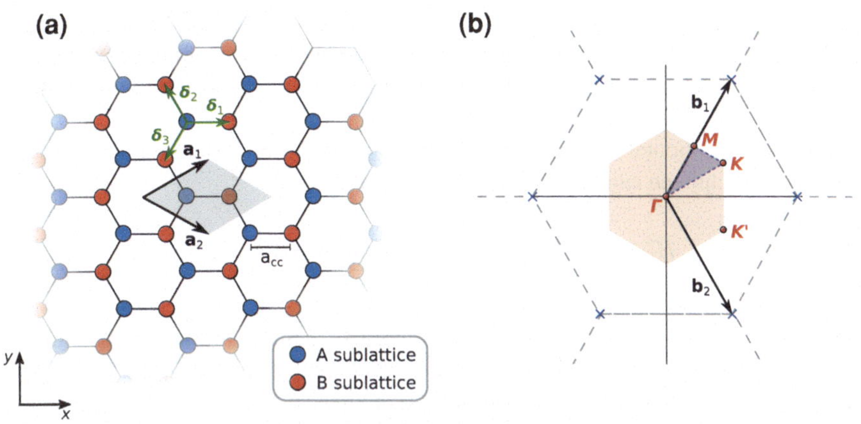

Fig. 3.1 Crystal structure and reciprocal lattice of one-atom-thick graphene. **a** Direct (real-space) graphene's lattice, with the atoms corresponding to the A and B sublattices depicted in blue and red color, respectively. The two lattice vectors $\mathbf{a}_1$ and $\mathbf{a}_2$ are also shown, along with the crystal's unit cell (shaded in gray). The nearest-neighbor vectors $\boldsymbol{\delta}_m$ (where $m \in \{1, 2, 3\}$) are also represented. The distance between neighboring carbon atoms is $a_{CC} \simeq 1.42$ Å. **b** Illustration of graphene's reciprocal lattice (lattice points marked by the blue crosses), spanned by the reciprocal lattice vectors $\mathbf{b}_1$ and $\mathbf{b}_2$. The first Brillouin zone is indicated by the shaded hexagon. Also shown are the high-symmetry points $\boldsymbol{\Gamma}$, $\boldsymbol{M}$, $\boldsymbol{K}$, and $\boldsymbol{K}'$

3.1.2 Tight-Binding Description

The tight-binding model is one of the most common theoretical tools used in condensed matter physics for computing the electronic structure of solids [11–13]. In this regard, graphene is no exception; in fact, the tight-binding technique has been a workhorse for describing the physics of graphene in numerous settings [1, 3, 14]. This fact becomes even more impressive in the light of the recent widespread use of computational methods such as density functional theory (DFT) for solving the electronic structure problem [15, 16]. The main motivation for using the tight-binding model over more complex methods is that, despite its simplicity, the tight-binding approach can still successfully account for the main features of the system's Hamiltonian, which constitutes a stepping-stone toward the calculation of many of the system's properties (e.g., electronic bandstructure, optical properties, electronic transport, etc).

Specifically, graphene's electronic structure is well described by the simple nearest-neighbor tight-binding Hamiltonian,[3] which contains only one parameter describing interaction between nearest-neighbors [1, 3, 14, 17, 22, 23]. One can of course go further, and also include a next-nearest-neighbors hopping term (and so

[3]Curiously, the first theoretical description of graphene was actually under the tight-binding approximation, developed by Wallace in 1947 [17] in the context of the band theory of graphite, long before the isolation of monolayer graphene in 2004(–2005) [18–21].

on) [1, 3, 17]. However, more often than not, the economical nearest-neighbor tight-binding model gives an excellent account of graphene's low-energy spectrum (i.e., around the Fermi level), which reflects the good agreement between the tight-binding description and the *ab initio* calculation [24].

The aforementioned nearest-neighbor tight-binding Hamiltonian governing $2p_z$ electrons in graphene can be written as[4] [1, 2, 14]

$$H_{\text{TB}} = -t \sum_{n,m} \left\{ |A, \mathbf{R}_n\rangle\langle B, \mathbf{R}_n + \boldsymbol{\delta}_m| + \text{H.c.} \right\}, \tag{3.5}$$

where H.c. stands for the Hermitian conjugate, $t \approx 2.8\,\text{eV}$ is the nearest-neighbor hopping energy [1], and $|A, \mathbf{R}_n\rangle$ denotes the Wannier state associated with a site belonging to the sublattice A, at position $\mathbf{R}_n$ (i.e., in the n-th unit cell). An equivalent definition holds for $|B, \mathbf{R}_n + \boldsymbol{\delta}_m\rangle$, where $\boldsymbol{\delta}_m$ (with $m = \{1, 2, 3\}$) refers to the nearest-neighbor vectors described earlier [cf. Eq. (3.2)]. Introducing the following Fourier representation of the Wannier states in terms of Bloch states of wavevector $\mathbf{k}$,

$$|A, \mathbf{R}_n\rangle = \frac{1}{\sqrt{N}} \sum_{\mathbf{k}} e^{-i\mathbf{k}\cdot\mathbf{R}_n} |A, \mathbf{k}\rangle \qquad \text{(and similarly for states in the } B \text{ sublattice)}, \tag{3.6}$$

where N is the number of unit cells of the crystal, ones finds

$$H_{\text{TB}} = -t \sum_{\mathbf{k}} \left\{ f(\mathbf{k}) |A, \mathbf{k}\rangle\langle B, \mathbf{k}| + \text{H.c.} \right\} \qquad \text{with} \qquad f(\mathbf{k}) = \sum_{m=1}^{3} e^{i\mathbf{k}\cdot\boldsymbol{\delta}_m}. \tag{3.7}$$

Equivalently, one may write the tight-binding Hamiltonian as $H_{\text{TB}} = \sum_{\mathbf{k}} \boldsymbol{\psi}_{\mathbf{k}}^{\dagger} \cdot \boldsymbol{H}_{\mathbf{k}} \cdot \boldsymbol{\psi}_{\mathbf{k}}$, where we have introduced $\boldsymbol{\psi}_{\mathbf{k}}^{\dagger} = \left[|A, \mathbf{k}\rangle, |B, \mathbf{k}\rangle \right]$ and

$$\boldsymbol{H}_{\mathbf{k}} = -t \begin{bmatrix} 0 & f(\mathbf{k}) \\ f^*(\mathbf{k}) & 0 \end{bmatrix}. \tag{3.8}$$

From here, the electronic spectrum of graphene is readily obtained:

$$\begin{aligned} E_{\mathbf{k},s} &= s\,|f(\mathbf{k})| \\ &= s \sqrt{3 + 2\cos\left(\sqrt{3}k_y a_{\text{CC}}\right) + 4\cos\left(\frac{3}{2}k_x a_{\text{CC}}\right)\cos\left(\frac{\sqrt{3}}{2}k_y a_{\text{CC}}\right)}, \end{aligned} \tag{3.9}$$

where $s = \pm 1$ is the band-index. Strikingly, the valence band ($s = -1$) and the conduction band ($s = 1$) touch at the so-called Dirac points $\boldsymbol{K}$ and $\boldsymbol{K}'$, rendering pristine graphene gapless. Because of this, together with the fact that, in neutral graphene, the

[4] The spin degree of freedom has been omitted here since spin-orbit effects are weak in graphene [25, 26].

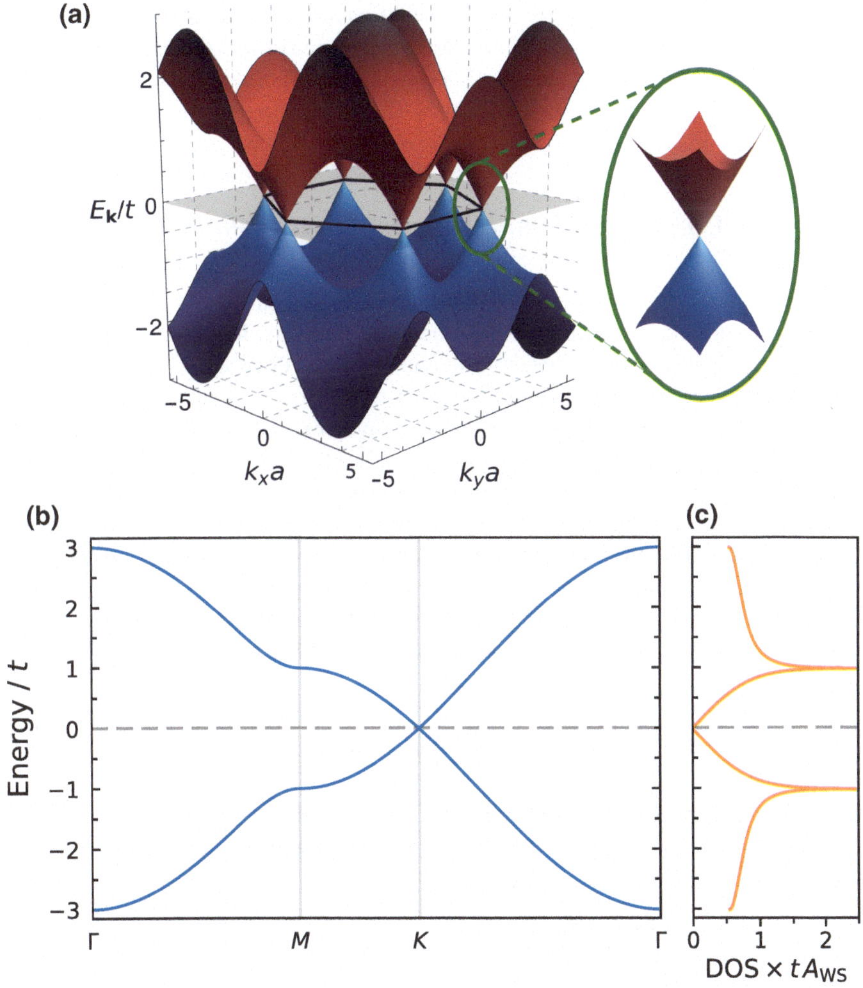

Fig. 3.2 Electronic structure of graphene within the tight-binding approximation. **a** Energy-momentum relation for electrons in graphene, obtained from Eq. (3.9). The inset shows a close-up of graphene's bandstructure in the vicinity of one of the six Dirac cones [i.e., near one the $\boldsymbol{K}(\boldsymbol{K}')$ points], where the dispersion relation is approximately linear. **b** Band diagram along the $\Gamma M K \Gamma$ path. Again, notice the linear regime around the K point. **c** Density of states (DOS) in graphene; A_{WS} is the area of the Wigner–Seitz unit cell. The sharp peaks at energies $\pm t$ correspond to Van Hove singularities [1] associated with the M point (saddle-point)

valence band is completely filled and the conduction band is empty, graphene is classified as a semi-metal. The electronic structure of monolayer graphene is detailed in Fig. 3.2, where the semi-metallic behavior, as well as the famous conical dispersion in the vicinity of the $\boldsymbol{K}(\boldsymbol{K}')$ point can be distinctly seen.

Crucially, in the vast majority of the situations of interest (including the ones studied in this thesis), the physics of graphene is governed by its low-energy electronic spectrum, that is, in the close neighborhood of the Dirac points. In the following, we explore in detail some of the iconic features of graphene in this regime.

3.1.3 The Continuum Limit: Massless Dirac Hamiltonian

Although in the previous section we have derived the full bandstructure (in the sense that Eq. (3.9) provides us with the bandstructure in the entire **k**-space), the prevailing processes associated with the electronic and optical properties of solids typically occur around specific points (usually of high-symmetry) in their bandstructures—namely, points associated with band extrema, in semiconductors,[5] and around the Fermi level, in metals. Similarly, the physics of graphene is dominated by the bandstructure of the π electrons in the vicinity of the $\boldsymbol{K}$ and $\boldsymbol{K}'$ points, that is, the Dirac points (the reason for this designation will become clear shortly).

Hence, motivated by this fact, we perform an expansion around the $\boldsymbol{K}$ point by introducing the shifted wavevector $\mathbf{q} = \mathbf{k} - \boldsymbol{K}$, where $|\mathbf{q}| \ll |\boldsymbol{K}|$ (i.e., the continuum limit, $|\mathbf{q}|\, a_{\mathrm{CC}} \ll 1$); then, to first order in momentum, we obtain $f(\mathbf{k}) \simeq -\frac{3}{2} a_{\mathrm{CC}}\, e^{i\pi/6} \left(q_x - i q_y\right)$. Therefore, the Hamiltonian (3.8) becomes[6]

$$\mathcal{H}_{\boldsymbol{K}} \equiv \boldsymbol{H}_{\boldsymbol{K}+\mathbf{q}} = \hbar v_F \begin{bmatrix} 0 & q_x - i q_y \\ q_x + i q_y & 0 \end{bmatrix}, \tag{3.10}$$

where $v_F \equiv 3 t a_{\mathrm{CC}}/(2\hbar) \approx c/300 = 1 \times 10^6$ m/s is the Fermi velocity of charge carriers in graphene [1]. Applying the same procedure around $\boldsymbol{K}'$ point, we get

$$\mathcal{H}_{\boldsymbol{K}'} \equiv \boldsymbol{H}_{\boldsymbol{K}'+\mathbf{q}} = \hbar v_F \begin{bmatrix} 0 & q_x + i q_y \\ q_x - i q_y & 0 \end{bmatrix}. \tag{3.11}$$

The previous effective Hamiltonians can be written more compactly in terms of Pauli matrices, that is,

$$\mathcal{H}_{\boldsymbol{K}} = \hbar v_F \boldsymbol{\sigma} \cdot \mathbf{q}, \tag{3.12a}$$

$$\mathcal{H}_{\boldsymbol{K}'} = \hbar v_F \boldsymbol{\sigma}^* \cdot \mathbf{q}, \tag{3.12b}$$

[5]This observation has contributed to the development of $\mathbf{k} \cdot \mathbf{p}$ method. See, for instance, Ref. [27].

[6]While also performing a unitary transformation of the basis to in order to exclude the $e^{\pm i\pi/6}$ phase factors (for convenience alone).

with $\boldsymbol{\sigma} = (\sigma_x, \sigma_y)$. In the present shape, it becomes clear that these Hamiltonians have the form of the ones associated with a relativistic massless 2D Dirac equation[7] [1, 3, 28, 29]. It is for this reason that the corners of the first Brillouin zone—the $\boldsymbol{K}$ and $\boldsymbol{K'}$ points (or valleys)—are designated as Dirac points in the graphene literature [1].

The eigenenergies of the above Dirac Hamiltonians are straightforward to obtain, reading

$$E_{\mathbf{q},s} = s\hbar v_F q, \tag{3.13}$$

where, as before, $s = \pm 1$ is the band-index, and where $q \equiv |\mathbf{q}| = \sqrt{q_x^2 + q_y^2}$. This result formally shows that the energy-momentum dispersion of charge carriers in graphene is indeed linear near the Dirac points (as we have anticipated, though heuristically, in Fig. 3.2). This unusual dispersion is responsible for many of graphene's distinct properties [1, 2, 30]. Furthermore, the eigenstates of the Dirac Hamiltonians (3.12) take the form[8]

$$\Psi^{\boldsymbol{K}}_{\mathbf{q},s} = \frac{1}{\sqrt{2}} \begin{bmatrix} e^{-i\theta_{\mathbf{q}}/2} \\ s\, e^{i\theta_{\mathbf{q}}/2} \end{bmatrix} \quad \text{and} \quad \Psi^{\boldsymbol{K'}}_{\mathbf{q},s} = \frac{1}{\sqrt{2}} \begin{bmatrix} e^{i\theta_{\mathbf{q}}/2} \\ s\, e^{-i\theta_{\mathbf{q}}/2} \end{bmatrix}, \tag{3.14}$$

for each valley, where $\theta_{\mathbf{q}} = \arctan(q_y/q_x)$. In fact, the full effective Hamiltonian describing graphene's massless Dirac Fermions is a 4×4 matrix with a block structure. Specifically, it reads

$$\mathcal{H}_{\mathrm{D}} = \begin{bmatrix} \mathcal{H}_K & \mathbf{0} \\ \mathbf{0} & \mathcal{H}_{K'} \end{bmatrix}, \tag{3.15}$$

along with the corresponding four-component wavefunctions (akin to a four-spinor) $\Psi_{\mathrm{D}} = \left[\Psi^{\boldsymbol{K}}_A, \Psi^{\boldsymbol{K}}_B, \Psi^{\boldsymbol{K'}}_A, \Psi^{\boldsymbol{K'}}_B\right]^{\mathrm{T}}$. Together, they take into account the two sublattices and the two inequivalent Dirac cones (or valleys). The Dirac Hamiltonian (3.15) can also be written in a more condensed form, namely

$$\mathcal{H}_{\mathrm{D}} = \hbar v_F \left(\tau_0 \otimes \sigma_x\, q_x + \tau_z \otimes \sigma_y\, q_y\right), \tag{3.16}$$

where the τ_j matrices are also Pauli matrices (with τ_0 denoting the identity matrix), but acting on the subspace associated with the valley degree of freedom. This notation is introduced here in order to better distinguish from the Pauli matrices σ_j, which are assigned to the sublattice subspace instead.

---

[7]The Dirac equation with vanishing mass is sometimes also known as the Weyl equation or the Dirac–Weyl equation.

[8]This can be easily seen by writing the Dirac Hamiltonians in polar form; for instance, for the $\boldsymbol{K}$ valley, Eq. (3.10) [or Eq. (3.12a)], transforms to:

$$\mathcal{H}_K = \hbar v_F q \begin{bmatrix} 0 & e^{-i\theta_{\mathbf{q}}} \\ e^{i\theta_{\mathbf{q}}} & 0 \end{bmatrix}.$$

Density of states of massless Dirac fermions in graphene

The unconventional linear dispersion exhibited by charge carriers in graphene naturally makes its electronic properties different from traditional two-dimensional electron gases (2DEGs) with parabolic dispersion. Such an example is the electronic density of states (DOS). This quantity is given by

$$\mathrm{DOS}(E) = \frac{g}{A}\sum_{\mathbf{k},s}\delta(E - E_{\mathbf{k},s}) \xrightarrow[\text{limit}]{\text{continuum}} g\int \frac{d\mathbf{k}}{(2\pi)^2}\delta(E - E_{\mathbf{k},s})$$
$$\Rightarrow \quad \mathrm{DOS}(E) = \frac{2\,|E|}{\pi(\hbar v_F)^2}, \tag{3.17}$$

where g is a degeneracy factor which, for graphene, amounts to $g = g_s g_v = 4$ owing to the spin ($g_s = 2$) and valley ($g_v = 2$) degeneracies. We stress that this is in stark contrast with the constant $\mathrm{DOS} = \frac{m^*}{\pi\hbar^2}$ displayed by traditional 2DEGs. Moreover, notice that although graphene is a semi-metal—i.e., it has zero bandgap—its density of states is nevertheless zero for neutral graphene. Finally, this peculiar DOS leads to a different dependence of the Fermi energy with carrier density. In particular, the electronic density is calculated by integrating the DOS up to the Fermi level, $n_e = \int_0^\infty \mathrm{DOS}(E) f(E) dE$, where $f(E)$ is the Fermi–Dirac distribution. For graphene at zero temperature (or for $E_F \gg k_B T$), this leads to

$$n_e = \frac{E_F^2}{\pi(\hbar v_F)^2} \qquad \Leftrightarrow \qquad n_e = \frac{k_F^2}{\pi}, \tag{3.18}$$

and thus the Fermi wavevector is related to the carrier density via $k_F = \sqrt{\pi n_e}$.

We end this section by comparing the Dirac approximation to the tight-binding results obtained in Sect. 3.1.2. This is done in Fig. 3.3. Clearly, the low-energy Dirac model provides an excellent description of graphene's electronic properties for energies up to $\lesssim 0.6t$ ($\lesssim 1.7$ eV). Throughout this thesis—and in the vast majority of the

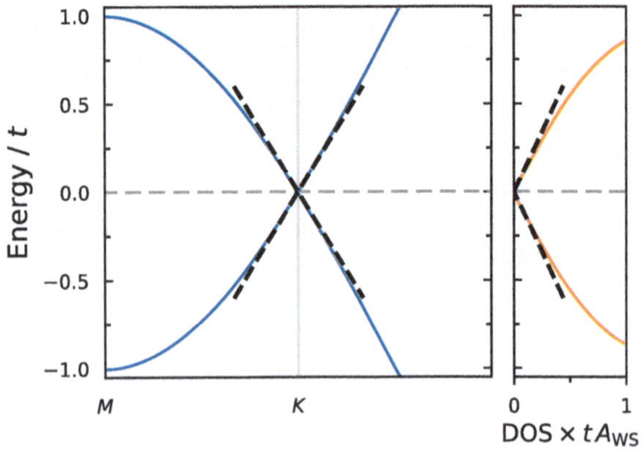

Fig. 3.3 Electronic structure of graphene within the continuum Dirac approximation, shown by the dashed black lines superimposed to the tight-binding calculations (colored lines; they are a close-up of the results presented in Fig. 3.2b–c)

cases considered in the literature—we will always work well within this regime. In particular, we will be able to make use of the Dirac approach for computing the optical response of monolayer graphene. For energies above this threshold, deviations from conical Dirac spectrum arise, an effect that is commonly known as trigonal warping [1, 31].

3.2 Optical Properties of Graphene

The response of a given system to external stimuli can be fully determined upon calculating its appropriate response function(s). Prominent examples of the latter are the dielectric function and the conductivity, although other quantities, such as the density-density response function, are equally notorious. Evidently, all these response functions are interlinked, and therefore the utilization of one or the other is solely a matter of choice or convenience.

In this section we introduce the conductivity of graphene (which, for the two-dimensional graphene, it is a *surface* conductivity) within the framework of the linear-response theory in the Dirac limit. We start by presenting the dynamical—i.e., frequency-dependent—conductivity of single-layer graphene in the local response approximation, specifying both the intraband and interband contributions that make up the overall optical conductivity. Next, we relax the assumption of a local response and consider the full nonlocal conductivity: we do so by summarizing the main features arising from the nonlocal Kubo formalism, which is formally equivalent to the nonlocal dielectric function in the so-called random-phase approximation (RPA).

It should be emphasized that the knowledge of graphene's conductivity constitutes a central cornerstone in graphene photonics and plasmonics, since it embodies all the information about material's electrodynamics, including the physics governing the interaction between light and graphene, and the properties of the plasmonic excitations supported by it.

3.2.1 Dynamical Conductivity of Graphene

The optical conductivity of graphene in the local-response approximation can be split into two distinct contributions: one describing intraband processes, characterized by transitions within the same band (the conduction band for electrons or the valence band for holes), and another describing vertical interband transitions (cf. Fig. 3.4). In this vein, we thus write:

$$\sigma(\omega) = \sigma_{\text{intra}}(\omega) + \sigma_{\text{inter}}(\omega). \tag{3.19}$$

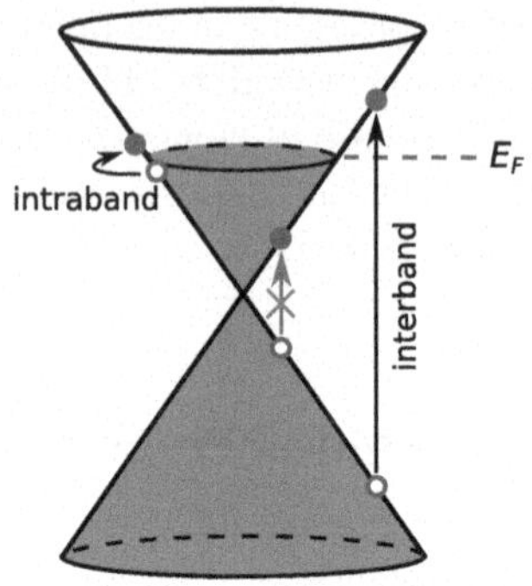

Fig. 3.4 Pictorial illustration of both intraband processes and vertical interband transitions contributing to the dynamical conductivity of doped graphene (at zero temperature). The crossed out transition is representative of a process prevented by Pauli blocking

This separation, though unnecessary, is useful as we will see in what follows. Both the intra- and interband terms arise naturally from the calculation of graphene's conductivity using the Kubo formalism, from which one obtains [2, 32–34]

$$\sigma_{\text{intra}}(\omega) = \frac{\sigma_0}{\pi}\frac{4iE_F}{\hbar\omega + i\hbar\gamma}\left[1 + \frac{2k_BT}{E_F}\ln\left(1 + e^{-E_F/(k_BT)}\right)\right], \tag{3.20a}$$

for the intraband part, and

$$\sigma_{\text{inter}}(\omega) = \sigma_0\left[G(\hbar\omega/2) + i\frac{4\hbar\omega}{\pi}\int_0^\infty dE\,\frac{G(E) - G(\hbar\omega/2)}{(\hbar\omega)^2 - 4E^2}\right], \tag{3.20b}$$

for the interband contribution, where

$$G(E) = \frac{\sinh\left[E/(k_BT)\right]}{\cosh\left[E_F/(k_BT)\right] + \cosh\left[E/(k_BT)\right]}.$$

In the previous expressions, we have introduced the universal conductivity of graphene $\sigma_0 = e^2/(4\hbar)$ [2, 14, 35], whereas k_B denotes the Boltzmann constant. Moreover, E_F is the Fermi energy of graphene[9] (relative to the Dirac point) and γ is a phenomenological scattering rate, here considered in the spirit of the relaxation-time approximation.

In the low-temperature limit, that is, for $E_F \gg k_BT$, Eqs. (3.20) simplify considerably, and reduce to

$$\sigma_{\text{intra}}(\omega) = \frac{\sigma_0}{\pi}\frac{4iE_F}{\hbar\omega + i\hbar\gamma}, \tag{3.21a}$$

$$\sigma_{\text{inter}}(\omega) = \sigma_0\left[\Theta(\hbar\omega - 2E_F) + \frac{i}{\pi}\ln\left|\frac{\hbar\omega - 2E_F}{\hbar\omega + 2E_F}\right|\right], \tag{3.21b}$$

[9]For hole-doped graphene, the expressions presented throughout this section are nevertheless valid upon performing the replacement $E_F \to |E_F|$.

where $\Theta(x)$ stands for the Heaviside step function. These simple analytical expressions often constitute an excellent approximation for moderately doped graphene—even at room temperature—since $k_B T \simeq 26$ meV under ambient conditions, a value that is typically at least one order of magnitude smaller than E_F in most graphene plasmonics experiments. Now, focusing for a moment on the interband term alone (and on the zero temperature result, for the sake of simplicity), notice that there is no absorption[10] for frequencies below $2E_F$, a behavior that is consequence of the so-called Pauli blocking (named after the Pauli exclusion principle, which is the reason for it)—see Fig. 3.4. Accordingly, long-lived graphene plasmons [2] can only exist for $\hbar\omega \lesssim 2E_F$; we shall revisit this point in more detail in Chap. 4.

Lastly, in the low-frequency regime, i.e., for $\hbar\omega \ll 2E_F$, which typically encompasses the terahertz (THz) window up to the mid-infrared (mid-IR) region of the electromagnetic spectrum, the dynamical conductivity of graphene is essentially dominated by its intraband term, thereby taking the familiar Drude-like form:

$$\sigma(\omega) \approx \frac{\sigma_0}{\pi} \frac{4iE_F}{\hbar\omega + i\hbar\gamma}. \tag{3.22}$$

Figure 3.5 depicts the dynamical conductivity of monolayer graphene arising from the expressions outlined in the preceding paragraphs. Notably, both the real and imaginary parts of graphene's conductivity exhibit a weak temperature dependence across a broad range of frequencies (because $E_F \gg k_B T$), the only exception being in the spectral range around the onset of vertical interband transitions (i.e., near $2E_F$). The latter behavior simply reflects the smearing out of the Pauli blocking threshold. Furthermore, in Fig. 3.5b we show empirically that in the THz regime the optical conductivity of graphene is well described by the Drude-type result (3.22).

3.2.2 *Nonlocal Conductivity of Graphene and the Nonlocal RPA Dielectric Function*

The dynamical conductivity of graphene in the local-response approximation corresponds to the longwavelength limit of the more general nonlocal conductivity of graphene $\sigma(q, \omega)$, which is both wavevector- and frequency-dependent. In the following, we present a minimal description of the main steps towards the computation of the noninteracting (retarded) density-density correlation function,[11][12] $\chi_0(q, \omega)$,

---

[10]Recall that absorption is encoded in the real part of the system's conductivity, Re σ.

[11]The noninteracting density-density correlation function $\chi_0(q, \omega)$ also goes by the name of noninteracting (or bare) density-density response function, or polarizability, Lindhard function [36], or even as the bare pair-bubble diagram in the language of Feynman diagrammatics [2, 37, 38].

[12]The detailed derivation of the density-density response function for independent electrons is somewhat lengthy and therefore here we simply outline the main steps and results. A thorough derivation of the density-density response function for the noninteracting homogeneous 3D electron gas can be found in a number of textbooks in condensed matter theory, see, for instance, Refs. [37–41]. The

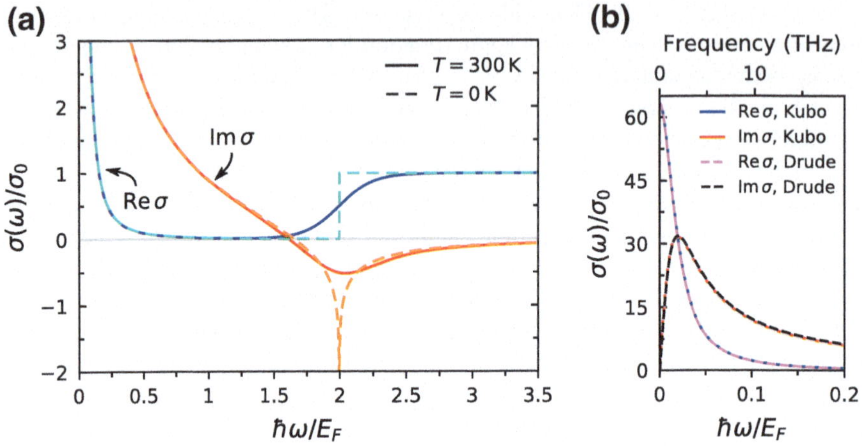

Fig. 3.5 Optical conductivity of graphene in the local response approximation. **a** Real and imaginary parts of the dynamical conductivity of extended graphene obtained from the Kubo formula in the longwavelength limit [i.e., $\sigma(\omega) \equiv \sigma(q \to 0, \omega)$]. The solid lines show the conductivity at room temperature ($T = 300$ K) whereas the dashed lines represent the zero temperature result. **b** Graphene's dynamical conductivity in the THz spectral range, calculated using the Kubo formula at room temperature [Eqs. (3.20), solid lines] and using the Drude-type expression [Eqs. (3.22), dashed lines]. Material parameters: in both panels we have used $E_F = 0.4$ eV and $\hbar\gamma = 8$ meV

from which the nonlocal conductivity of graphene and its nonlocal RPA dielectric function can be straightforwardly obtained.

Specifically, for massless Dirac fermions in extended, pristine graphene the non-interacting density-density correlation function is given by [2, 42, 43]

$$\chi_0(q, \omega) = \frac{g}{(2\pi)^2} \int d\mathbf{k} \sum_{s,s'} \frac{f(E_{\mathbf{k},s}) - f(E_{\mathbf{k}+\mathbf{q},s'})}{E_{\mathbf{k},s} - E_{\mathbf{k}+\mathbf{q},s'} + \hbar(\omega + i\eta)} F_{s,s'}(\mathbf{k}, \mathbf{q}), \tag{3.23}$$

where $g = g_s g_v = 4$ is the overall degeneracy factor including both the spin and valley degeneracies, $f(E_{\mathbf{k},s}) = \{\exp[(E_{\mathbf{k},s} - E_F)/(k_B T)] + 1\}^{-1}$ is the Fermi–Dirac distribution at energy $E_{\mathbf{k},s}$, and $F_{s,s'}(\mathbf{k}, \mathbf{q})$ is a form factor that comes from the overlap between Dirac wavefunctions, and that amounts to [2, 42, 43]

$$F_{s,s'}(\mathbf{k}, \mathbf{q}) = \frac{1}{2}\left[1 + ss' \frac{k^2 + \mathbf{k} \cdot \mathbf{q}}{k|\mathbf{k} + \mathbf{q}|}\right]. \tag{3.24}$$

version for a homogeneous 2D electron gas with parabolic dispersion is outlined in Refs. [39, 40], whereas the derivation of the same quantity for graphene is presented in Refs. [2, 42, 43].

At zero temperature the Fermi–Dirac distributions reduce to step functions, thereby simplifying (to some extent) the integral (3.23). The explicit evaluation of the density-density correlation function then leads to the result[13] [2, 42, 43]

$$\chi_0(x, y) = -\frac{2k_F}{\pi \hbar v_F} + \frac{1}{4\pi}\frac{k_F}{\hbar v_F}\frac{x^2}{\sqrt{y^2 - x^2}}\left\{-i\pi + \tilde{C}\left(\frac{y+2}{x}\right)\right.$$
$$\left. - \Theta\left(\frac{2-y}{x} - 1\right)\left[\tilde{C}\left(\frac{2-y}{x}\right) - i\pi\right] - \Theta\left(\frac{y-2}{x} + 1\right)\tilde{C}\left(\frac{y-2}{x}\right)\right\}, \tag{3.25}$$

where we have introduced the dimensionless wavevector and energy, $x = q/k_F$ and $y = \hbar\omega/E_F$, respectively, and the auxiliary complex function

$$\tilde{C} = x\sqrt{x^2 - 1} - \ln\left(x + \sqrt{x^2 - 1}\right). \tag{3.26}$$

At this point, the expression for the nonlocal density-density response function of graphene (3.25) may not look particularly elucidative. Nevertheless, as we shall see in a moment, it is a pivotal ingredient for the description of the nonlocal optical response of graphene. For now, it is instructive to separate the density-density response function of doped graphene into its real and imaginary parts, i.e., $\chi_0(q, \omega) = \text{Re}\,\chi_0(q, \omega) + \text{Im}\,\chi_0(q, \omega)$, and analyze the different regions of the (q, ω)-phase-space where χ_0 exhibits different behaviors—see Box 3.1 and Fig. 3.6 for details.

Box 3.1—Graphene's density-density correlation function $\chi_0(q, \omega)$

The real and imaginary parts of $\chi_0(x, y)$ can also be written as piecewise functions in terms of real functions alone [2], namely

$$\text{Re}\,\chi_0(x, y) = \begin{cases} -\frac{2k_F}{\pi\hbar v_F} + \frac{1}{4\pi}\frac{k_F}{\hbar v_F}\frac{x^2}{\sqrt{y^2-x^2}}\left[C_\text{h}\left(\frac{y+2}{x}\right) - C_\text{h}\left(\frac{2-y}{x}\right)\right] & \text{, in 1B} \\ -\frac{2k_F}{\pi\hbar v_F} & \text{, in 1A} \\ -\frac{2k_F}{\pi\hbar v_F} + \frac{1}{4\pi}\frac{k_F}{\hbar v_F}\frac{x^2}{\sqrt{y^2-x^2}}C_\text{h}\left(\frac{y+2}{x}\right) & \text{, in 2B} \\ -\frac{2k_F}{\pi\hbar v_F} + \frac{1}{4\pi}\frac{k_F}{\hbar v_F}\frac{x^2}{\sqrt{x^2-y^2}}C\left(\frac{2-y}{x}\right) & \text{, in 2A} \\ -\frac{2k_F}{\pi\hbar v_F} + \frac{1}{4\pi}\frac{k_F}{\hbar v_F}\frac{x^2}{\sqrt{y^2-x^2}}\left[C_\text{h}\left(\frac{y+2}{x}\right) - C_\text{h}\left(\frac{y-2}{x}\right)\right] & \text{, in 3B} \\ -\frac{2k_F}{\pi\hbar v_F} + \frac{1}{4\pi}\frac{k_F}{\hbar v_F}\frac{x^2}{\sqrt{x^2-y^2}}\left[C\left(\frac{y+2}{x}\right) + C\left(\frac{2-y}{x}\right)\right] & \text{, in 3A} \end{cases} \tag{3.27a}$$

[13] The derivation of this result can be found in Refs. [2, 42, 43], and it relies on the use of standard techniques of complex analysis, such as, for instance, the application of the Sokhotski–Plemelj formula: $\lim_{\eta\to 0^+}\int_{-\infty}^{\infty}\frac{f(x)}{x \pm i\eta}dx = \mathcal{P}\int_{-\infty}^{\infty}\frac{f(x)}{x}dx \mp i\pi f(0)$, where $\mathcal{P}$ denotes the Cauchy principal value.

for the real part of graphene's 2D polarizability, and

$$\mathrm{Im}\,\chi_0(x,y) = \begin{cases} 0 & \text{, in 1B} \\ \frac{1}{4\pi}\frac{k_F}{\hbar v_F}\frac{x^2}{\sqrt{x^2-y^2}}\left[C_\mathrm{h}\left(\frac{2-y}{x}\right) - C_\mathrm{h}\left(\frac{y+2}{x}\right)\right] & \text{, in 1A} \\ \frac{1}{4\pi}\frac{k_F}{\hbar v_F}\frac{x^2}{\sqrt{y^2-x^2}}C\left(\frac{2-y}{x}\right) & \text{, in 2B} \\ -\frac{1}{4\pi}\frac{k_F}{\hbar v_F}\frac{x^2}{\sqrt{x^2-y^2}}C_\mathrm{h}\left(\frac{y+2}{x}\right) & \text{, in 2A} \\ -\frac{1}{4}\frac{k_F}{\hbar v_F}\frac{x^2}{\sqrt{y^2-x^2}} & \text{, in 3B} \\ 0 & \text{, in 3A} \end{cases} \tag{3.27b}$$

for the corresponding imaginary part. The different regions akin to each branch of the above piecewise functions are defined schematically in Fig. 3.6. In the previous expressions, as in Eq. (3.25), we have defined $x = q/k_F$ and $y = \hbar\omega/E_F$. Here, we have also introduced the following auxiliary (real) functions:

$$C_\mathrm{h}(z) = z\sqrt{z^2-1} - \mathrm{arccosh}(z)\,, \tag{3.28a}$$

$$C(z) = z\sqrt{1-z^2} - \arccos(z)\,. \tag{3.28b}$$

Of particular importance for our considerations on graphene plasmonics is the 1B region of the (q,ω)-phase-space (cf. Fig. 3.6), where $\mathrm{Im}\,\chi_0 = 0$ and therefore in

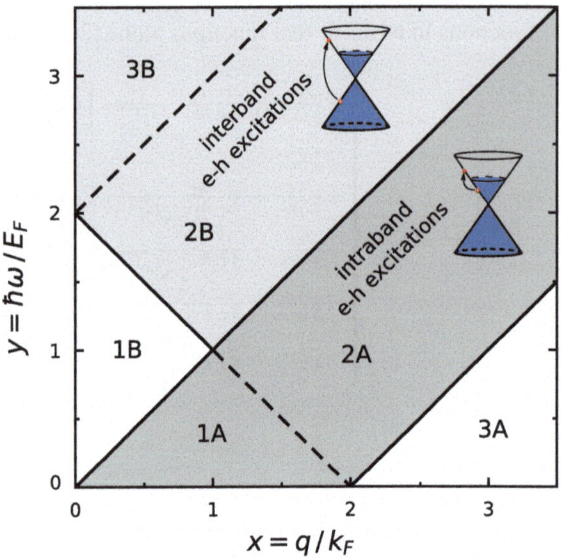

Fig. 3.6 Regions of the polarization function of graphene corresponding to the piecewise partition of the real and imaginary parts of χ_0 as described by Eqs. (3.27a) and (3.27b), respectively. The two different shades of gray indicate the domains of the single-particle continuum where where intra- and interband electron-hole excitations (broadly known as Landau damping) occur, and are associated with regions where $\mathrm{Im}\,\chi_0 \neq 0$ [see Eqs. (3.27a) and (3.27b)]

this region plasmons in graphene can attain large life-times [2, 44, 45]. In essence, this "damping-proofed" domain corresponds to the nonlocal extension of the Pauli blocking mechanism previously mentioned when we considered the local optical response of graphene.[14] Outside this region, plasmon damping becomes non-zero because Im $\chi_0 \neq 0$, even for pristine graphene at zero temperature, due to emergence of a new decay channel through which graphene plasmons can decay into electron-hole pair excitations [2, 42, 43]. This mechanism of plasmon decay is referred to as Landau damping in the literature [2, 37, 38, 43, 44, 46].

In passing, we mention that the above-noted zero temperature result(s) for the noninteracting density-density response function can be generalized for an arbitrary finite temperature by making use of Maldague's result [39, 47]:

$$\chi_0(q, \omega, T) = \int_0^\infty dE \, \frac{\chi_0(q, \omega, T = 0)|_{E_F = E}}{4k_B T \cosh^2\left(\frac{E - \mu(T)}{2k_B T}\right)} \tag{3.29}$$

where $\mu(T)$ designates the system's chemical potential at temperature T.

Nonlocal conductivity of graphene. Equipped with the expression for the non-interacting density-density response function of graphene—see Eq. (3.25) or Eqs. (3.27)—the nonlocal conductivity of the material is readily obtainable from the relation [2, 37]

$$\sigma(q, \omega) = ie^2 \frac{\omega}{q^2} \chi_0(q, \omega), \tag{3.30}$$

which is derivable by making use of the continuity equation [2].

The local conductivity of graphene (3.21) corresponds to the longwavelength limit of Eq. (3.30), that is, $\sigma(\omega) = ie^2 \frac{\omega}{q^2} \chi_0(q \to 0, \omega)$, where the limit needs to be taken judiciously while maintaining $\omega > v_F q$ (see Refs. [39, 42]).

Nonlocal RPA dielectric function of graphene. The random-phase approximation (RPA) is one of the workhorses for describing the optical properties of solids [37, 39–41]. Historically, the linear-response RPA result has been derived in a number of ways[15] [2, 36, 37, 39–41, 60], for instance, via the self-consistent field method [41, 60] or through the diagrammatic approach for the screened interaction [37, 38, 41]. Naturally, both descriptions are ultimately equivalent.

---

[14] Clearly, in the longwavelength limit, i.e., $q \to 0$ (vertical transitions only), the local version of Pauli blocking encountered in Eqs. (3.21) and Fig. 3.4 is reinstated.

[15] In this context, besides the works referenced in the text, it is also just and appropriate to mention other seminal contributions made by a number of people during the 1950s, including, for instance, Bohm and Pines [48–53], Brout and Sawada [54–56], and Nozières and Pines [57–59]. Indeed, these works established the existence of collective plasma oscillations due to the long-range Coulomb interaction between electrons in a homogeneous electron gas, and are often considered as the foundational bedrock for the quantum theory of plasmons. The term "plasmon" was introduced by Pines [53] as the quantum of elementary excitation associated with such collective oscillation.

A homogeneous electron gas subjected to an external potential, ϕ_{ext}, develops an induced density, ρ_{ind}, that corresponds to the deviation from its original equilibrium density. In turn, this perturbation creates an additional induced potential, ϕ_{ind}, in accordance with [37, 38, 41]

$$\phi_{\text{ind}}(\mathbf{r}, t) = \int \frac{\rho_{\text{ind}}(\mathbf{r}', t)}{4\pi\epsilon_0|\mathbf{r} - \mathbf{r}'|} d\mathbf{r}' \quad \rightarrow \quad \phi_{\text{ind}}(\mathbf{q}, \omega) = \frac{1}{e^2} v_{\mathbf{q}}\, \rho_{\text{ind}}(\mathbf{q}, \omega), \tag{3.31}$$

where $v_{\mathbf{q}}$ is the Fourier transform of the bare Coulomb potential, and it is given by

$$v_{\mathbf{q}} = \begin{cases} \frac{e^2}{\epsilon_0 q^2} & \text{, in 3D} \\ \frac{e^2}{2\epsilon_0 q} & \text{, in 2D} \end{cases}, \tag{3.32}$$

in three and two dimensions, respectively. The main notion behind the RPA dielectric function is that an electron gas, when perturbed by an external potential ϕ_{ext}, responds to an effective total (or self-consistent) potential comprised by both ϕ_{ext} and ϕ_{ind}, that is, $\phi_{\text{tot}} = \phi_{\text{ext}} + \phi_{\text{ind}}$; then, the dielectric function is the quantity that relates the total and external potentials, and it is defined by [37, 38, 40, 41]

$$\epsilon(\mathbf{q}, \omega) = \frac{\phi_{\text{ext}}(\mathbf{q}, \omega)}{\phi_{\text{tot}}(\mathbf{q}, \omega)}. \tag{3.33}$$

Furthermore, the *proper* density-density response function $\chi(\mathbf{q}, \omega)$ establishes a relation between the induced charge density ρ_{ind} and the external potential ϕ_{ext} via [37, 38, 40, 41]

$$\rho_{\text{ind}}(\mathbf{q}, \omega) = e^2 \chi(\mathbf{q}, \omega) \phi_{\text{ext}}(\mathbf{q}, \omega). \tag{3.34}$$

Therefore, combining Eqs. (3.31) and (3.34) yields

$$\phi_{\text{ind}}(\mathbf{q}, \omega) = v_{\mathbf{q}} \chi(\mathbf{q}, \omega) \phi_{\text{ext}}(\mathbf{q}, \omega), \tag{3.35}$$

and thus one can write

$$\begin{aligned} \phi_{\text{tot}}(\mathbf{q}, \omega) &= \phi_{\text{ext}}(\mathbf{q}, \omega) + \phi_{\text{ind}}(\mathbf{q}, \omega) \\ &= \left[1 + v_{\mathbf{q}}\, \chi(\mathbf{q}, \omega)\right] \phi_{\text{ext}}(\mathbf{q}, \omega). \end{aligned} \tag{3.36}$$

From this expression, and recalling the definition (3.33), the dielectric function can be readily identified as [37, 38, 40, 41]

$$\epsilon^{-1}(\mathbf{q}, \omega) = 1 + v_{\mathbf{q}}\, \chi(\mathbf{q}, \omega). \tag{3.37}$$

Notice that in the previous expression $\chi(\mathbf{q}, \omega)$ is the *proper* density-density correlation function. However, and rather unfortunately, its explicit calculation is, in practice, an onerous task. In the RPA, the full density-density correlation function $\chi(\mathbf{q}, \omega)$ is approximated by $\chi^{\text{RPA}}(\mathbf{q}, \omega)$ [37, 38, 41],

$$\chi^{\text{RPA}}(\mathbf{q}, \omega) = \frac{\chi_0(\mathbf{q}, \omega)}{1 - v_{\mathbf{q}}\, \chi_0(\mathbf{q}, \omega)}, \tag{3.38}$$

which can be formulated in terms of the *noninteracting* density-density correlation function $\chi_0(\mathbf{q}, \omega)$ determined before [recall Eq. (3.25) or Eqs. (3.27)]. We note that in the RPA explicit electron-electron interactions are neglected, with the exception of screening (at the single-particle level) [41]. As such, it is expected that its applicability is limited to the high-density limit and to weakly correlated systems [39]. Some corrections beyond the RPA will be addressed in Sect. 7.2.

In this vein, by replacing the full density-density response function in Eq. (3.37) by its RPA version (3.38), one finally obtains the nonlocal RPA dielectric function [2, 37, 38, 40, 41]:

$$\epsilon^{\text{RPA}}(\mathbf{q}, \omega) = 1 - v_{\mathbf{q}}\, \chi_0(\mathbf{q}, \omega)\,. \tag{3.39}$$

Specifically, for two-dimensional graphene, one has [2]

$$\boxed{\epsilon^{\text{RPA}}(\mathbf{q}, \omega) = 1 - \frac{e^2}{2\epsilon_0 q}\chi_0(\mathbf{q}, \omega)}\,. \tag{3.40}$$

Here, we note that self-sustained plasmon excitations are given by the zeros of the RPA dielectric function. This can also be seen by inspecting of Eq. (3.33) [from which it is clear that $\phi_{\text{tot}}(\mathbf{q}, \omega) = \frac{\phi_{\text{ext}}(\mathbf{q}, \omega)}{\epsilon(\mathbf{q}, \omega)}$], since it suggests that—under such conditions—a finite field can persist even in the absence of an (or, rather, of an *infinitesimally small*) external field.

Lastly, the energy-loss function of the system is typically defined by [2, 11, 61]

$$\text{L}(\mathbf{q}, \omega) = -\operatorname{Im}\left\{\frac{1}{\epsilon^{\text{RPA}}(\mathbf{q}, \omega)}\right\}, \tag{3.41}$$

which, for instance, is closely related to the electron energy-loss spectrum [61, 62].

Density-density response function in the relaxation-time approximation

So far in our discussion of the noninteracting density-density response function we have peremptorily neglected electron scattering processes. A rigorous description of these is, in general, not trivial. Hence, we present an extension of the previous collisionless result were the aforementioned difficulty is bypassed by making use of the relaxation-time approximation (RTA). It should be noted, however, that although one may feel tempted to perform the familiar prescription $\omega \to \omega + i\gamma$ in order to introduce collisions within the spirit of the RTA, this naïve substitution is flawed since it fails to conserve the number of particles. As shown by Mermin [63] (see also

Ref. [64]), the correct approach is to consider that collisions relax the electronic density to a *local* equilibrium (with a *local* chemical potential) rather than to a uniform (*global*) equilibrium. Using this argument, Mermin showed that the appropriate form of the noninteracting density-density response function in the RTA can be written as [2, 39, 63]

$$\chi_0^\tau(q,\omega) = \frac{(1+i\gamma/\omega)\,\chi_0(q,\omega+i\gamma)}{1+(i\gamma/\omega)\left[\chi_0(q,\omega+i\gamma)/\chi_0(q,0)\right]}\,. \tag{3.42}$$

Evidently, the corresponding nonlocal conductivity and nonlocal RPA dielectric function in the RTA follow from Eqs. (3.30) and (3.39) simply by performing the replacement $\chi_0(q,\omega) \to \chi_0^\tau(q,\omega)$.

References

1. Castro Neto AH, Guinea F, Peres NMR, Novoselov KS, Geim AK (2009) Rev Mod Phys 81:109. https://doi.org/10.1103/RevModPhys.81.109
2. Gonçalves PAD, Peres NMR (2016) An introduction to graphene plasmonics, 1st edn. World Scientific, Singapore. http://www.worldscientific.com/worldscibooks/10.1142/9948, https://doi.org/10.1142/9948
3. Katsnelson MI (2012) Graphene: carbon in two dimensions. Cambridge University Press. https://doi.org/10.1017/CBO9781139031080
4. Pauling L (1931) J Am Chem Soc 53(4):1367. https://doi.org/10.1021/ja01355a027
5. Pauling L (1960) The nature of the chemical bond, 3rd edn. Cornell University Press, Ithaca, New York
6. Pierson HO (1993) Handbook of carbon, graphite, diamonds and fullerenes. William Andrew Publishing, Oxford
7. Sofo JO, Chaudhari AS, Barber GD (2007) Phys Rev B 75:153401. https://doi.org/10.1103/PhysRevB.75.153401
8. Elias DC, Nair RR, Mohiuddin TMG, Morozov SV, Blake P, Halsall MP, Ferrari AC, Boukhvalov DW, Katsnelson MI, Geim AK, Novoselov KS (2009) Science 323(5914):610. https://science.sciencemag.org/content/323/5914/610, https://doi.org/10.1126/science.1167130
9. Nair RR, Ren W, Jalil R, Riaz I, Kravets VG, Britnell L, Blake P, Schedin F, Mayorov AS, Yuan S, Katsnelson MI, Cheng HM, Strupinski W, Bulusheva LG, Okotrub AV, Grigorieva IV, Grigorenko AN, Novoselov KS, Geim AK (2010) Small 6(24):2877. https://onlinelibrary.wiley.com/doi/abs/10.1002/smll.201001555, https://doi.org/10.1002/smll.201001555
10. Samarakoon DK, Chen Z, Nicolas C, Wang XQ (2011) Small 7(7):965. https://onlinelibrary.wiley.com/doi/abs/10.1002/smll.201002058, https://doi.org/10.1002/smll.201002058
11. Grosso G, Parravicini GP (2013) Solid state physics, 2nd edn. Academic Press
12. Ashcroft NW, Mermin ND (1976) Solid state physics. Harcourt College Publishers, New York
13. Singleton J (2010) Band theory and electronic properties of solids, 1st edn. Oxford Master Series in Physics. Oxford University Press
14. Peres NMR (2010) Rev Mod Phys 82:2673. https://doi.org/10.1103/RevModPhys.82.2673
15. Fiolhais C, Marques MA, Ullrich CA, Nogueira F (eds) (2003) A primer in density functional theory. Lecture notes in physics. Springer, New York
16. Kohanoff J (2006) Electronic structure calculations for solids and molecules: theory and computational methods. Cambridge University Press. https://doi.org/10.1017/CBO9780511755613
17. Wallace PR (1947) Phys Rev 71:622. https://doi.org/10.1103/PhysRev.71.622

18. Novoselov KS, Geim AK, Morozov SV, Jiang D, Zhang Y, Dubonos SV, Grigorieva IV, Firsov AA (2004) Science 306(5696):666. https://science.sciencemag.org/content/306/5696/666, https://doi.org/10.1126/science.1102896
19. Novoselov KS, Geim AK, Morozov SV, Jiang D, Katsnelson MI, Grigorieva IV, Dubonos SV, Firsov AA (2005) Nature 438:197. https://www.nature.com/articles/nature04233, https://doi.org/10.1038/nature04233
20. Zhang Y, Tan YW, Stormer HL, Kim P (2005) Nature 438:201. https://www.nature.com/articles/nature04235, https://doi.org/10.1038/nature04235
21. Novoselov KS, Jiang D, Schedin F, Booth TJ, Khotkevich VV, Morozov SV, Geim AK (2005) Proc Natl Acad Sci USA 102(30):10451. https://www.pnas.org/content/102/30/10451, https://doi.org/10.1073/pnas.0502848102
22. Saito R, Dresselhaus G, Dresselhaus MS (1998) Physical properties of carbon nanotubes. Imperial College Press
23. McCann E (eds) (2012) Graphene nanoelectronics: metrology, synthesis, properties and applications. Springer, pp. 237–275. https://doi.org/10.1007/978-3-642-22984-8_8
24. Reich S, Maultzsch J, Thomsen C, Ordejón P (2002) Phys Rev B 66:035412. https://doi.org/10.1103/PhysRevB.66.035412
25. Min H, Hill JE, Sinitsyn NA, Sahu BR, Kleinman L, MacDonald AH (2006) Phys Rev B 74:165310. https://doi.org/10.1103/PhysRevB.74.165310
26. Yao Y, Ye F, Qi XL, Zhang SC, Fang Z (2007) Phys Rev B 75:041401. https://doi.org/10.1103/PhysRevB.75.041401
27. Voon LCLY, Willatzen M (2009) The $k \cdot p$ method. Springer
28. Semenoff GW (1984) Phys Rev Lett 53:2449. https://doi.org/10.1103/PhysRevLett.53.2449
29. Ando T (2005) J Phys Soc Jpn 74(3):777. https://doi.org/10.1143/JPSJ.74.777
30. Abergel DSL, Apalkov V, Berashevich J, Ziegler K, Chakraborty T (2010) Adv Phys 59(4):261. https://doi.org/10.1080/00018732.2010.487978
31. Ando T, Nakanishi T, Saito R (1998) J Phys Soc Jpn 67(8):2857. https://doi.org/10.1143/JPSJ.67.2857
32. Falkovsky LA, Varlamov AA (2007) Eur Phys J B 56(4):281. https://doi.org/10.1140/epjb/e2007-00142-3
33. Falkovsky LA, Pershoguba SS (2007) Phys Rev B 76:153410. https://doi.org/10.1103/PhysRevB.76.153410
34. Stauber T, Peres NMR, Geim AK (2008) Phys Rev B 78:085432. https://doi.org/10.1103/PhysRevB.78.085432
35. Nair RR, Blake P, Grigorenko AN, Novoselov KS, Booth TJ, Stauber T, Peres NMR, Geim AK (2008) Science 320(5881):1308. https://science.sciencemag.org/content/320/5881/1308, https://doi.org/10.1126/science.1156965
36. Lindhard J, Danske Videnskab KGL (1954) Selskab Mat-Fys Medd 28
37. Bruus H, Flensberg K (2004) Many-body quantum theory in condensed matter physics: an introduction. Oxford Graduate Texts. Oxford University Press
38. Jishi RA (2013) Feynman diagram techniques in condensed matter physics. Cambridge University Press. https://doi.org/10.1017/CBO9781139177771
39. Giuliani G, Vignale G (2005) Quantum theory of the electron liquid. Cambridge University Press. https://doi.org/10.1017/CBO9780511619915
40. Sólyom J (2010) Fundamentals of the physics of solids. Springer
41. Mahan GD (2000) Many-particle physics, 3rd edn. Springer, New York
42. Wunsch B, Stauber T, Sols F, Guinea F (2006) New J Phys 8(12):318. https://doi.org/10.1088/1367-2630/8/12/318
43. Hwang EH, Das Sarma S (2007) Phys Rev B 75:205418. https://doi.org/10.1103/PhysRevB.75.205418
44. García de Abajo FJ (2014) ACS Photon 1(3):135. https://doi.org/10.1021/ph400147y
45. Woessner A, Lundeberg MB, Gao Y, Principi A, Alonso-González P, Carrega M, Watanabe K, Taniguchi T, Vignale G, Polini M, Hone J, Hillenbrand R, Koppens FHL (2015) Nat Mater 14:421. https://www.nature.com/articles/nmat4169, https://doi.org/10.1038/nmat4169

46. Yan H, Low T, Zhu W, Wu Y, Freitag M, Li X, Guinea F, Avouris P, Xia F (2013) Nat Photon 7(5):394. http://www.doi.org/10.1038/nphoton.2013.57, https://doi.org/10.1038/nphoton.2013.57
47. Maldague PF (1978) Surf Sci 73:296. http://www.sciencedirect.com/science/article/pii/0039602878905071, https://doi.org/10.1016/0039-6028(78)90507-1
48. Bohm D, Pines D (1950) Phys Rev 80:903. https://doi.org/10.1103/PhysRev.80.903.2
49. Bohm D, Pines D (1951) Phys Rev 82:625. https://doi.org/10.1103/PhysRev.82.625
50. Pines D, Bohm D (1952) Phys Rev 85:338. https://doi.org/10.1103/PhysRev.85.338
51. Bohm D, Pines D (1953) Phys Rev 92:609. https://doi.org/10.1103/PhysRev.92.609
52. Pines D (1953) Phys Rev 92:626. https://doi.org/10.1103/PhysRev.92.626
53. Pines D (1956) Rev Mod Phys 28:184. https://doi.org/10.1103/RevModPhys.28.184
54. Brout R (1957) Phys Rev 108:515. https://doi.org/10.1103/PhysRev.108.515
55. Sawada K (1957) Phys Rev 106:372. https://doi.org/10.1103/PhysRev.106.372
56. Sawada K, Brueckner KA, Fukuda N, Brout R (1957) Phys Rev 108:507. https://doi.org/10.1103/PhysRev.108.507
57. Nozières P, Pines D (1958) Phys Rev 109:741. https://doi.org/10.1103/PhysRev.109.741
58. Nozières P, Pines D (1958) Phys Rev 109:762. https://doi.org/10.1103/PhysRev.109.762
59. Nozières P, Pines D (1958) Phys Rev 109:1062. https://doi.org/10.1103/PhysRev.109.1062
60. Ehrenreich H, Cohen MH (1959) Phys Rev 115:786. https://doi.org/10.1103/PhysRev.115.786
61. Egerton RF (2011) Electron energy-loss spectroscopy in the electron microscope, 3rd edn. Springer
62. García de Abajo FJ (2010) Rev Mod Phys 82:209. https://doi.org/10.1103/RevModPhys.82.209
63. Mermin ND (1970) Phys Rev B 1:2362. https://doi.org/10.1103/PhysRevB.1.2362
64. Greene MP, Lee HJ, Quinn JJ, Rodriguez S (1969) Phys Rev 177:1019. https://doi.org/10.1103/PhysRev.177.1019

Chapter 4
Fundamentals of Graphene Plasmonics

Doped graphene supports plasmonic excitations in the terahertz (THz) and mid-infrared (mid-IR) regions of the electromagnetic spectrum[1] [4–7], and it is often regarded as an excellent plasmonic material [8–10]. Some of the alluring properties of graphene plasmons [4–7] are direct consequence of graphene's tantalizing electronic and optical properties [4, 11], while some others are related to its two-dimensionality alone [4, 6, 12, 13]. The latter are therefore not unique to graphene and thus may be observed in other two-dimensional electron systems [14, 15]. Nevertheless, the unique electronic structure of graphene is of paramount importance for many of its captivating plasmonic features [4–7, 16]. Furthermore, and on a more a practical level, the mechanical stability the graphene together with its intrinsic atomic thickness makes it particularly appealing also from a technological standpoint.

A key advantage of graphene plasmons—specially when compared with traditional surface plasmons supported by three-dimensional metals (3D)—is the ability of actively control the frequency of plasmon resonances in graphene simply by controlling its Fermi level,[2] which can be routinely achieved by means of electrostatic gating [17–19] or by chemical doping [20]. This ability is in stark contrast to conventional plasmonics based on 3D metals, and hence this fact constitutes a significant advantage of graphene plasmonics. In addition, another noteworthy advantage of graphene plasmonics is the fact that plasmons in graphene can propagate while experiencing relatively low-losses when compared with traditional plasmonic

[1] In principle, plasmons in graphene could potentially be pushed towards the near-infrared or even the visible provided that sufficiently high doping can be achieved (this is because, as we have seen in Sect. 3.2, for frequencies above the interband threshold, interband Landau damping prevents the existence of long-lived graphene plasmons). Some encouraging steps have been taken in this direction [1–3], but reaching doping levels corresponding to Fermi energies $E_F \gtrsim 1$ eV remains experimentally challenging.

[2] The relation between graphene's Fermi energy, E_F, and its electronic density, n_e, is given by $E_F = \hbar v_F \sqrt{\pi n_e}$ [recall Eq. (3.18)]. Typical values for the carrier density in doped graphene are $n_e = 10^{11} - 10^{13}\ \mathrm{cm}^{-2}$ [17–19], though values as large as $n_e = 4 \times 10^{14}\ \mathrm{cm}^{-2}$ have been reported [1].

P. A. D. Gonçalves, *Plasmonics and Light–Matter Interactions in Two-Dimensional Materials and in Metal Nanostructures*, Springer Theses,
https://doi.org/10.1007/978-3-030-38291-9_4

materials (this is particularly true for graphene encapsulated in hexagonal boron nitride (hBN)) [8–10].

In this chapter, we present the elementary theory describing plasmonic excitations in graphene and in graphene-based systems. We start by considering plasmons supported by extended, continuous graphene sheets, and then move on to the treatment of plasmons in finite-sized graphene nanostructures and in patterned graphene. Along the way, we comment on several key features of graphene plasmons. Finally, we note that although here particular emphasis is given to plasmons in graphene, the theoretical description set forth in this chapter can be swiftly applied to many other polaritonic excitations [8, 21] in the ever-increasing number[3] of two-dimensional (2D) and quasi-2D materials.

4.1 Plasmons in Extended Graphene

4.1.1 Plasmons in Monolayer Graphene

We consider a graphene monolayer placed at $z = 0$ and sandwiched between two dielectric media, characterized by the relative permittivities ϵ_1 and ϵ_2, as illustrated in Fig. 4.1. The derivation of the dispersion relation of plasmons propagating along an extended graphene sheet essentially follows the same guidelines as in the derivation of the classical SPP spectrum in a dielectric–metal interface (detailed in Sect. 2.2.1.1). The crucial difference, however, is that in the present case the presence of graphene is taken into account by introducing a nonvanishing *surface* current density, $\mathbf{J}_{2D}$. For a uniform graphene layer—i.e., unstrained, unpatterned, pristine graphene—the induced surface current due to an electric field is given by $\mathbf{J}_{2D} = \sigma \mathbf{E}_\parallel$, where $\sigma \equiv \sigma(q, \omega)$ is the (in general, nonlocal) conductivity of graphene that we have studied in Sect. 3.2, and $\mathbf{E}_\parallel$ is the total in-plane electric field evaluated at the position of the graphene sheet (i.e., at $z = 0$). Crucially, all the steps leading to Eqs. (2.19)–(2.21) are basically unaltered; the exception are the boundary conditions, which now read [recall Eqs. (2.9)]:

$$\begin{array}{l} \hat{\mathbf{n}}_{12} \times (\mathbf{E}_2 - \mathbf{E}_1)|_{z=0} = 0 \\ \hat{\mathbf{n}}_{12} \times (\mathbf{H}_2 - \mathbf{H}_1)|_{z=0} = \mathbf{J}_{2D} \end{array} \quad \Rightarrow \quad \begin{array}{l} E_{x,2} - E_{x,1} = 0 \\ H_{y,2} - H_{y,1} = \sigma E_{x,1} \end{array} . \tag{4.1}$$

Using the relations (2.20) between field components, the determinantal solution associated with the previous algebraic system of equations produces the following implicit condition for the dispersion relation of graphene plasmons [4]:

[3]At the time of writing, several hundredths of 2D materials have been identified. Most of them and their properties can be found in "2D materials databases" such as the ones provided in Refs. [22, 23].

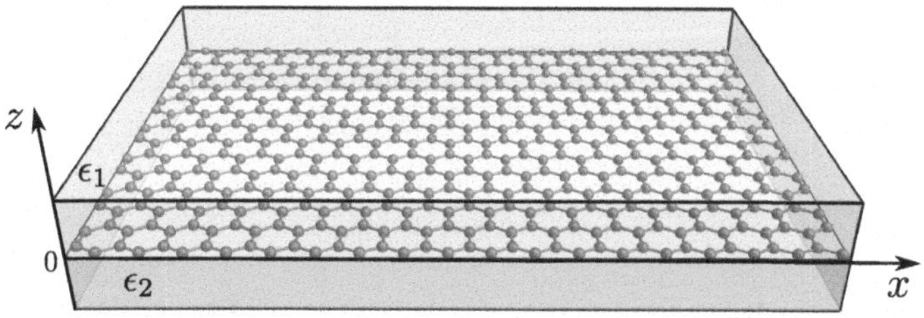

Fig. 4.1 Illustration of a sheet of monolayer graphene sandwiched between two semi-infinite media with relative permittivities ϵ_1 and ϵ_2, respectively, for $z > 0$ and $z < 0$. The graphene layer is located in the $z = 0$ plane

$$\frac{\epsilon_1}{\sqrt{q^2 - k_0^2\epsilon_1}} + \frac{\epsilon_2}{\sqrt{q^2 - k_0^2\epsilon_2}} + \frac{i\sigma(q,\omega)}{\omega\epsilon_0} = 0\,, \tag{4.2}$$

where q is the graphene plasmon's wavevector and $k_0 = \omega/c$. In general, this dispersion relation has to be solved by numerical means. Furthermore, by analyzing Eq. (4.2), it is clear that confined TM graphene plasmons can only exist in regions of the phase-space where $\operatorname{Im}\sigma > 0$; in the longwavelength limit, this is fulfilled for energies $\hbar\omega/E_F \lesssim 1.667$ [4].

Box 4.1—Scattering coefficients for a planar interface in the presence of graphene

The scattering coefficients associated with p-polarized, or transverse magnetic (TM), and s-polarized, or transverse electric (TE), waves are straightforwardly derived in the same way as the traditional Fresnel coefficients between two nonmagnetic media [4, 24], but now augmented with the boundaries conditions for a nonvanishing surface current density [cf. Eqs. (4.1)]. Specifically, for a system like the one portrayed in Fig. 4.1, the scattering coefficients read [4]:

$$r_p = \frac{\epsilon_2 k_{z,1} - \epsilon_1 k_{z,2} + k_{z,1}k_{z,2}\,\sigma/(\omega\epsilon_0)}{\epsilon_2 k_{z,1} + \epsilon_1 k_{z,2} + k_{z,1}k_{z,2}\,\sigma/(\omega\epsilon_0)}\,, \tag{4.3a}$$

$$t_p = \sqrt{\frac{\epsilon_1}{\epsilon_2}}\frac{2\epsilon_2 k_{z,1}}{\epsilon_2 k_{z,1} + \epsilon_1 k_{z,2} + k_{z,1}k_{z,2}\,\sigma/(\omega\epsilon_0)}\,, \tag{4.3b}$$

for p-polarized waves, where $k_{z,j} = \sqrt{\epsilon_j k_0^2 - q^2}$ (with $\operatorname{Im} k_{z,j} > 0$), and [4]

$$r_s = \frac{k_{z,1} - k_{z,2} - \mu_0 \omega \sigma}{k_{z,1} + k_{z,2} + \mu_0 \omega \sigma}, \tag{4.4a}$$

$$t_s = \frac{2k_{z,1}}{k_{z,1} + k_{z,2} + \mu_0 \omega \sigma}, \tag{4.4b}$$

for s-polarized waves. The impinging plane-wave is assumed to come from medium 1, and q denotes the in-plane (i.e., parallel to the interface) wavevector component.

As in the case of SPPs supported by a dielectric–metal interface (see Sect. 2.2.1.1 and Box 2.1), the dispersion relation for TM graphene plasmons[4] could also be determined by inspecting the poles of the system's reflection coefficient for p-polarized waves, r_p. This can be immediately realized from Eq. (4.3a) in Box 4.1, while noting that $k_{z,j} = i\kappa_j \equiv i\sqrt{q^2 - \epsilon_j k_0^2}$ for confined surface waves.

The dispersion relation of plasmons in doped graphene is shown in Fig. 4.2, where it has been calculated using three different models for the conductivity of graphene. The simplest of these is the Drude-like expression for the conductivity of graphene [Eq. (3.22)], which predicts a square-root-type dispersion, $\omega_{\text{GP}} \propto \sqrt{q}$, for graphene plasmons (GPs). The next step in the complexity ladder is the inclusion of both intraband and interband contributions in graphene's conductivity within the framework of the local RPA [Eq. (3.21)]; here, the presence of interband effects effectively restricts the GPs' dispersion to energies $\hbar\omega/E_F \lesssim 1.667$, as discussed in the text after Eq. (4.2). Finally, we also show the dispersion of GPs obtained by employing the more sophisticated full nonlocal conductivity of graphene obtained within the formalism of the nonlocal RPA [Eq. (3.30)] (together with Mermin's particle-conserving relaxation-time approximation). The nonlocal model imparts significant deviations of the GPs' dispersion from its local prediction, particularly as q/k_F increases. Naturally, this deviation is more pronounced for large wavevectors (and correspondingly larger frequencies). Another important feature that can only be rigorously described using the full nonlocal model is the substantial change of the GPs' dispersion when it enters the region associated with interband Landau damping, where GPs become short-lived due to their ability to decay rapidly into electron-hole pairs. This aspect becomes especially transparent through the observation of the disappearance of a well defined plasmon branch in the color plot of Fig. 4.2, immediately after entering the interband electron-hole continuum.

[4] In passing—and mostly for the sake of curiosity alone—we note that, unlike surface plasmon polaritons at single dielectric–metal interfaces [25, 26], graphene is also capable of supporting TE surface waves [4, 27]. These are, however, weakly bound modes situated very close to the light line and therefore are of little fundamental or technological interest. For this reason, they are often disregarded in the literature; likewise, throughout this work (with the exception of this footnote), we consider only TM graphene plasmons.

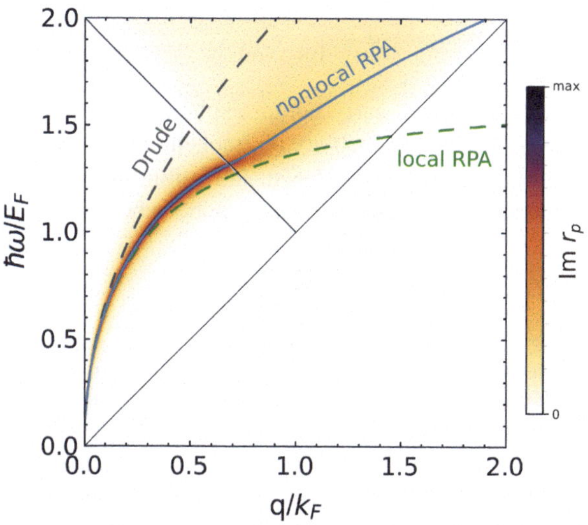

Fig. 4.2 Dispersion relation of graphene plasmons in extended graphene. The different curves represent the graphene plasmon's dispersion (4.2) (here plotted as $\hbar\omega$ vs $\operatorname{Re} q$) calculated using different models for the conductivity of graphene (in the zero-temperature limit), namely, the Drude-type conductivity [Eq. (3.22); gray dashed line], the local RPA result [green dashed line; Eq. (3.21)], and the full nonlocal RPA conductivity of graphene [blue solid line; Eq. (3.30)] in the relaxation-time approximation using Mermin's prescription [see Eq. (3.42)]. The colored intensity plot in the background expresses the loss function via the imaginary part of the TM reflection coefficient, $\operatorname{Im} r_p$, obtained within the full nonlocal RPA framework (here plotted as $\log_{100}[1 + \operatorname{Im} r_p(q, \omega)]$, for presentation purposes alone). Material parameters: $\epsilon_1 = \epsilon_2 = 1$, $E_F = 0.4$ eV, and $\hbar\gamma = 8$ meV

Nonretarded regime. Due to the extreme subwavelength confinement promoted by GPs [4, 5, 7, 9, 10, 28], it transpires that the nonretarded limit (defined by the absence of retardation effects) can be taken without any loss of accuracy in virtually all the relevant scenarios. In fact, treating plasmons in graphene within a nonretarded approach often constitutes an excellent approximation, as demonstrated in a number of works [4, 5, 9, 10, 29–33]. In this spirit, we take the nonretarded limit (i.e., $q \gg k_0$) of Eq. (4.2), thereby obtaining the nonretarded dispersion relation of graphene plasmons[5]:

$$q = \frac{2i\omega\epsilon_0\bar{\epsilon}}{\sigma(q,\omega)}, \tag{4.5}$$

where we have introduced the quantity $\bar{\epsilon} = (\epsilon_1 + \epsilon_2)/2$ without loss of generality. Notice that the previous equation is still an implicit condition for the GPs' nonretarded spectrum, due to the explicit dependence of the conductivity on both q and ω. In the local limit, and for low-loss, one can obtain a closed-form expression for the (real part) of the GPs' wavevector as a function of frequency, that is,

[5] In fact, this expression is not only valid for plasmons in graphene, but it also holds for *any* confined excitation supported by a 2D material sitting in an interface between two semi-infinite media.

$$q = \frac{2\omega\epsilon_0\bar{\epsilon}}{\mathrm{Im}\,\sigma(\omega)}. \tag{4.6}$$

At this point, it is instructive to carry out an explicit calculation by adopting the Drude-like expression for the conductivity of graphene (3.22), which for negligible losses takes the form $\sigma_D(\omega) \simeq \frac{ie^2}{\pi\hbar}\frac{E_F}{\hbar\omega}$. Hence, and within these approximations, the dispersion for graphene plasmons in extended graphene becomes[6] [4]

$$\boxed{\hbar\omega_{GP} = \sqrt{\frac{2\alpha}{\bar{\epsilon}}\, E_F\, \hbar c\, q}}\,, \tag{4.7}$$

where $\alpha = \frac{e^2}{4\pi\epsilon_0\hbar c} \simeq 1/137$ denotes the fine-structure constant. Chiefly, this result predicts that plasmons in graphene exhibit a $\omega_{GP} \propto \sqrt{q}$ dispersion in the local, intraband-dominated conductivity regime. Incidentally, such square-root dependence of the plasmon dispersion with wavevector is a manifestation of the system's two-dimensionality, as this behavior is also observed in conventional 2D electron gases with a parabolic band [4, 6, 13, 34]. However, the dependence of the plasmon energy with the carrier density is different between graphene and traditional 2DEGs. In particular, graphene plasmons exhibit a $\omega_{GP} \propto \sqrt{q}\, n_e^{1/4}$ dependence, whereas plasmons in conventional 2DEGs exhibit a $\omega_p^{2DEG} \propto \sqrt{q}\, n_e^{1/2}$ scaling instead[7] [4]. Evidently, this difference is ascribed to the unique linear energy-momentum dispersion that is characteristic of graphene's massless Dirac fermions.

In closing, let us provide a rough estimate of the maximum confinement of the electromagnetic fields that can be achieved using plasmons in an extended, continuous graphene sheet. With the help of Eq. (4.7), it is easy to see that the ratio between

---

[6]The generalization of this result for the case of a Drude-type conductivity with finite (but weak) damping is straightforwardly given by $\omega_{GP} = \frac{1}{\hbar}\sqrt{\frac{2\alpha}{\bar{\epsilon}}\, E_F\, \hbar c\, q} - i\frac{\gamma}{2}$.

[7]The nonretarded dispersion relation for 2D plasmons supported by 2D electron systems can be generically written as [cf. Eq. (4.6)]

$$\omega_{\text{2D-plasmon}} = \sqrt{\frac{\mathcal{D}}{2\pi\epsilon_0\bar{\epsilon}}\, q}\,,$$

where $\mathcal{D}$ is the so-called Drude weight. This quantity is defined via the Drude conductivity for an arbitrary 2D electron system through $\sigma(\omega) = \frac{i}{\pi}\frac{\mathcal{D}}{\omega + i\gamma}$. Since the Drude weight of graphene, $\mathcal{D}_G = e^2 E_F/\hbar^2 = e^2\hbar^{-1} v_F\sqrt{\pi n_e}$, is different than that of a parabolic 2DEG, $\mathcal{D}_{2DEG} = e^2\pi n_e/m^*$, this leads to 2D plasmons whose dispersions exhibit distinct dependences as a function of the electronic density, namely

$$\omega_{\text{2D-plasmon}} = \begin{cases} C_G\, q^{1/2} n_e^{1/4} & \text{, for plasmons in graphene} \\ C_{2DEG}\,(q\, n_e)^{1/2} & \text{, for plasmons in a conventional 2DEG} \end{cases},$$

where, for the sake of completeness, the constants in the previous expressions are $C_G = \sqrt{\frac{e^2}{2\sqrt{\pi}\epsilon_0\bar{\epsilon}}\, v_F}$ and $C_{2DEG} = \sqrt{\frac{e^2}{2\epsilon_0\bar{\epsilon} m^*}}$.

the GP's wavelength, $\lambda_{\mathrm{GP}} = 2\pi/q_{\mathrm{GP}}$, and that of photon in free-space, $\lambda_0 = 2\pi c/\omega$, with the same frequency ω, is given by

$$\frac{\lambda_{\mathrm{GP}}}{\lambda_0} = \frac{2\alpha}{\bar{\epsilon}} \frac{E_F}{\hbar\omega}. \tag{4.8}$$

For graphene on a typical dielectric substrate, and for frequencies near $\omega \approx E_F/\hbar$ (so that one is sufficiently away from interband effects), which typically falls in the mid-IR, one obtains $\lambda_{\mathrm{GP}} \approx \alpha\lambda_0 \approx 10^{-2}\lambda_0$, that is, the GP's wavelength can be up to 100 times smaller than that of light propagating in free-space.

4.1.2 Plasmons in Double-Layer Graphene

We now consider two doped graphene sheets separated by a distance d. The interlayer medium is assumed to be a dielectric medium with relative permittivity ϵ_2, and the whole structure is encapsulated by a substrate characterized by ϵ_3 and a superstrate described by ϵ_1, as represented schematically in Fig. 4.3. Each graphene layer is described by a conductivity as shown in the figure. The computation of the plasmon dispersion for a graphene double-layer can be performed using the same principles described in Sect. 2.2.1.2, and a detailed derivation can be found in Ref. [4].

In particular, the dispersion relation for coupled graphene plasmons in double-layer graphene stems from the solutions of [4]

$$\left[\frac{\epsilon_1}{\kappa_1} + \frac{\epsilon_2}{\kappa_2} + \frac{i\sigma_{12}(q,\omega)}{\omega\epsilon_0}\right]\left[\frac{\epsilon_3}{\kappa_3} + \frac{\epsilon_2}{\kappa_2} + \frac{i\sigma_{23}(q,\omega)}{\omega\epsilon_0}\right] e^{\kappa_2 d} = \left[\frac{\epsilon_1}{\kappa_1} - \frac{\epsilon_2}{\kappa_2} + \frac{i\sigma_{12}(q,\omega)}{\omega\epsilon_0}\right]\left[\frac{\epsilon_3}{\kappa_3} - \frac{\epsilon_2}{\kappa_2} + \frac{i\sigma_{23}(q,\omega)}{\omega\epsilon_0}\right] e^{-\kappa_2 d}, \tag{4.9}$$

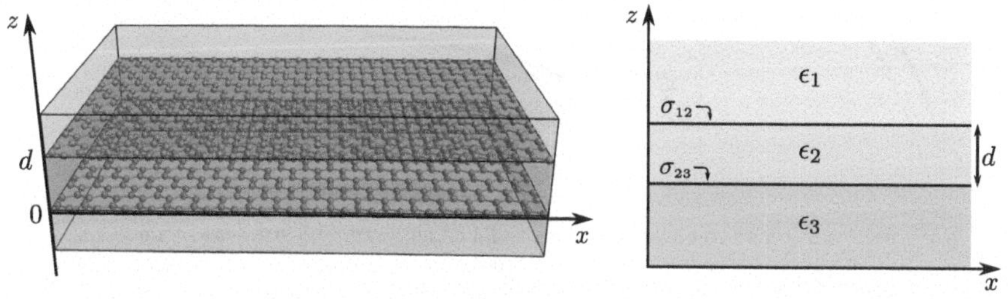

Fig. 4.3 Representation of a graphene double-layer structure where the two graphene sheets are distanced by an inner slab of thickness d and dielectric constant ϵ_2. The upper and lower media, with relative permittivities ϵ_1 and ϵ_3, respectively, are assumed to be semi-infinite for simplicity. The two individual graphene layers are placed at the interfaces defined by the $z = d$ and $z = 0$ planes, and are characterized by the surface conductivities σ_{12} and σ_{23}, correspondingly

where $\kappa_j \equiv \kappa_j(q,\omega) = \sqrt{q^2 - \epsilon_j k_0^2}$ with $j \in \{1, 2, 3\}$. From this expression, it is clear that for small interlayer separations (i.e., such that $\kappa_2 d \ll 1$), the GPs in each graphene layer can couple via the Coulomb interaction, which, as we shall see below, leads to the formation of two hybrid GP modes of the whole system [4, 35–37]. Conversely, for very large separations $d \to \infty$ (or large momenta such that $\kappa_2 d \gg 1$) the two interfaces decouple and one recovers the two dispersion relations akin to GPs in each one of the individual sheets.

A case of special interest—and also one that is particularly elucidative—is that of a double-layer graphene structure in which the two graphene sheets possess the same doping, and hence the same conductivity $\sigma \equiv \sigma_{12} = \sigma_{23}$. Moreover, for a symmetric dielectric environment $\epsilon_1 = \epsilon_3$ (and thus $\kappa_1 = \kappa_3$), Eq. (4.9) can be substantially simplified. Specifically, under this assumption Eq. (4.9) yields two coupled GP modes of opposite parity, namely [4]

$$\frac{\epsilon_2}{\kappa_2} \tanh\left(\frac{\kappa_2 d}{2}\right) + \frac{\epsilon_1}{\kappa_1} + \frac{i\sigma(q,\omega)}{\omega\epsilon_0} = 0\,, \tag{4.10a}$$

for the optical plasmon mode, where the charge density in both graphene sheets oscillate in-phase, and

$$\frac{\epsilon_2}{\kappa_2} \coth\left(\frac{\kappa_2 d}{2}\right) + \frac{\epsilon_1}{\kappa_1} + \frac{i\sigma(q,\omega)}{\omega\epsilon_0} = 0\,, \tag{4.10b}$$

for the acoustic plasmon mode, in which the charge density in both graphene sheets oscillate with opposite phase.

The dispersion diagram of the hybridized GP modes of a graphene double-layer is displayed in Fig. 4.4. The figure shows the plasmon spectrum of the system calculated under different approximations (in the same vein of Fig. 4.2 corresponding to mono-

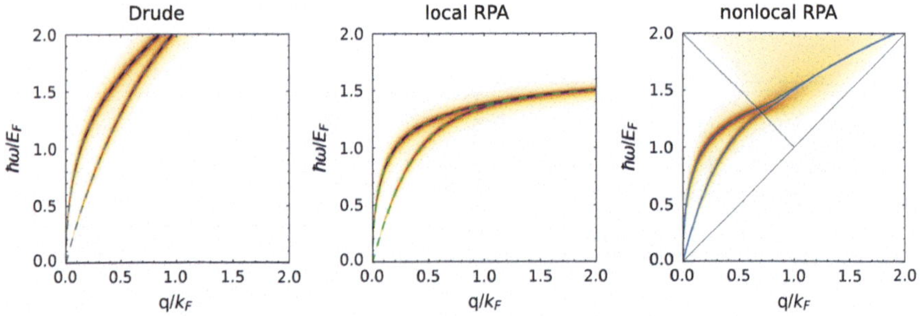

Fig. 4.4 Spectrum of graphene plasmons in double-layer graphene obtained using different models for the optical response of graphene. In all panels, the graphene sheets are assumed to be separated by a distance of $d = 5$ nm. The colormap in the background of each plot expresses the loss function via the imaginary part of the TM reflection coefficient, Im r_p^{DLG} (corresponding to each conductivity model). Material parameters: $\epsilon_1 = \epsilon_2 = \epsilon_3 = 1$, $E_F^{12} = E_F^{23} = 0.4$ eV, and $\hbar\gamma = 8$ meV

layer graphene). Two distinct modes can be clearly identified: an upper plasmonic branch corresponding to the optical GP, and a lower plasmonic branch corresponding to the acoustic GP. Furthermore, as in the single-layer case considered in Sect. 4.1.1, it can also be seen how the plasmon dispersion is successively affected—from the simplest to the most sophisticated model—by the onset of vertical interband transitions, and then by nonlocal effects (that manifest themselves as a change in the plasmon dispersion curve and in plasmon damping due to Landau damping).

We now turn our attention to the dependence of the coupled GP's dispersion on the interlayer separation d. For the sake of clarity, we assume that both graphene layers have the Fermi energy ($E_F^{12} = E_F^{23}$) and that they are embedded in a homogeneous dielectric medium (here assumed to be vacuum). As Fig. 4.5 plainly shows, the coupling between GPs living in each individual graphene sheet splits the plasmon dispersion of single-layer graphene (SLG) into an upper and a lower branch, associated with the optical and acoustic plasmon, respectively. Predominantly, the aforementioned splitting becomes successively more pronounced upon decreasing interlayer separation d, owing to the correspondingly stronger interlayer Coulomb coupling. In the $d \to 0$ limit, the optical plasmon branch becomes indistinguishable from that of a graphene monolayer with twice the Fermi energy, while the linear acoustic branch becomes indistinguishable from the $\omega = v_F q$ line (meaning that the GP's velocity slows down to the Fermi velocity of electrons in graphene; we shall revisit this point in Sect. 7.2).

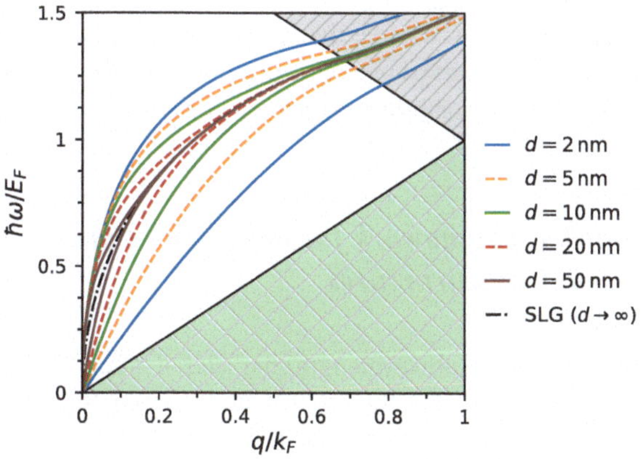

Fig. 4.5 Dispersion relation of plasmons in double-layer graphene structures with different interlayer separations d. The optical response of graphene is modeled within the nonlocal RPA with the Mermin-corrected relaxation-time approximation. The hatched regions denote the single-particle continuum where Landau damping occurs. Material parameters: $\epsilon_1 = \epsilon_2 = \epsilon_3 = 1$, $E_F^{12} = E_F^{23} = 0.3\,\text{eV}$, and $\hbar\gamma = 5\,\text{meV}$

4.2 Plasmons in Nanostructured Graphene

The studies carried out in the previous section on the basic properties of propagating graphene plasmons in extended graphene systems have enabled us to establish the fundamental theory and the elementary concepts behind plasmonic excitations in graphene. Notwithstanding, the vibrant field of graphene plasmonics is considerably wealthier than that, going well beyond the restricted subgroup of extended graphene structures. As such, the prime goal of the present section is to expand our previous considerations by discussing localized plasmons in graphene nanostructures [4, 32]. The latter are typically fabricated by patterning an otherwise continuous graphene sheet, and arguably constitute the most abundant subset of graphene-based plasmonic structures considered in the literature. A primary reason for this has to do with the fact that plasmons in nanostructured graphene can couple to light directly without the necessity to employ special coupling techniques (a feature that they share with LSPs in finite-sized metal particles). Examples of graphene nanostructures—considered either in isolation or in periodic arrangements—that have been used in plasmonics are graphene ribbons [17, 38–44], disks [3, 45–50], rings [46, 48], triangles [51, 52], or anti-dots [50, 53–55], just to mention a few.

In the present section, we provide a cursory overview of plasmonic excitations in nanostructured graphene. First, we introduce a general nonretarded formalism for modeling plasmons in two-dimensional nanostructures of arbitrary geometries [4, 32]. We then focus on two archetypal structures for graphene plasmonics—graphene nanoribbons and nanodisks—and compute and characterize the plasmonic resonances supported by them. Finally, we consider the interaction and ensuing hybridization of 2D plasmons in 2D nanoslits carved in atomically-thin crystals [31], and interpret the corresponding results and waveguiding properties of such structures.

4.2.1 *Nonretarded Framework for Graphene Plasmons in Generic Nanostructures*

As mentioned above, the deep subwavelength character of graphene plasmons renders retardation effects negligible, and therefore a nonretarded treatment provides an excellent description of plasmons in graphene nanostructures in nearly all relevant scenarios. Consequently, instead of solving the full vectorial Maxwell's equations, one can afford to solve the corresponding electrostatic problem, whereby plasmon excitations are governed by the Poisson equation for the (scalar) electric potential.

In what follows, we consider a generic 2D nanostructure lying in the $z = 0$ plane and encapsulated by a superstrate and a substrate with relative permittivities ϵ_1 and ϵ_2, respectively. Hence, one may write Poisson's equation as

$$\nabla^2 \Phi(\mathbf{r}) = -\frac{\rho(\mathbf{r})}{\epsilon_0 \bar{\epsilon}}, \tag{4.11}$$

where, as previously, $\bar{\epsilon} = (\epsilon_1 + \epsilon_2)/2$ is simply the average dielectric constant[8] of the media on either side of the 2D nanostructure. Both the electrostatic potential and the charge density are herein assumed to have a harmonic time dependence of the form $\Phi(\mathbf{r}, t) = \Phi(\mathbf{r})e^{-i\omega t}$ and $\rho(\mathbf{r}, t) = \rho(\mathbf{r})e^{-i\omega t}$. In turn, the charge density is restricted to the xy-plane and thus one may write $\rho(\mathbf{r}) = \rho_{2D}(\mathbf{r}_\parallel)\delta(z)$, where ρ_{2D} is a *surface* charge density and the in-plane vector $\mathbf{r}_\parallel$ is given by $\mathbf{r}_\parallel = x\,\hat{\mathbf{x}} + y\,\hat{\mathbf{y}}$. Furthermore, the induced surface charge density can be written in terms of the in-plane electrostatic potential $\phi(\mathbf{r}_\parallel)$ [with $\phi(\mathbf{r}_\parallel) \equiv \Phi(\mathbf{r}_\parallel, z = 0)$] by combining Ohm's law together with the continuity equation, yielding the following relation[9]

$$\rho_{2D}(\mathbf{r}_\parallel) = i\omega^{-1}\sigma(\omega)\nabla_\parallel \cdot \left[f(\mathbf{r}_\parallel)\nabla_\parallel\phi(\mathbf{r}_\parallel)\right], \tag{4.12}$$

where $\nabla_\parallel \equiv \hat{\mathbf{x}}\frac{\partial}{\partial x} + \hat{\mathbf{y}}\frac{\partial}{\partial y}$ is the in-plane 2D nabla operator and $f(\mathbf{r}_\parallel)$ is an envelope function that takes into account the geometry of the 2D nanostructure. Specifically, it can be defined through $\sigma(\mathbf{r}_\parallel, \omega) \equiv \sigma(\omega)f(\mathbf{r}_\parallel)$, where $f(\mathbf{r}_\parallel) = 1$ for $\mathbf{r}_\parallel$ within 2D nanostructure and $f(\mathbf{r}_\parallel) = 0$ otherwise.[10] At this point, the nonretarded optical response can be determined in a self-consistent fashion by solving the coupled equations (4.11) and (4.12).

We can make this more transparent by considering Poisson's equation (4.11) in integral form instead, namely

$$\begin{aligned}\Phi(\mathbf{r}) &= \frac{1}{\epsilon_0\bar{\epsilon}}\int d\mathbf{r}' G(\mathbf{r}, \mathbf{r}')\rho(\mathbf{r}') \\ &= \frac{1}{\epsilon_0\bar{\epsilon}}\int d\mathbf{r}'_\parallel\, G(\mathbf{r}_\parallel, \mathbf{r}'_\parallel; z, z' = 0)\rho_{2D}(\mathbf{r}'_\parallel) \\ &= \frac{i\sigma(\omega)}{\omega\epsilon_0\bar{\epsilon}}\int d\mathbf{r}'_\parallel\, G(\mathbf{r}_\parallel, \mathbf{r}'_\parallel; z, z' = 0)\left\{\nabla'_\parallel \cdot \left[f(\mathbf{r}'_\parallel)\nabla'_\parallel\phi(\mathbf{r}'_\parallel)\right]\right\},\end{aligned} \tag{4.13}$$

---

[8] We emphasize that this result is not a naïve "effective medium" approximation, but it is in fact an exact description of dielectric screening for a planar interface within the nonretarded regime.

[9] This can be shown explicitly by noting that the 2D version of the continuity equation $\nabla_\parallel \cdot \mathbf{J}_{2D} + \frac{\partial}{\partial t}\rho_{2D} = 0$ can be cast as $\nabla_\parallel \cdot \mathbf{J}_{2D} = i\omega\rho_{2D}$, and thus we have

$$\begin{aligned}\rho_{2D}(\mathbf{r}_\parallel) &= (i\omega)^{-1}\nabla_\parallel \cdot \mathbf{J}_{2D} \\ &\stackrel{\text{i}}{=} (i\omega)^{-1}\nabla_\parallel \cdot \left[\sigma(\mathbf{r}_\parallel, \omega)\mathbf{E}_\parallel(\mathbf{r}_\parallel)\right] \\ &= (i\omega)^{-1}\sigma(\omega)\nabla_\parallel \cdot \left[f(\mathbf{r}_\parallel)\mathbf{E}_\parallel(\mathbf{r}_\parallel)\right] \\ &\stackrel{\text{ii}}{=} i\omega^{-1}\sigma(\omega)\nabla_\parallel \cdot \left[f(\mathbf{r}_\parallel)\nabla_\parallel\phi(\mathbf{r}_\parallel)\right],\end{aligned}$$

where, in particular, Ohm's law $\mathbf{J}_{2D} = \sigma\mathbf{E}_\parallel$ and the relation $\mathbf{E}_\parallel = -\nabla\Phi(\mathbf{r}_\parallel, z = 0) \equiv -\nabla\phi(\mathbf{r}_\parallel)$ have been used, respectively, in steps (i) and (ii).

[10] In this way, it is implicitly assumed that the 2D structure is homogeneously doped. Nevertheless, in general, $f(\mathbf{r}_\parallel)$ does not have to be defined in such a step-like fashion; in fact, that assumption can be relaxed so that the formalism considered here can be straightforwardly extended to inhomogeneously doped 2D nanostructures as well (see, for instance, Refs. [56, 57] for examples).

where $G(\mathbf{r}, \mathbf{r}')$ is the Green's function that satisfies the differential equation associated with Eq. (4.11), that is, $\nabla^2 G(\mathbf{r}, \mathbf{r}') = -\delta(\mathbf{r} - \mathbf{r}')$. Crucially, notice that the previous equation tells us that the potential in the *entire space* can be deduced provided that the *in-plane* potential within the 2D nanostructure is known (this is the only contribution since outside the material $f(\mathbf{r}_\parallel) = 0$). To that end, we now evaluate the previous equality at $z = 0$, so that one obtains the following self-consistent integro-differential equation for the potential within the 2D nanostructure:

$$\boxed{\phi(\mathbf{r}_\parallel) = \frac{i\sigma(\omega)}{\omega\epsilon_0\bar{\epsilon}} \int d\mathbf{r}'_\parallel\, g(\mathbf{r}_\parallel, \mathbf{r}'_\parallel) \left\{\nabla'_\parallel \cdot \left[f(\mathbf{r}'_\parallel)\nabla'_\parallel \phi(\mathbf{r}'_\parallel)\right]\right\}}\,, \tag{4.14}$$

with $g(\mathbf{r}_\parallel, \mathbf{r}'_\parallel) \equiv G(\mathbf{r}_\parallel, \mathbf{r}'_\parallel; z = 0, z' = 0)$. Equation (4.14) becomes particularly elucidative (and useful) by writing it in terms of dimensionless quantities: introducing $\tilde{\mathbf{r}}_\parallel = \mathbf{r}_\parallel / L$ and $\tilde{\nabla}_\parallel = L\nabla_\parallel$, where L is some appropriate characteristic length of the structure under consideration, the above equation can be recast as

$$\Lambda\,\phi(\tilde{\mathbf{r}}_\parallel) = -2 \int d\tilde{\mathbf{r}}'_\parallel\, g(\tilde{\mathbf{r}}_\parallel, \tilde{\mathbf{r}}'_\parallel) \left\{\tilde{\nabla}'_\parallel \cdot \left[f(\tilde{\mathbf{r}}'_\parallel)\tilde{\nabla}'_\parallel \phi(\tilde{\mathbf{r}}'_\parallel)\right]\right\}\,, \tag{4.15a}$$

$$\text{where} \quad \Lambda = \frac{2i\omega\epsilon_0\bar{\epsilon}L}{\sigma(\omega)}\,, \tag{4.15b}$$

which is now clearly scale-invariant. This is because the integral has been stripped of any reference to the structure's size upon introducing the above-noted dimensionless spatial variables. Moreover, there is no frequency or any material dependencies left in the right-hand side of Eq. (4.15a), and thus that term has a *purely geometrical meaning*. Consequently, this formalism can in fact be applied to *any 2D material* that is capable of supporting plasmonic resonances (or other collective modes, e.g., phonon polaritons or exciton polaritons).

Naturally, the usefulness of the general scale-independent integro-differential equation epitomized by Eq. (4.15a) relies on one's ability to solve it. In general, this cannot be done analytically and hence one has to rely on numerical methods. In spite of this, it is still possible to solve Eq. (4.15a) using semi-analytical techniques [4, 32]. Broadly speaking, the idea behind such semi-analytical methods is typically to expand the electric potential $\phi(\tilde{\mathbf{r}}_\parallel)$ using a suitable set of basis functions. Throughout this section, we choose to base our semi-analytical framework on an expansion using a basis containing orthogonal polynomials[11] [4, 29, 31, 56, 58–62]. The specific type of orthogonal polynomials [63] is intimately related to the particular geometry of the nanostructure under consideration. In any case, by performing such an expansion of the potential figuring in Eq. (4.15a) followed by the use of the

[11] See Ref. [4] for an all-encompassing introduction on the use of this method for calculating plasmonic excitations in several graphene nanostructures.

appropriate orthogonality relation [63], the integro-differential equation (4.15a) can be transformed into a standard matrix eigenvalue problem

$$\sum_{m=0}^{\infty} U_{nm}^{(\nu)} c_m^{(\nu)} = \Lambda_\nu c_n^{(\nu)} \qquad \Rightarrow \qquad \mathbf{U}_\nu \mathbf{c}_\nu = \Lambda_\nu \mathbf{c}_\nu \,, \tag{4.16}$$

where the index ν categorizes the plasmon resonance. Lastly, the diagonalization of the square-matrix $\mathbf{U}^{(\nu)}$ yields the system's plasmon eigenfrequencies via

$$\Lambda_\nu(\omega_\nu) = \lambda_\nu \qquad \Leftrightarrow \qquad \frac{2i\omega_\nu \epsilon_0 \bar{\epsilon} L}{\sigma(\omega_\nu)} = \lambda_\nu \,, \tag{4.17}$$

where λ_ν denotes the eigenvalues of $\mathbf{U}^{(\nu)}$. On the other end, the corresponding eigenvectors allow the construction of the in-plane potential in the 2D nanostructure and thus the mode profile associated with the plasmon eigenfrequency ω_ν. From here, one can also compute the induced charge density within the 2D material using Eq. (4.12), or to determine the potential in the entire three-dimensional space through Eq. (4.13) [the corresponding electric field then simply follows from $\mathbf{E}(\mathbf{r}) = -\nabla\Phi(\mathbf{r})$]. Notice that either one of those calculations does not require a significant amount of labor, since they only contain derivatives or integrals of functions involving polynomials and simple weighting functions (typically either an exponential, a Gaussian, or a polynomial).

Before applying this formalism to a particular 2D plasmonic structure, it is enlightening to substitute graphene's Drude conductivity in Eq. (4.17) and thereby establish an explicit connection between the eigenfrequencies ω_ν and the eigenvalues λ_ν. Hence, assuming low-loss (i.e., $\omega_\nu \ll \gamma$), one finds the relation

$$\omega_\nu \simeq \frac{1}{\hbar}\sqrt{\frac{2\alpha}{\bar{\epsilon}} E_F \hbar c \frac{\lambda_\nu}{L}} - i\frac{\gamma}{2} \,, \tag{4.18}$$

which bears close resemblance to the dispersion relation of GPs in extended graphene [cf. Eq. (4.7)], where the role previously played by the wavevector is now transferred to $\lambda_\nu L^{-1}$. Notably, this result formally validates the intuitively expected behavior $\omega \propto \sqrt{1/L}$, that here is now appropriately weighted by the eigenvalue λ_ν that accounts for the particular geometry of the graphene nanostructure.

Plasmons in two-dimensional nanoribbons. In the context of graphene plasmonics, studies using the ribbon geometry proliferate in the literature [4, 17, 38–44], thereby making it a prototypical system to study plasmons in two dimensions. Below, we provide an economical overview of the application of the nonretarded framework outlined above to the specific case of the ribbon geometry.

In what follows, we consider a 2D nanoribbon of finite width $W = 2a$ (along the x-axis) and that is translational invariant along the y-direction. Hence, the electric potential in such a system can be written as $\Phi(\mathbf{r}) = \Phi(x, z)e^{ik_y y}$ (and similarly for the charge density); thus the wavevector parallel to the ribbon's edges, k_y, can be used to parameterize the plasmon modes of the ribbon. Concretely, in the language of Eq. (4.15), the in-plane potential $\phi(\tilde{\mathbf{r}}_\parallel)$ and the function $g(\tilde{\mathbf{r}}_\parallel, \tilde{\mathbf{r}}'_\parallel)$ are found to be given by $\phi(\tilde{\mathbf{r}}_\parallel) = \phi(\tilde{x})\, e^{i\beta\tilde{y}}$ and $g(\tilde{x}, \tilde{x}') = (2\pi)^{-1} K_0(\beta|\tilde{x} - \tilde{x}'|)$ [4], where K_0 is the zero-th order modified Bessel function of the second kind, and where we have introduced the dimensionless variables $(\tilde{x}, \tilde{y}) = (x/a, y/a)$ and normalized wavevector $\beta = k_y a$. Likewise, since we have assigned $L \equiv a$ (i.e., half of the ribbon's width), the dimensionless parameter (4.15b) is then $\Lambda = 2i\omega\epsilon_0\bar{\epsilon}a/\sigma(\omega)$. Having specified all the ingredients entering in the integro-differential equation (4.15), the next step is to expand the potential $\phi(\tilde{x})$ using a suitable basis in terms of orthogonal polynomials. For the ribbon geometry, such an expansion takes the form $\phi(\tilde{x}) = \sum_{n=0}^{\infty} c_n P_n(\tilde{x})$, where $P_n(\tilde{x})$ denotes the Legendre polynomials. Finally, substituting the aforementioned expansion into Eq. (4.15a) and exploiting the orthogonality relation of the Legendre polynomials[12] yields an eigenproblem in the familiar form of Eq. (4.16).

The plasmon spectrum of 2D nanoribbons calculated using the method described in the previous paragraph is shown in Fig. 4.6. As depicted in Fig. 4.6a, the eigenvalues $\lambda_\beta = \{\lambda_\beta^{[0]}, \lambda_\beta^{[1]}, \lambda_\beta^{[2]}, ...\}$ are parameterized by β and form a manifold of plasmon bands indexed by n, where $n = 0$ corresponds to the monopolar eigenmode, $n = 1$ to the dipolar eigenmode, and so on. Each band therefore describes a ribbon plasmon propagating along the y-direction with (normalized) wavevector β and exhibiting quantization (for $n \geq 1$) across the ribbon's width.[13]

In addition to our semi-analytical results (shown as colored solid lines), we have also superimposed analytical fits (represented by the black dashed lines) to the corresponding data, and whose fitting parameters are listed in Table 4.1. Similarly, Fig. 4.6b shows the same results, but where a Drude-like conductivity has been assumed, thereby allowing us to write the plasmon dispersion diagram as $\omega_\beta^{[n]} = \omega_a \sqrt{\lambda_\beta^{[n]}}$ [cf. Eq. (4.18)], where $\omega_a = \hbar^{-1}\sqrt{(2\alpha/\bar{\epsilon})E_F\, \hbar c\, a^{-1}}$ for the case of a graphene nanoribbon.

Figure 4.6c–f portray the calculated charge density profiles that are associated with the four lowest-frequency eigenmodes, and empirically endorse the standing-wave-like picture mentioned previously. Notwithstanding, in the limit of large $\beta = k_y a$, the plasmon bands indicated in Fig. 4.6a–b can be divided into two groups based on their asymptotic behavior: one containing the two lowest-frequency eigenmodes, characterized by $\lambda_\beta^{[n\leq 1]}$, and another containing the modes defined by $\lambda_\beta^{[n\geq 2]}$. In particular, the dispersion curves associated with the monopolar and dipolar ribbon eigenmodes become degenerate and asymptotically approach the dispersion of quasi-1D edge plasmons supported by a half-sheet, specifically $\lim_{\beta\to\infty} \lambda_\beta^{[n\leq 1]} = \lambda_{\text{hs}}\,\beta$,

[12]Explicitly: $\int_{-1}^{1} d\tilde{x}\, P_n(\tilde{x}) P_m(\tilde{x}) = 2\delta_{nm}/(2n+1)$.

[13]This standing-wave-like, quantized behavior is thus suggestive of the condition $k_x^n W = n\pi$. It turns out, however, that the computed spectrum is only reasonably well described if one augments this elementary picture by allowing a n-dependent "reflection phase-shift", i.e., $k_x^n W + \varphi_n = n\pi$. This kind of analysis has been explored in some works [64, 65], but caution should be taken when attempting to attribute a direct physical origin for such fitted phase-shifts.

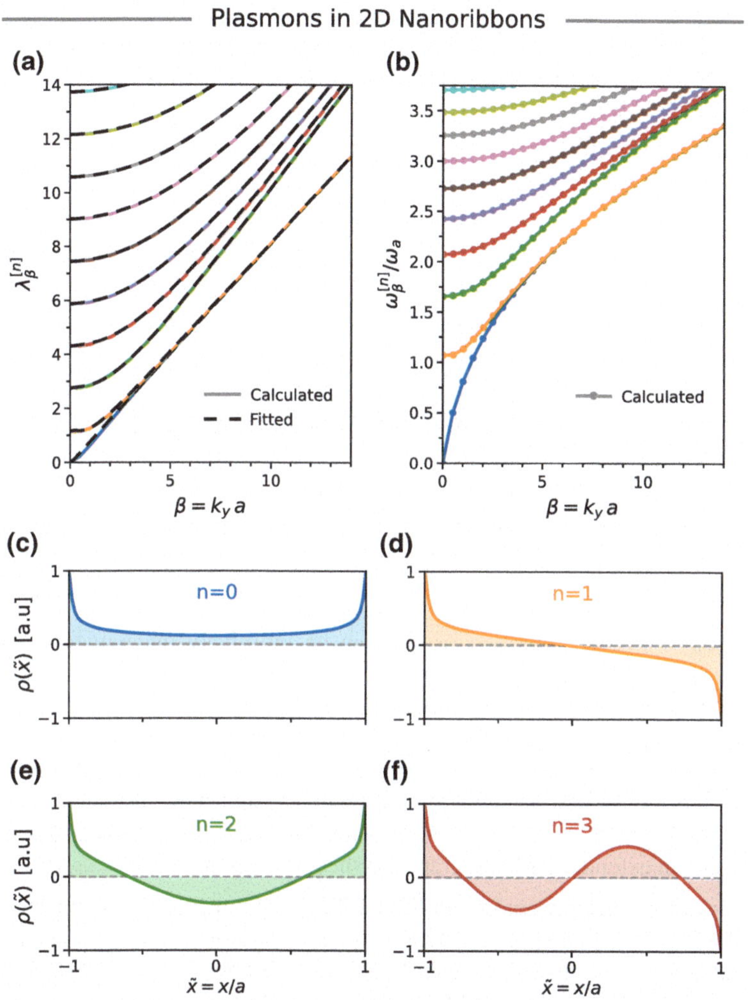

Fig. 4.6 Plasmons in an individual 2D nanoribbon of width $W = 2a$. **a** Eigenvalues associated with plasmons of a 2D nanoribbon calculated using the semi-analytical method described in the text (colored solid lines) and corresponding fits (black dashed lines). **b** Dispersion relation of the plasmonic eigenmodes supported by 2D nanoribbons, obtained using the calculated eigenvalues and assuming a Drude-type conductivity [see Eqs. (4.17) and (4.18)]. Here, we have defined $\omega_a = \sqrt{\frac{\mathcal{D}}{2\pi\epsilon_0\bar{\epsilon}}a^{-1}}$, where $\mathcal{D}$ is the Drude weight of the 2D electron system ($\mathcal{D}_{\mathrm{G}} = e^2 E_F/\hbar^2$ for graphene, and $\mathcal{D}_{\mathrm{2DEG}} = e^2\pi n_e/m^*$ for a conventional 2DEG). **c–f** Spatial profiles of the charge density across the ribbon associated with the four lowest-frequency plasmon eigenmodes, for a fixed $\beta = k_y a = 1.5$. The quantities in each panel are normalized with respect to their absolute maxima. All results **(a–f)** were obtained by truncating the eigensystem with $N = 20$ [cf. Eq. (4.16)], which provides converged results for the depicted modes

Table 4.1 Fits to the eigenvalues describing the spectrum of plasmons in a 2D nanoribbon. Analytical fitting function [66] (top row) written as a linear term plus a Padé approximant of the form [1/2], followed by the corresponding fitting constants obtained by fitting the analytical formula to the calculated data presented in Fig. 4.6

$$\lambda_\beta^{[n]} = c_1^n \beta + \frac{c_2^n + c_3^n \beta}{1 + c_4^n \beta + c_5^n \beta^2}$$

n	c_1^n	c_2^n	c_3^n	c_4^n	c_5^n
0	0.8216	0	−12.397	233.56	−9.6187
1	0.8106	1.1541	0.2760	0.5905	1.5169
2	1	2.7402	0.1302	0.3411	0.1928
3	1	4.2980	0.2383	0.2633	0.0667
4	1	5.8731	0.2856	0.2069	0.0336
5	1	7.4389	0.3055	0.1683	0.0200
6	1	9.0120	0.3138	0.1413	0.0131
7	1	10.579	0.3151	0.1211	0.0092
8	1	12.152	0.3119	0.1057	0.0068
9	1	13.732	0.3044	0.0933	0.0052

where[14] $\lambda_{\text{hs}} \approx 0.8216$. This observation can be intuitively understood by noting that these two modes are highly—and solely—confined to the ribbon's edges. On the other hand, the dispersion curves corresponding to eigenmodes affiliated with the second group (containing the set of higher-order modes with $n \geq 2$) pile-up at the line defined by $\lambda_\beta^{[n\geq 2]} = \beta$, when $\beta \to \infty$. As such, they acquire a successively more bulk-like behavior with growing β, and their dispersion becomes $\omega_\beta^{[n\geq 2]} \to \omega_{k_y}^{[n\geq 2]} = \hbar^{-1}\sqrt{(2\alpha/\bar{\epsilon})E_F \hbar c\, k_y}$, and thus indistinguishable from that of plasmons in an extended, infinite graphene sheet [cf. Eq.(4.7)].

Plasmons in two-dimensional nanodisks. Like the ribbon, the disk geometry is also one of the most prominent nanostructures for investigating plasmons in nanostructured graphene, a fact that is reflected by the vast number of works devoted to this configuration [3, 4, 45–50]. In the following, we outline the key points for applying the nonretarded formalism introduced in this section to describe plasmon resonances supported by a conductive 2D nanodisk (e.g., a doped graphene nanodisk).

Let us consider a single 2D disk of radius R made from an atomically-thin material. In the present case, by virtue of axial symmetry, we can express the potential as $\Phi(\mathbf{r}) = \Phi(x, z)e^{il\theta}$, where an analogous relation holds for the corresponding induced charge density. Therefore, we anticipate that the plasmon eigenmodes can be classified in terms of their angular momentum l, each containing a further subset of radial eigenindexes n that allocate the mode's type of radial confinement. Moreover, since the natural length scale of a disk is its radius, we adopt $L \equiv R$, so that we have

[14]The value $\lambda_{\text{hs}} \approx 0.82155866$ stems from the application of the Wiener–Hopf technique for calculating the dispersion of edge plasmons in a half-sheet. Explicitly, it is given by the root of the equation $\int_0^{\pi/2} \ln\left([\lambda_{\text{hs}} \sin(x)]^{-1} - 1\right) dx = 0$ [67].

$\Lambda = 2i\omega\epsilon_0\bar{\epsilon}R/\sigma(\omega)$ and the in-plane potential becomes $\phi(\tilde{\mathbf{r}}_\parallel) = \phi(\tilde{r})e^{il\theta}$, where $\tilde{r} = r/R$. Additionally, the dimensionless function $g(\tilde{\mathbf{r}}_\parallel, \tilde{\mathbf{r}}'_\parallel)$ admits a representation of the form $g(\tilde{r}, \tilde{r}') = \frac{1}{2}\int_0^\infty dp J_l(p\tilde{r})J_l(p\tilde{r}')$, where J_l is the Bessel function of the first kind of order l. Next, we expand the radial potential inside the disk—associated with an angular momentum l—using a basis in terms of Jacobi polynomials, that is, specifically, $\phi(\tilde{r}) = \sum_{n=0}^\infty c_n \tilde{r}^l P_n^{(l,0)}(1 - 2\tilde{r}^2)$, where $P_n^{(l,0)}$ denotes the Jacobi polynomials. Lastly, using the orthogonality relation of Jacobi polynomials,[15] the integro-differential equation for the potential (4.15) can be converted into a standard eigenvalue problem as in Eq. (4.16). Curiously—and perhaps rather surprisingly—in the present case the matrix associated with this eigenproblem possesses *analytical* matrix elements.[16]

The eigenvalues $\lambda_l^{[n]}$ characterizing the spectrum of plasmon resonances in 2D nanodisks are depicted in Fig. 4.7a–b as a function of the angular momentum l (where the $l = 0$ modes have been omitted, and where the color is used to indicate eigenmodes with the same radial eigenindex n). Like in the ribbon configuration, the spectrum shows a manifold of eigenmodes whose energies increase with increasing eigenindex n; however, and contrasting the ribbon geometry, these do not form "bands", courtesy of the azimuthal quantization l that has no equivalent in the ribbon structure. The nature of the disk's eigenmodes with distinct combinations of $\{l, n\}$ pairs becomes readily apparent upon inspecting the corresponding induced charge density profiles, as shown in Fig. 4.7c. The associated eigenvalues $\lambda_l^{[n]}$ are referenced in Table 4.2 for the reader's convenience. Unsurprisingly, the charge density profiles are naturally reminiscent of the displacement ascribed to the normal modes of a vibrating circular membrane, where the "quantum numbers" l and n classify the quantization along the azimuthal and radial directions, respectively.

The disk's eigenmodes of dipole character, i.e., with $l = 1$, are of particular importance in what regards their excitation and interaction with plane-waves, since they can couple strongly to these. Evidently, in the nonretarded limit the fundamental dipole mode $\{1, 1\}$ dominates the optical response of the system under plane-wave illumination, and whose absorption cross section has been shown to be quite substantial [32, 47].

4.2.2 Hybridized Plasmons in Two-Dimensional Nanoslits

The nonretarded framework set forth in Sect. 4.2.1 constitutes a versatile theoretical tool to describe plasmons in graphene nanostructures, or, more generally, collective

[15]Explicitly: $\int_0^1 d\tilde{r}\, \tilde{r}^{2l+1} P_n^{(l,0)}(1 - 2\tilde{r}^2) P_m^{(l,0)}(1 - 2\tilde{r}^2) = \frac{1}{2}\frac{\delta_{nm}}{2n+l+1}$.

[16]This fact was noted by Fetter in Ref. [59], where he used the same method for expanding the charge density instead of expanding the potential. Naturally, such a formulation is tantamount—and ultimately equivalent—to the one described here. In any case, and perhaps more important for practical matters, we stress that the approach we follow here also admits a matrix with analytical matrix elements.

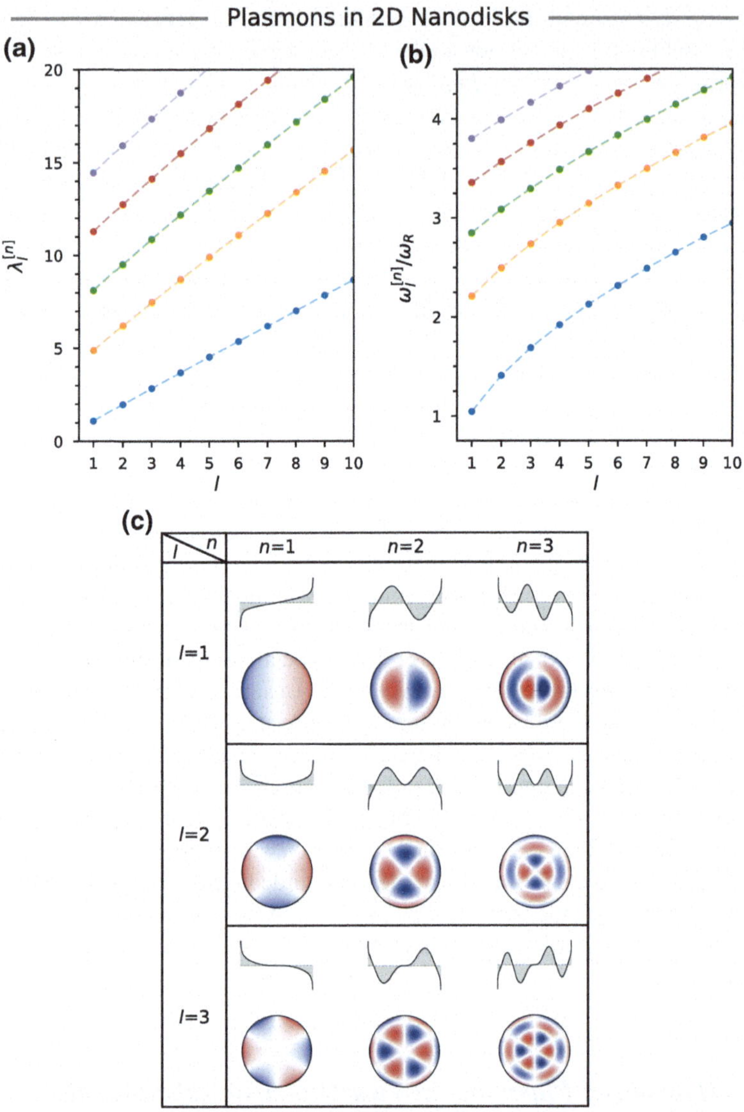

Fig. 4.7 Plasmons in an individual 2D nanordisk of radius R. **a** Eigenvalues for $l \geq 1$ associated with plasmon resonances of a 2D nanodisk obtained using the nonretarded framework described in the text (colored circles; the dashed lines connecting the circles are only for visual guidance). **b** Corresponding dispersion curves in the case where the 2D material's optical response is well described by a conductivity of the Drude kind [see Eqs. (4.17) and (4.18)]. Here, $\omega_R = \sqrt{\frac{\mathcal{D}}{2\pi\epsilon_0\bar{\epsilon}} R^{-1}}$, where $\mathcal{D}$ is the Drude weight of the 2D electron system. **c** Spatial profiles of the normalized induced charge density for eigenmodes with different $\{l, n\}$ pairs. All results **(a–c)** were obtained by truncating the expansion to $N = 100$

Table 4.2 Computed eigenvalues $\lambda_l^{[n]}$ corresponding to different $\{l, n\}$ eigenmodes that classify the plasmon resonances in individual 2D nanodisks (cf. Fig. 4.7)

	$n = 1$	$n = 2$	$n = 3$
$\lambda_{l=1}^{[n]}$	1.0978	4.9141	8.1338
$\lambda_{l=2}^{[n]}$	1.9943	6.2456	9.5456
$\lambda_{l=3}^{[n]}$	2.8557	7.5125	10.899

modes supported by *any* nanostructured atomically-thin material. As demonstrated in the examples given above, this semi-analytical technique provides an accurate and reliable description of the spectral features of such self-sustaining collective modes (e.g., plasmons), along with the corresponding spatial distributions of the potential, charge density, and electric field. However, it should be noted that the two prototypical nanosystems considered in the aforementioned section constitute only two species (albeit particularly distinguished ones) of a vast zoo of plasmonic nanostructures. Additionally, by combining different structures one can further augment the set of available plasmon resonances through plasmon hybridization [31, 68, 69], which results in new hybrid modes of the mingled system. Motivated by this, in Publication J [31] we have theoretically investigated plasmon coupling and hybridization in 2D nanoslits—obtained by removing a ribbon of an otherwise continuous 2D material—and extend our previous semi-analytical treatment also to anisotropic 2D crystals. The goal of the present section if therefore to provide an overview of the results reported in that publication [31].

Theoretical background. The setup under consideration—i.e., a 2D nanoslit carved out of an arbitrary atomically-thin material—is sketched in Fig. 4.8. Notice that a 2D nanoslit of width $W = 2a$ can be thought of as two co-planar, parallel half-sheets separated by that distance. It should be stressed that although in the following we focus on plasmonic excitations, the formalism presented herein is also readily applicable to other polaritonic modes [8, 21], such as exciton polaritons in transition metal dichalcogenides (TMDCs) or phonon polaritons in hexagonal boron nitride (hBN).

As mentioned above, we describe the governing physics using the nonretarded formalism outlined in Sect. 4.2.1 adapted to the present geometry (and, later on, generalized also to anisotropic 2D media). Owing to the systems' translational invariance along the y-direction, we can express the electrostatic potential as $\Phi(\mathbf{r}) = \Phi(x, z)e^{ik_y y}$ [and the charge density admits a similar decomposition, i.e., $\rho(\mathbf{r}) = \rho(x, z)e^{ik_y y}$], and thus Poisson's equation takes the form

$$\left[\frac{\partial^2}{\partial x^2} + \frac{\partial^2}{\partial z^2} - k_y^2\right] \Phi(x, z) = -\frac{\rho_{\text{2D}}(x)}{\epsilon_0 \bar{\epsilon}} \delta(z) , \tag{4.19}$$

where the fact that $\rho(x, z) = \rho_{\text{2D}}(x)\delta(z)$ has been used. Then, in the spirit of Sect. 4.2.1, one can work out an integro-differential equation for the potential in the

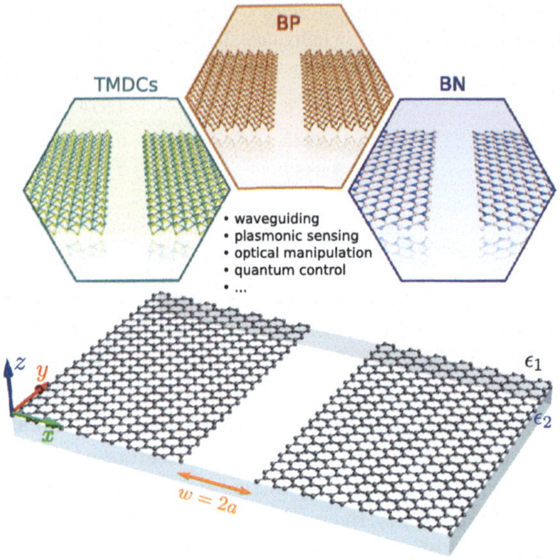

Fig. 4.8 Illustration of two-dimensional nanoslits of width $W = 2a$ made from different 2D crystals, including doped graphene, transition metal dichalcogenides (TMDCs), black phosphorus (BP), and hexagonal boron nitride (hBN). For instance, doped graphene and BP support plasmon polaritons, TMDCs can sustain exciton polaritons, and hBN supports phonon polaritons. Although only monolayers are portrayed, our framework is also applicable to their few-layer counterparts as long as a 2D conductivity can be attributed to them

form of Eq. (4.15). Before explicitly doing so, we first note that it is advantageous to exploit the mirror symmetry of our structure with respect to the plane bisecting the slit (defined by $x = 0$). In particular, we can divide the solutions into even and odd eigensolutions and then solve for the eigenpotential in one side of the system alone, say, for $x > 0$. With this in mind, one arrives at the following self-consistent integro-differential equation[17]:

$$\underbrace{\frac{2i\omega\epsilon_0\bar{\epsilon}\, a}{\sigma(\omega)}}_{\Lambda}\phi^+(\tilde{x}) = -\frac{1}{\pi}\left\{\int_1^\infty d\tilde{x}'\,\mathcal{K}_\eta(\beta;\tilde{x},\tilde{x}')\left[\frac{\partial^2\phi^+(\tilde{x}')}{\partial\tilde{x}'^2} - \beta^2\phi^+(\tilde{x}')\right] + \left.\frac{\partial\phi^+(\tilde{x})}{\partial\tilde{x}}\right|_{x=1}\mathcal{K}_\eta(\beta;\tilde{x},1)\right\}, \tag{4.20}$$

with the integral Kernel $\mathcal{K}_\eta(\beta;\tilde{x},\tilde{x}') = K_0(\beta|\tilde{x}-\tilde{x}'|) + \eta K_0(\beta|\tilde{x}+\tilde{x}'|)$, where $\eta = 1$ for even eigensolutions and $\eta = -1$ for odd eigensolutions. As in the ribbon case, we have defined the dimensionless quantities $\tilde{x} = x/a$, $\tilde{x}' = x'/a$, and $\beta = k_y a$. We then proceed by expanding the potential $\phi^+(\tilde{x})$ using a basis containing Laguerre polynomials, i.e., $\phi^+(\tilde{x}) = e^{-\beta(\tilde{x}-1)}\sum_{n=0}^{\infty} c_n L_n(2\beta[\tilde{x}-1])$, where the aforementioned polynomials are denoted by L_n, and satisfy the orthogonality relation $\int_0^\infty e^{-t}L_n(t)L_m(t)dt = \delta_{nm}$. Again, such a procedure then leads to an eigen-

[17]See Supporting Information of Publication J for further details.

value problem [cf. Eq. (4.16)], where the specific matrix elements are described in the Supporting Information of Publication J[18].

Plasmon dispersion. The dispersion of the eigenmodes propagating along the 2D nanoslit follows from the implicit condition

$$\frac{2i\omega_\nu\epsilon_0\bar{\epsilon}a}{\sigma(\omega_\nu)} = \lambda_\nu \,, \tag{4.21}$$

where λ_ν, with $\nu = \{\beta, \eta\}$, refers to the eigenvalues[19] stemming from the solution of the eigensystem obtained using Eq. (4.20) together with the 0expansion just described in the preceding paragraph.

Assuming, for purposes of definiteness alone, that the 2D nanoslit is made of graphene with a frequency-dependent conductivity well described by the Drude model, namely, $\sigma(\omega) = \frac{ie^2 E_F}{\pi\hbar^2\omega}$ (assuming negligible losses), then the dispersion relation of the system's eigenmodes is given by:

$$\omega(\beta) = \omega_a\sqrt{\lambda_{\beta,\eta}} \,, \tag{4.22}$$

which—for a given β—yields a pair of eigenfrequencies, each ascribed to $\eta = \pm$. Here, ω_a is the nonretarded dispersion of plasmons propagating in extended, continuous graphene with wavevector $k_y = a^{-1}$, that is, $\omega_a = \hbar^{-1}\sqrt{(2\alpha/\bar{\epsilon})E_F\hbar c\, a^{-1}}$. For a traditional 2DEG, the formula (4.22) is unchanged upon employing $\omega_a = \sqrt{(2\pi\alpha/\bar{\epsilon})(n_e/m^*)\hbar c\, a^{-1}}$ instead, where n_e and m^* stand for the carrier density and the effective mass, respectively.

The reader should appreciate that, despite its simple and compact form, Eq. (4.22) (or the more general Eq. (4.21) for that matter) entails a comprehensive description of the effect of the nanoslit's width in the coupling and ensuing hybridization of the modes sustained at opposite edges of the structure. Such information is naturally embodied in the eigenvalues $\lambda_{\beta,\eta}$, which essentially depend on the dimensionless parameter $\beta = k_y a$, or, in other words, on the slit width to plasmon wavelength ratio. Therefore, the complete knowledge of the plasmonic spectrum can be fetched by diagonalizing the matrix $\mathbf{U}$ for a set of β-values, and then inserting the determined eigenvalues into Eq. (4.22). The outcome of that operation is shown in Fig. 4.9. As a guide to the eye, we have also included the dispersion of edge plasmons supported by an individual 2D half-sheet (green dashed line), as well as the plasmon dispersion of the 2D plasmon in the unpatterned, pristine 2D host material (black dashed-dotted line). Due to the electrostatic scaling law epitomized by Eq. (4.21), we stress that the results presented in Fig. 4.9a are valid for an arbitrary dielectric environment,

---

[18] Here we note an unfortunate typographic error in the Supporting Information of Publication J: in Eq. (S14) the minus sign is a *global* one, i.e., it should appear in both terms of the said equation [also, notice that the definition of the matrix elements U_{nm} in this thesis differs from the one in Publication J by a factor of $2\beta/\pi$ (this is simply due to a different choice of Λ)].

[19] We warn the reader than their definition here is slightly different from the one we used in Publication J [mainly to be consistent with the one adopted when writing Eqs. (4.15)–(4.16)]. The results in the end are naturally fully equivalent.

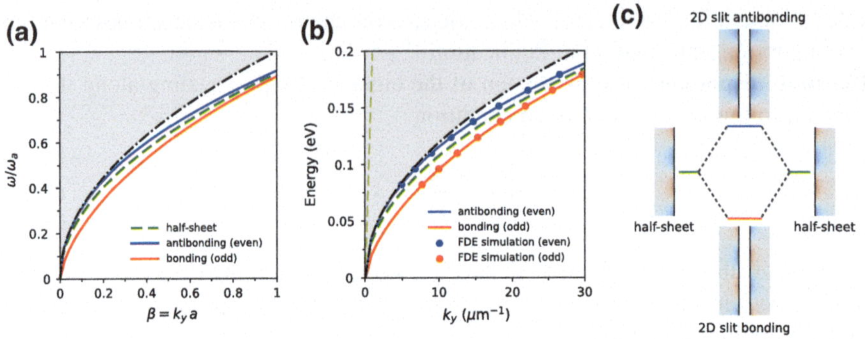

Fig. 4.9 Dispersion relation of hybrid edge plasmons supported by a 2D nanoslit. The spectrum contains two eigenmodes of opposite parity, corresponding to a bonding (red solid line) and an antibonding (blue solid line) mode arising from the interaction between edge plasmons in opposite 2D half-sheets. The black dot-dashed line represents a bulk 2D plasmon propagating along the unpatterned, continuous 2D material, whereas the green dashed line represents the dispersion of bare edge plasmons propagating along the edge of a single half-sheet. **a** Spectrum of coupled plasmons in a 2D nanoslit, obtained via Eq. (4.22). **b** Plasmon dispersion for a representative $W = 2a = 50$ nm graphene nanoslit calculated via Eq. (4.21), where we have employed the local Kubo formula at $T = 300$ K for the conductivity of graphene (other parameters are $E_F = 0.5$ eV, $\hbar\gamma = 3.7$ meV, and $\epsilon_1 = \epsilon_2 = 1$). The colored circles correspond to data points obtained from full-wave numerical calculations based on FDE analysis (in frequency domain) [70]. The results obtained using the semi-analytical technique are show by the red and blue solid lines (and by the green dashed line for the single, uncoupled half-sheet). **c** Plasmon hybridization scheme and calculated induced charge densities of the hybridized plasmon modes, obtained using our semi-analytical formalism (for $\beta = 0.4$). For all calculations, we have truncated the expansion to $N = 20$, which we have found to be sufficient in order to get fully converged solutions with high precision

nanoslit width, and irrespective of the host (isotropic) 2D crystal (upon choosing the appropriate ω_a, as discussed above). Crucially, the figure plainly shows that the hybridization between the edge modes of the two half-planes results in a splitting of the unperturbed half-sheet edge plasmon into a pair of new hybrid eigenmodes. These arise from antisymmetric and symmetric hybridizations of the bare edge plasmons, giving rise to a bonding and an antibonding branch, lying below and above, respectively, the plasmon band of a single half-sheet—see Fig. 4.9a–b. As the name suggests, in the case of the bonding mode the induced charge density in opposite half-planes oscillates in anti-phase (odd symmetry), whereas for the antibonding mode such oscillations are in-phase (even symmetry). Naturally, these resonances occur at different frequencies (for a fixed propagation constant), hence giving rise to the aforementioned energy splitting. We further illustrate this by drawing a plasmon hybridization scheme—see Fig. 4.9c—where the calculated eigendensities are also depicted.

Figure 4.9b shows the solution of the condition (4.21) for a specific structure: a 50 nm-wide graphene nanoslit where the optical response of graphene has been modeled using Kubo's formula for the conductivity at finite temperature [cf. Eqs. (3.20)]. In addition, we compare the results of our semi-analytical theory against data obtained from numerical full-wave electrodynamic simulations using a commercially available finite-difference eigenmode (FDE) solver [70]. The observed agreement between both techniques is outstanding. Such a fact unambiguously demonstrates the ability of our nonretarded semi-analytical framework to rigorously and accurately describe plasmonic excitations in 2D nanoslits. All of this with the added advantage a semi-analytical method provides in portraying a clear and intuitive picture of the underlying physics, and, importantly, without the necessity of relying on often time-consuming numerical simulations.

It is also worth noting that for large wavevectors (in relation to the momentum-scale introduced by a^{-1}, i.e., for $k_y a \gg 1$) the dispersion associated with both eigensolutions asymptotically converge to that of an edge plasmon in an individual half-plane (cf. Fig. 4.9a). In this limit, the Coulomb interaction between the neighboring edges falls off rapidly and therefore the two half-sheets become effectively decoupled. Indeed, in that regime, we recover the dispersion of the edge plasmon of a single half-sheet, due to the fact that $\lim_{\beta\to\infty} \lambda_{\beta,\eta} = \lambda_{\text{hs}}\, \beta \approx 0.8216\, \beta$ (cf. footnote 14 and related discussion; a slightly different[20]—and more concrete—analysis in also given in Publication J).

Finally, we portray in Fig. 4.10 the semi-analytically calculated electric field distributions for the coupled modes supported by a 2D nanoslit. The distinct nature of the hybridized plasmon modes is striking, with a dipole-like bonding mode and a monopole-like antibonding mode clearly visible. Also worthy of notice is the high degree of field confinement delivered by the eigenmodes of the 2D nanoslit—particularly in the vicinity of the slit's edges—, which, for that reason, makes this geometry specially well-suited for deep subwavelength waveguiding [4, 29, 30, 39] and plasmonic circuitry [71], or else for plasmonic sensing [42, 44] and surface-enhanced Raman spectroscopy (SERS) [72]. Other potential applications also include the control of the decay rate of quantum emitters (via the Purcell enhancement) [73–76] or quantum information processing [77–79].

Nanoslits made from anisotropic 2D materials. So far, we have implicitly assumed that the conductivity of the 2D material was isotropic. This is indeed the case for unstrained graphene and group-VI TMDCs such as MoS_2, WS_2, or $MoSe_2$. Nevertheless, that remains a particular case of a broader picture. Notably, there has been a growing interest in the plasmonics of anisotropic 2D materials, either as a platform to enhance and tune their inherent optical birefringence [80–82] or in the context of hyperbolic nanophotonics [83, 84]. Examples of anisotropic 2D materials include black phosphorus (BP) [85, 86], trichalcogenides like TiS_3 [82, 87], and group-VII TMDCs (for instance, ReS_2) [88]. Among these, few-layer black phosphorus and its monolayer version—dubbed as phosphorene—have been the subject of remarkable attention from the nanophotonics community, owing both to its high carrier mobility and attractive optical properties [86]. For these reasons, in the following we inves-

[20] Specifically, in Publication J we define the parameter Λ and the matrix $\mathbf{U}$ in such a way that in the $\beta \to \infty$ limit the corresponding eigenvalue becomes a constant [31] (as opposed to becoming linearly dependent on β). Evidently, both approaches are fully equivalent.

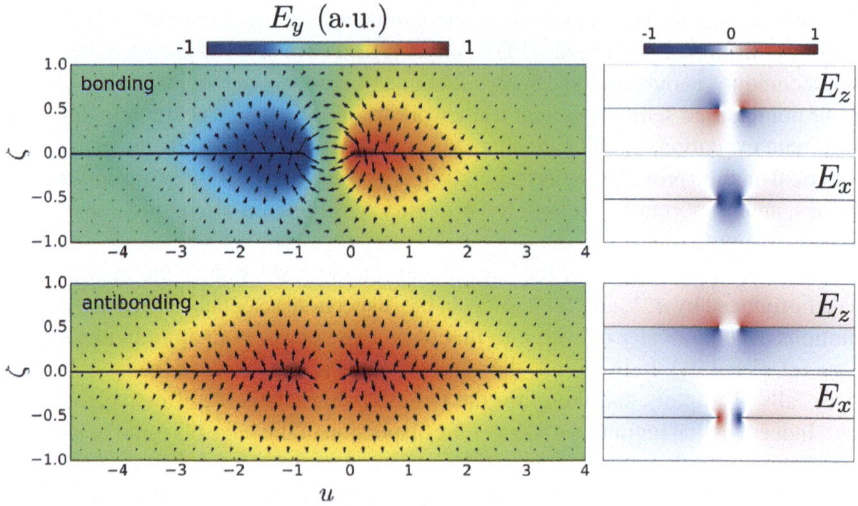

Fig. 4.10 Electric field distributions in the xz- ($u\zeta$)-plane corresponding to the two hybrid edge plasmon modes of a 2D nanoslit, computed using the semi-analytic theory described in the text (for $\beta = k_y a = 0.4$). Here, the dimensionless spatial variables read: $u = \beta(\tilde{x} - 1)$ and $\zeta = k_y z = \beta\tilde{z}$. The black solid line indicates the 2D plasmonic material. The three uppermost panels display the electric field of the bonding mode, whereas the ones at the bottom depict the same quantities for the antibonding mode. The vector plots superimposed onto the two main panels illustrate the full 2D vector field, $\mathbf{E}(u, \zeta) = E_x(u, \zeta)\mathbf{u}_x + E_z(u, \zeta)\mathbf{u}_z$, while the background, in rainbow colors, shows the y-component of the electric field (i.e., the component parallel to the edges of the nanoslit). The length of the arrows is proportional to the norm of the electric-field vector at that point, $|\mathbf{E}(u, \zeta)|$, in logarithmic scale. Each individual panel is normalized with respect to their own maximum values, and the region depicted in the smaller panels has the same dimensions as the main plots

tigate 2D nanoslits made from anisotropic 2D crystals, and, subsequently, we focus and discuss the case of doped phosphorene (which is a semiconductor).

The key aspect differentiating this case from the isotropic scenario considered above, lies in the fact that the surface conductivity of the atomically-thin material is now a tensor; specifically, it can be written as[21]

$$\boldsymbol{\sigma}_{\mathrm{BP}}(\omega) = \sigma_{\mathrm{iso}}(\omega) \begin{pmatrix} m_{xx}^{-1} & 0 \\ 0 & m_{yy}^{-1} \end{pmatrix} , \tag{4.23}$$

where $m_{xx} \equiv m_x^*/m_0$ and $m_{yy} \equiv m_y^*/m_0$ are, respectively, the effective masses of the charge carriers along the x- and y-direction (that will be oriented along the crystallographic axis of high-symmetry). Clearly, the anisotropy of the optical response will be controlled by the different effective masses. Additionally, in the above formula,

[21] The low-energy bandstructure of monolayer BP can be approximated by that of an ordinary parabolic 2D semiconductor, whose conductivity can be constructed in terms of the carrier effective masses along the high-symmetry directions (here, the zigzag and armchair directions) [83, 85, 86].

$\sigma_{\rm iso}(\omega) = \frac{ie^2}{\omega}\frac{n_e}{m_0}$ is the conductivity of an isotropic 2DEG with parabolic dispersion, where n_e stands the carrier density and m_0 is the electron rest mass.

Conveniently, one can still profit from the work performed earlier in the isotropic setting by implementing a few basic modifications in order to contemplate the medium's anisotropy. We shall refrain ourselves from enumerating the mathematical details here, but they are provided in the Supporting Information of Publication J. The corollary of such procedure is an eigenvalue problem resembling the one encountered previously, but where now the matrix that would be equivalent to $\mathbf{U}$ is supplemented by the addition of another matrix that accounts for the system's anisotropy. Naturally, as before, the dispersion of the plasmon eigenmodes of the anisotropic 2D nanoslit are determined by the eigenvalues of the total matrix. Denoting these by ξ, we find that the spectrum of the guided modes in a doped phosphorene nanoslit follows from

$$\omega_{\rm BP}(\beta) = \omega_a \sqrt{\xi_{\beta,\eta}}\,, \tag{4.24}$$

which is entirely analogous to Eq. (4.22) but with distinct eigenvalues. All the details of the anisotropy and specificities of the 2D crystal are therefore contained in such "anisotropic eigenvalues", ξ.

Figure 4.11 shows the dispersion relation of anisotropic plasmons supported by an electron-doped phosphorene nanoslit. Two different cases are considered: one where the edges of the slit run along the zigzag direction (left panel), and another in which the nanoslit runs along the armchair direction (right panel)—see the figure's insets. Although it is clear that the qualitative features already observed in the isotropic case remain—namely the existence of a bonding and an antibonding mode, respectively below and above the dispersion curve of the half-sheet plasmon—the spectrum in the anisotropic case is quantitatively very different depending on the orientation of the 2D nanoslit with respect to the phosphorene's crystal axes. The dramatic contrast between the plasmon dispersion and hybridization in the two cases depicted in Fig. 4.11 reflects the strong anisotropy of black phosphorus (which in turn stems from its puckered honeycomb lattice). Indeed, while the dispersion curves associated with the hybrid modes of a zigzag slit are barely indistinguishable from each other (and also from the half-sheet and bulk phosphorene plasmon), the splitting between the bonding and antibonding modes of an armchair phosphorene nanoslit is substantial. This arises because, for instance in the latter case, the charge carrier effective mass is significantly smaller along the slit's armchair edges and it is larger in the direction perpendicular to it (evidently, the situation is reversed in the case of a zigzag nanoslit).

We remark that the semi-analytical results presented in Fig. 4.11 were benchmarked and subsequently validated by rigorous electrodynamic numerical calculations based on the finite-element method (a comparison is shown in the Supporting Information of Publication J) [89]. Therefore, in this manner, we have demonstrated that our semi-analytical model is extensible to nanoslits made from anisotropic 2D materials as well. This further emphasizes the applicability and versatility of our method to describe plasmonic excitations in a wide variety of nanostructured 2D plasmonic materials, both with and without anisotropy [90]. In this regard, the same

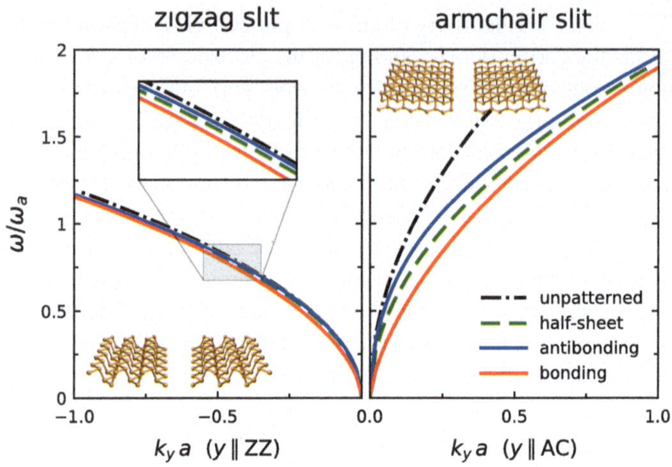

Fig. 4.11 Dispersion of coupled, hybridized plasmon modes propagating along an electron-doped phosphorene nanoslit. Both the case of a nanoslit patterned along the zigzag and the armchair directions is shown. We have assigned negative values of k_y to represent the hybrid plasmonic modes sustained at a zigzag slit, and positive values of k_y to identify the eigenmodes of an armchair nanoslit. In our calculations, we take the following parameters for the (anisotropic) electron effective masses: $m_{\mathrm{ZZ}} = 0.7m_0$ and $m_{\mathrm{AC}} = 0.15m_0$ [85, 86], for the electronic bands along the zigzag and armchair directions, respectively

formalism used for here for phosphorene call also be straightforwardly applied to other anisotropic materials, or even otherwise isotropic materials under the application of uniaxial strain, which effectively breaks the isotropy.

We believe that this investigation contributes with a new building block—a one-atom-thick nanoslit—to the 2D toolkit of hybrid plasmon resonances, thereby expanding our freedom and capabilities for designing new tunable plasmonic systems based on flatland plasmonics.

References

1. Efetov DK, Kim P (2010) Phys Rev Lett 105:256805. https://doi.org/10.1103/PhysRevLett.105.256805
2. Bezares FJ, Sanctis AD, Saavedra JRM, Woessner A, Alonso-González P, Amenabar I, Chen J, Bointon TH, Dai S, Fogler MM, Basov DN, Hillenbrand R, Craciun MF, García de Abajo FJ, Russo S, Koppens FHL (2017) Nano Lett 17(10):5908. https://doi.org/10.1021/acs.nanolett.7b01603
3. Wang Z, Li T, Almdal K, Mortensen NA, Xiao S, Ndoni S (2016) Opt Lett 41(22):5345. http://ol.osa.org/abstract.cfm?URI=ol-41-22-5345, https://doi.org/10.1364/OL.41.005345
4. Gonçalves PAD, Peres NMR (2016) An introduction to graphene plasmonics, 1st edn (World Scientific, Singapore. https://doi.org/10.1142/9948
5. García de Abajo FJ (2014) ACS Photonics 1(3):135. https://doi.org/10.1021/ph400147y

6. Grigorenko AN, Polini M, Novoselov KS (2012) Nat Photonics 6:749. https://doi.org/10.1038/nphoton.2012.262
7. Low T, Avouris P (2014) ACS Nano 8(2):1086. https://doi.org/10.1021/nn406627u
8. Low T, Chaves A, Caldwell JD, Kumar A, Fang NX, Avouris P, Heinz TF, Guinea F, Martin-Moreno L, Koppens F (2017) Nat Mater 16:182. https://www.nature.com/articles/nmat4792, https://doi.org/10.1038/nmat4792
9. Woessner A, Lundeberg MB, Gao Y, Principi A, Alonso-González P, Carrega M, Watanabe K, Taniguchi T, Vignale G, Polini M, Hone J, Hillenbrand R, Koppens FHL (2015) Nat Mater 14:421. https://www.nature.com/articles/nmat4169, https://doi.org/10.1038/nmat4169
10. Ni GX, McLeod AS, Sun Z, Wang L, Xiong L, Post KW, Sunku SS, Jiang BY, Hone J, Dean CR, Fogler MM, Basov DN (2018) Nature 557:530. https://doi.org/10.1038/s41586-018-0136-9
11. Castro Neto AH, Guinea F, Peres NMR, Novoselov KS, Geim AK (2009) Rev Mod Phys 81:109. https://doi.org/10.1103/RevModPhys.81.109
12. Santoyo BM, del Castillo-Mussot M (1993) Rev Mex Fis 39:640
13. Linghua K, Baorong Y, Xiwei H (2007) Plasma Sci Technol 9(5):519. https://doi.org/10.1088/1009-0630/9/5/01
14. Ando T, Fowler AB, Stern F (1982) Rev Mod Phys 54:437. https://doi.org/10.1103/RevModPhys.54.437
15. Grimes CC, Adams G (1976) Phys Rev Lett 36:145. https://doi.org/10.1103/PhysRevLett.36.145
16. Koppens FHL, Chang DE, García de Abajo FJ (2011) Nano Lett 11(8):3370. https://doi.org/10.1021/nl201771h
17. Ju L, Geng B, Horng J, Girit C, Martin M, Hao Z, Bechtel HA, Liang X, Zettl A, Shen YR, Wang F (2011) Nat Nanotechnol 6:630. https://doi.org/10.1038/nnano.2011.146
18. Fei Z, Rodin AS, Andreev GO, Bao W, McLeod AS, Wagner M, Zhang LM, Zhao Z, Thiemens M, Dominguez G, Fogler MM, Neto AHC, Lau CN, Keilmann F, Basov DN (2012) Nature 487:82. https://doi.org/10.1038/nature11253
19. Chen J, Badioli M, Alonso-González P, Thongrattanasiri S, Huth F, Osmond J, Spasenović M, Centeno A, Pesquera A, Godignon P, Elorza AZ, Camara N, García de Abajo FJ, Hillenbrand R, Koppens FHL (2012) Nature 487:77. https://doi.org/10.1038/nature11254
20. Liu H, Liu Y, Zhu D (2011) J Mater Chem 21:3335. https://doi.org/10.1039/C0JM02922J
21. Basov DN, Fogler MM, García de Abajo FJ (2016) Science 354:6309. https://science.sciencemag.org/content/354/6309/aag1992, https://doi.org/10.1126/science.aag1992
22. Haastrup S, Strange M, Pandey M, Deilmann T, Schmidt PS, Hinsche NF, Gjerding MN, Torelli D, Larsen PM, Riis-Jensen AC, Gath J, Jacobsen KW, Mortensen JJ, Olsen T, Thygesen KS (2018) 2D Mater 5(4):042002. http://c2db.fysik.dtu.dk, https://doi.org/10.1088/2053-1583/aacfc1
23. Mounet N, Gibertini M, Schwaller P, Campi D, Merkys A, Marrazzo A, Sohier T, Castelli IE, Cepellotti A, Pizzi G, Marzari N (2018) Nat Nanotechnol 13:246. https://doi.org/10.24435/materialscloud:2017.0008/v3; https://doi.org/10.1038/s41565-017-0035-5
24. Novotny L, Hecht B (2012) Principles of nano-optics, 2nd edn. Cambridge University Press, Cambridge
25. Maradudin AA, Barnes WL, Sambles JR (eds) (2014) Modern plasmonics, 1st edn. Elsevier
26. Maier SA (2007) Plasmonics: fundamentals and applications. Springer, Berlin
27. Mikhailov SA, Ziegler K (2007) Phys Rev Lett 99:016803. https://doi.org/10.1103/PhysRevLett.99.016803
28. Alcaraz Iranzo D, Nanot S, Dias EJC, Epstein I, Peng C, Efetov DK, Lundeberg MB, Parret R, Osmond J, Hong JY, Kong J, Englund DR, Peres NMR, Koppens FHL (2018) Science 360(6386):291. https://science.sciencemag.org/content/360/6386/291, https://doi.org/10.1126/science.aar8438
29. Gonçalves PAD, Dias EJC, Xiao S, Vasilevskiy MI, Mortensen NA, Peres NMR (2016) ACS Photonics 3(11):2176. https://doi.org/10.1021/acsphotonics.6b00674
30. Gonçalves PAD, Bozhevolnyi SI, Mortensen NA, Peres NMR (2017) Optica 4(6):595. http://www.osapublishing.org/optica/abstract.cfm?URI=optica-4-6-595, https://doi.org/10.1364/OPTICA.4.000595

31. Gonçalves PAD, Xiao S, Peres NMR, Mortensen NA (2017) ACS Photonics 4(12):3045. https://doi.org/10.1021/acsphotonics.7b00558
32. Yu R, Cox JD, Saavedra JRM, García de Abajo FJ (2017) ACS Photonics 4(12):3106. https://doi.org/10.1021/acsphotonics.7b00740
33. Lundeberg MB, Gao Y, Asgari R, Tan C, Van Duppen B, Autore M, Alonso-González P, Woessner A, Watanabe K, Taniguchi T, Hillenbrand R, Hone J, Polini M, Koppens FHL (2017) Science 357(6347):187. http://science.sciencemag.org/content/357/6347/187, https://doi.org/10.1126/science.aan2735
34. Stern F (1967) Phys Rev Lett 18:546. https://doi.org/10.1103/PhysRevLett.18.546
35. Hwang EH, Das Sarma S (2009) Phys Rev B 80:205405. https://doi.org/10.1103/PhysRevB.80.205405
36. Stauber T, Gómez-Santos G (2012) Phys Rev B 85:075410. https://doi.org/10.1103/PhysRevB.85.075410
37. Zhu JJ, Badalyan SM, Peeters FM (2013) Phys Rev B 87:085401. https://doi.org/10.1103/PhysRevB.87.085401
38. Gonçalves PAD, Dias EJC, Bludov YV, Peres NMR (2016) Phys Rev B 94:195421. https://doi.org/10.1103/PhysRevB.94.195421
39. Christensen J, Manjavacas A, Thongrattanasiri S, Koppens FHL, García de Abajo FJ (2012) ACS Nano 6(1):431. https://doi.org/10.1021/nn2037626
40. Yan H, Low T, Zhu W, Wu Y, Freitag M, Li X, Guinea F, Avouris P, Xia F (2013) Nat Photonics 7(5):394. https://doi.org/10.1038/nphoton.2013.57
41. Luxmoore IJ, Gan CH, Liu PQ, Valmorra F, Li P, Faist J, Nash GR (2014) ACS Photonics 1(11):1151. https://doi.org/10.1021/ph500233s
42. Rodrigo D, Limaj O, Janner D, Etezadi D, García de Abajo FJ, Pruneri V, Altug H (2015) Science 349(6244):165. https://science.sciencemag.org/content/349/6244/165, https://doi.org/10.1126/science.aab2051
43. Sorger C, Preu S, Schmidt J, Winnerl S, Bludov YV, Peres NMR, Vasilevskiy MI, Weber HB (2015) New J Phys 17(5):053045. https://doi.org/10.1088/1367-2630/17/5/053045/meta
44. Farmer DB, Avouris P, Li Y, Heinz TF, Han SJ (2016) ACS Photonics 3(4):553. https://doi.org/10.1021/acsphotonics.6b00143
45. Yan H, Li X, Chandra B, Tulevski G, Wu Y, Freitag M, Zhu W, Avouris P, Xia F (2012) Nat Nanotechnol 7:330. https://doi.org/10.1038/nnano.2012.59
46. Yan H, Xia F, Li Z, Avouris P (2012) New J Phys 14(12):125001. https://doi.org/10.1088/1367-2630/14/12/125001/meta
47. Thongrattanasiri S, Koppens FHL, García de Abajo FJ (2012) Phys Rev Lett 108:047401. https://doi.org/10.1103/PhysRevLett.108.047401
48. Fang Z, Thongrattanasiri S, Schlather A, Liu Z, Ma L, Wang Y, Ajayan PM, Nordlander P, Halas NJ, García de Abajo FJ (2013) ACS Nano 7(3):2388. https://doi.org/10.1021/nn3055835
49. Fang Z, Wang Y, Schlather AE, Liu Z, Ajayan PM, García de Abajo FJ, Nordlander P, Zhu X, Halas NJ (2014) Nano Lett 14(1):299. https://doi.org/10.1021/nl404042h
50. Zhu X, Wang W, Yan W, Larsen MB, Bøggild P, Pedersen TG, Xiao S, Zi J, Mortensen NA (2014) Nano Lett 14(5):2907. https://doi.org/10.1021/nl500948p
51. Cox JD, García de Abajo FJ (2014) Nat. Commun. 5:5725. https://doi.org/10.1038/ncomms6725
52. Wang W, Christensen T, Jauho AP, Thygesen KS, Wubs M, Mortensen NA (2015) Sci Rep 5:9535. https://doi.org/10.1038/srep09535
53. Schultz MH, Jauho AP, Pedersen TG (2011) Phys Rev B 84:045428. https://doi.org/10.1103/PhysRevB.84.045428
54. Yeung KYM, Chee J, Yoon H, Song Y, Kong J, Ham D (2014) Nano Lett 14(5):2479. https://doi.org/10.1021/nl500158y
55. Liu PQ, Valmorra F, Maissen C, Faist J (2015) Optica 2(2):135. http://www.osapublishing.org/optica/abstract.cfm?URI=optica-2-2-135, https://doi.org/10.1364/OPTICA.2.000135
56. Wang W, Apell P, Kinaret J (2011) Phys Rev B 84:085423. https://doi.org/10.1103/PhysRevB.84.085423

57. Silveiro I, García de Abajo FJ (2014) Appl Phys Lett 104(13):131103. https://doi.org/10.1063/1.4870046
58. Fetter AL (1986) Phys Rev B 33:3717. https://doi.org/10.1103/PhysRevB.33.3717
59. Fetter AL (1986) Phys Rev B 33:5221. https://doi.org/10.1103/PhysRevB.33.5221
60. Xia X, Quinn JJ (1994) Phys. Rev. B 50:8032. https://doi.org/10.1103/PhysRevB.50.8032
61. Wang W, Apell SP, Kinaret JM (2012) Phys Rev B 86:125450. https://doi.org/10.1103/PhysRevB.86.125450
62. Wang W (2012) J Phys: Condens Matter 24(40):402202. https://doi.org/10.1088/0953-8984/24/40/402202
63. Abramowitz M, Stegun IA (1972) Handbook of mathematical functions: with formulas, graphs, and mathematical tables. Dover, New York. http://people.math.sfu.ca/cbm/aands/
64. Nikitin AY, Low T, Martin-Moreno L (2014) Phys Rev B 90:041407. https://doi.org/10.1103/PhysRevB.90.041407
65. Velizhanin KA (2015) Phys Rev B 91:125429. https://doi.org/10.1103/PhysRevB.91.125429
66. Christensen T (2015) From classical to quantum plasmonics in three and two dimensions. PhD thesis, Technical University of Denmark
67. Volkov VA, Mikhailov SA (1988) Sov Phys JETP 67(8):1639
68. Prodan E, Radloff C, Halas NJ, Nordlander P (2003) Science 302(5644):419. https://science.sciencemag.org/content/302/5644/419, https://doi.org/10.1126/science.1089171
69. Steele JM, Grady NK, Nordlander P, Halas NJ (2007) Surface Plasmon Nanophotonics. Springer, Berlin, pp 183–196. https://doi.org/10.1007/978-1-4020-4333-8_13
70. Lumerical Solutions Inc., MODE Solutions. http://www.lumerical.com/tcad-products/mode/
71. Bozhevolnyi S (2008) Plasmonic nanoguides and circuits, 1st edn. Pan Stanford
72. Kneipp K, Wang Y, Kneipp H, Perelman LT, Itzkan I, Dasari RR, Feld MS (1997) Phys Rev Lett 78:1667. https://doi.org/10.1103/PhysRevLett.78.1667
73. Russell KJ, Liu TL, Cui S, Hu EL (2012) Nat Photonics 6:459. https://www.nature.com/articles/nphoton.2012.112, https://doi.org/10.1038/nphoton.2012.112
74. Akselrod GM, Argyropoulos C, Hoang TB, Ciracì C, Fang C, Huang J, Smith DR, Mikkelsen MH (2014) Nat Photonics 8:835. https://www.nature.com/articles/nphoton.2014.228, https://doi.org/10.1038/nphoton.2014.228
75. Pelton M (2015) Nat Photonics 9:427. https://www.nature.com/articles/nphoton.2015.103, https://doi.org/10.1038/nphoton.2015.103
76. Gonçalves PAD, Christensen T, Rivera N, Jauho AP, Mortensen NA, Soljačić M (2020) Plasmon-emitter interactions at the nanoscale. Nat Commun 11:366. https://doi.org/10.1038/s41467-019-13820-z
77. Bozhevolnyi SI, Mortensen NA (2017) Nanophotonics 6:1185. https://doi.org/10.1515/nanoph-2016-0179
78. Chang DE, Sørensen AS, Hemmer PR, Lukin MD (2006) Phys Rev Lett 97:053002. https://doi.org/10.1103/PhysRevLett.97.053002
79. Calafell IA, Cox JD, Radonjić M, Saavedra JRM, García de Abajo FJ, Rozema LA, Walther P (2019) npj Quantum Inf 5:37. https://doi.org/10.1038/s41534-019-0150-2
80. Khaliji K, Fallahi A, Martin-Moreno L, Low T (2017) Phys Rev B 95:201401. https://doi.org/10.1103/PhysRevB.95.201401
81. Yang H, Jussila H, Autere A, Komsa HP, Ye G, Chen X, Hasan T, Sun Z (2017) ACS Photonics 4(12):3023. https://doi.org/10.1021/acsphotonics.7b00507
82. Silva-Guillén JA, Canadell E, Ordejón P, Guinea F, Roldán R (2017) 2D Mater 4(2):025085. https://doi.org/10.1088/2053-1583/aa6b92/meta
83. Nemilentsau A, Low T, Hanson G (2016) Phys Rev Lett 116:066804. https://doi.org/10.1103/PhysRevLett.116.066804
84. Gjerding MN, Petersen R, Pedersen TG, Mortensen NA, Thygesen KS (2017) Nat Commun 8:320. https://doi.org/10.1038/s41467-017-00412-y
85. Low T, Roldán R, Wang H, Xia F, Avouris P, Moreno LM, Guinea F (2014) Phys Rev Lett 113:106802. https://doi.org/10.1103/PhysRevLett.113.106802

86. Carvalho A, Wang M, Zhu X, Rodin AS, Su H, Neto AHC (2016) Nat Rev Mater 1:16061. https://doi.org/10.1038/natrevmats.2016.61
87. Dai J, Zeng XC (2015) Ang Chem 127(26):7682. https://doi.org/10.1002/ange.201502107
88. Rahman M, Davey K, Qiao SZ (2017) Adv Func Mater 27(10):1606129. https://doi.org/10.1002/adfm.201606129
89. COMSOL Multiphysics. www.comsol.com
90. Cupo A, Masih Das P, Chien CC, Danda G, Kharche N, Tristant D, Drndić M, Meunier V (2017) ACS Nano 11(7):7494. https://doi.org/10.1021/acsnano.7b04031

Chapter 5
Two-Dimensional Channel Plasmons in Nonplanar Geometries

The ability to effectively guide electromagnetic radiation below the diffraction limit is of the utmost importance in the prospect of all-optical circuitry. In this regard, a great deal of hope [1, 2] has been deposited in plasmonics due to the ability of surface plasmon polaritons (SPPs) to squeeze electromagnetic fields into sub-wavelength dimensions. Indeed, one of the most prominent applications of SPPs for optoelectronic technologies is the use of plasmonic waveguiding structures to route electromagnetic radiation below the diffraction limit [3–5]. An example of a class of such structures is the one involving the use of gap-SPPs (i.e., SPP modes sustained at dielectric gaps separating two metal surfaces), which are particularly appealing candidates for SPP-based waveguides due to their favorable balance between losses and field confinement [6–11]. A variation of this configuration is a V-shaped groove carved into a metallic substrate. Within this geometry, the corresponding propagating SPP modes are generally referred to as channel plasmon polaritons (CPPs) [7–9]. Over the last couple of decades, a plethora of fundamental explorations [12–14] and proof-of-concept experiments have been carried out demonstrating the usage of noble metal CPPs in plasmonic interferometers [15, 16], waveguides [17–21], ring-resonators [15], and for nanofocusing [22]. Additionally, in the context of quantum plasmonics [23], CPPs have also been explored for the control of the emission properties of quantum emitters [24].

On the other hand, the rise of graphene as a novel plasmonic material has attracted a great deal of attention owing to the prospect of long-lived, gate-tunable graphene plasmons (GPs) that are capable of producing large field confinements in the THz and mid-IR [25–28]. Nonplanar graphene plasmons have recently gained interest [25, 29–31], but the investigation of graphene channel plasmons still remains largely unexplored. Only very recently, research on the plasmonic properties of graphene wedges and grooves has been conducted [32–35].

In this chapter, we provide a comprehensive theoretical description of the salient features of graphene plasmons guided along the apex of a graphene-covered triangular wedge or groove. The structure under consideration is sketched in Fig. 5.1. An

P. A. D. Gonçalves, *Plasmonics and Light–Matter Interactions in Two-Dimensional Materials and in Metal Nanostructures*, Springer Theses,
https://doi.org/10.1007/978-3-030-38291-9_5

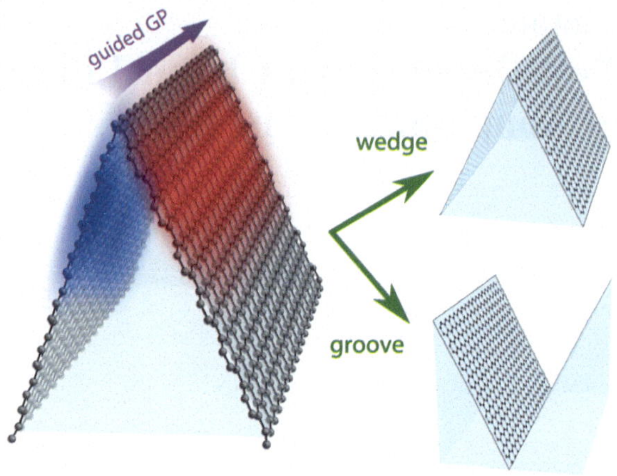

Fig. 5.1 Pictorial representation of a guided graphene plasmon (GP) mode propagating along a dielectric wedge. Herein, we denote the system as a *wedge* (*groove*) if the medium filling the inner side of the triangularly-shaped channel possesses a higher (lower) relative permittivity

important advantage of this configuration is that it does not involve any nanopatterning of the graphene sheet: it simply consists in depositing graphene onto a V-shaped wedge or groove previously sculpted in the receiving substrate or by controlling the folding of a 2D material using state-of-the art techniques.[1,2] In this way, graphene remains nearly pristine and free from fabrication-induced defects or edge-roughness (which are often present in other waveguiding graphene nanostructures, such as graphene ribbons [42]), both of which have a negative impact on the GPs propagation losses. Furthermore, by taking advantage of the third dimension one departs from the conventional flat-graphene geometries and effectively produces a one-dimensional channel which not only confines light in the vertical direction that bisects the channel, but is also capable of producing lateral confinement of the electromagnetic radiation.

In what follows, we determine the spectrum and ensuing plasmonic properties of guided GPs eigenmodes supported by a structure like the one illustrated in Fig. 5.1 (the theory described below is also applicable to other 2D materials capable of hosting collective resonances). We describe the underlying physics using two different theoretical approaches: (i) one that involves an extension of the semi-analytical orthogonal polynomial expansion technique previously introduced in Sect. 4.2.1, and (ii) a second one based on a generalization of the so-called effective-index method originally designed for noble-metal CPPs [8]. The entire content of these two methods have been published by the author in Publications N and K, respectively, for the

[1]The patterning of the substrate can be done, for instance, by employing the same techniques used to fabricate metallic grooves [7], followed by the deposition of graphene or, alternatively, even growing it on a pre-configured copper substrate [36]. Other possibilities include folding a graphene layer or by exploring the formation of wrinkles (either naturally occurring [37–39] or deliberately formed [36, 40, 41]).

[2]We bring to the reader's attention that the (at first sight idealized) triangularly-shaped structure proposed here (see Fig. 5.1) has actually been realized experimentally with graphene and other 2D materials in a different context [36, 38, 40, 41] (albeit those studies have focused primarily on electronic properties and strain-engineering).

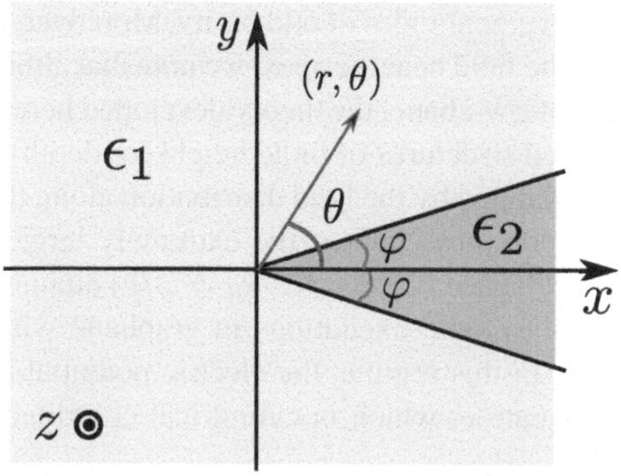

Fig. 5.2 Cross section of the triangular channel depicted in Fig. 5.1 with the specification of the chosen (cylindrical) coordinate system. The channel forms an angle of 2φ, where the 2D material lies *at* the two semi-infinite planes defined by $\theta = \pm\varphi$. The inner and outer media are characterized by the relative permittivities ϵ_2 (for $-\varphi \leq \theta \leq \varphi$) and ϵ_1 (for $\varphi \leq \theta \leq 2\pi - \varphi$), respectively

(i) and (ii) approaches. Hence, our aim here is to provide a succinct description of the main findings reported in those publications.

5.1 Graphene Plasmons in Triangular Channels: Wedge and Groove Configurations

We consider an idealized geometry in which a graphene monolayer is sandwiched between a triangular dielectric wedge with relative permittivity ϵ_2 and an outer dielectric material with relative permittivity ϵ_1, as depicted in Fig. 5.2. As it will become apparent later, our model is completely general irrespective of the specific values for the dielectric constants of the cladding insulators. However, for the sake of definiteness, we shall refer to a *wedge* whenever $\epsilon_2 > \epsilon_1$ and vice-versa to denote a *groove*.

Before outlining the details of the quasi-analytic theory developed to describe guided GPs in triangular wedges and grooves, we highlight that one can treat the cases of even and odd symmetry—defined after the potential or, equivalently, after the induced charge density, and in relation to the line bisecting the channel's cross-section—separately, as this makes the problem more amenable to handle. In particular, for the case of even symmetry (i.e., when the induced charges are symmetric in the graphene half-planes which constitute the V-shape), we have found that such even parity modes are not highly confined near the apex of the wedge/groove, with their dispersion being virtually the same as the one of GPs in a flat, planar graphene (see Fig. S2 in the Supporting Information of Publication N). Contrasting this, as it will become clear ahead, the corresponding odd eigensolutions exhibit strong field confinement near the apex of the wedge (or groove), and therefore hereafter we limit our analysis exclusively to eigenmodes of odd parity. Moreover, by virtue of the high degree of localization of the field near the apex, we note that although we assume (for simplicity) an infinitely long V-shape, the theory developed here remains adequate in the description of V-shaped structures of finite height (or depth) as long as their size is larger than the region spanned by the field distribution along the axis of symmetry.

As mentioned in the previous chapter, the extremely large wavevectors (when compared with light's free-space wavevector, $k_0 = \omega/c$) attained by graphene plasmons allow us to treat plasmonic excitations in graphene within the electrostatic limit with high accuracy. In this regime, the electric potential associated with GPs must satisfy Poisson's equation, which in cylindrical coordinates (cf. Fig. 5.2) can be written as

$$\left[\frac{\partial^2}{\partial r^2} + \frac{1}{r}\frac{\partial}{\partial r} + \frac{1}{r^2}\frac{\partial^2}{\partial \theta^2} - q^2\right]\Phi(r,\theta) = -\frac{\rho(r,\theta)}{\epsilon_0}, \tag{5.1}$$

where we have written the scalar potential as $\Phi(\mathbf{r}) = \Phi(r,\theta)e^{iqz}$, owing to the system's translational invariance along the z-axis (an implicit time-dependence of the form $e^{-i\omega t}$ is assumed herein). This effectively reduces our initial 3D problem into a 2D one, and will allow us to parameterize the dispersion relation of the guided modes in terms of the propagation constant q. Hence, the solutions of Eq. (5.1) essentially render guided wedge graphene plasmons (WGPs) or groove graphene plasmons (GGPs) that propagate along the longitudinal direction (i.e., the z-direction). The solution of this equation in the medium $j = \{1, 2\}$ can be formally written as[3]

$$\Phi(r,\theta) = \frac{i\sigma(\omega)}{\omega}\int_0^\infty dr' G_j(r,\theta;r',\varphi)\left[\frac{\partial^2}{\partial r'^2} - q^2\right]\Phi(r',\varphi), \tag{5.2}$$

where $\sigma(\omega)$ is the dynamical conductivity of graphene (or of an arbitrary atomically-thin medium), and $G_j(r,\theta;r',\varphi)$ is the Green's function associated with Eq. (5.1) in that medium. Its explicit form is somewhat cumbersome, but the interested reader may find it (along with its derivation) in the Supporting Information associated with Publication N. Also, we further note that when writing the preceding equation we have expressed the charge density as $\rho(r,\theta) = -en(r)\delta(\theta-\varphi)/r$, and where the 2D carrier density, $n(r)$, has been written in terms of the electrostatic potential by combining the continuity equation together with Ohm's law. Finally, notice that we only need to determine the potential in, say, the upper-half space (i.e., for $0 \le \theta \le \pi$), by exploiting the above-noted symmetry. In the spirit of Sect. 4.2.1, we set $\theta = \varphi$ in Eq. (5.2) and expand the potential *at* the graphene layer, $\phi(r) \equiv \Phi(r,\varphi)$, as $\phi(r) = \sum_n c_n L_n^{(0)}(qr)e^{-qr/2}$, with $L_n^{(0)}$ denoting the generalized Laguerre polynomials [43]. Making use of the appropriate orthogonality relations [43], it is then possible to translate the self-consistent integro-differential Eq. (5.2) into a standard linear algebra eigenproblem, namely

[3]We warn the reader in advance that throughout this section the definitions of the Green's function, the parameter Λ, etc, are slightly different that the ones adopted in Sect. 4.2.1.

$$\frac{i\omega}{q\sigma(\omega)} c_m = \sum_{n=0}^{\infty} U_{mn} c_n \,, \tag{5.3}$$

where the matrix elements U_{mn} read

$$U_{mn} = \int_0^\infty d\tilde{r} \int_0^\infty d\tilde{r}'\, G(\tilde{r}, \varphi; \tilde{r}', \varphi) e^{-\frac{\tilde{r}+\tilde{r}'}{2}} L_m^{(0)}(\tilde{r}) \times \left[\frac{3}{4} L_n^{(0)}(\tilde{r}') - L_{n-2}^{(2)}(\tilde{r}') - L_{n-1}^{(1)}(\tilde{r}') \right] . \tag{5.4}$$

We note that the double integration over the dimensionless variables $\tilde{r} = qr$ and $\tilde{r}' = qr'$ can be performed *analytically*, thereby making the computation of the matrix elements extremely fast.[4] Also, note that we have dropped the subscript j in the Green's function because the boundary condition at $\theta = \varphi$ enforces that $G_1(x, \varphi; y, \varphi) = G_2(x, \varphi; y, \varphi)$, and thus one can choose either Green's function arbitrarily without any loss of generality.

The eigenvalue problem (5.3) can be solved numerically, and, as before, the eigenvalues of $\mathbf{U}$, denoted below by $\tilde{\vartheta}$, determine the spectrum of WGPs (or GGPs) via

$$\frac{i\omega_\varphi}{q\sigma(\omega_\varphi)} = \tilde{\vartheta}_\varphi \quad \xrightarrow{\text{or, equivalently,}} \quad \frac{2i\omega_\varphi \epsilon_0 \bar{\epsilon}}{\sigma(\omega_\varphi)} \frac{1}{q} = \vartheta_\varphi \,, \tag{5.5}$$

where we have defined, for convenience, $\vartheta_\varphi \equiv 2\epsilon_0 \bar{\epsilon}\, \tilde{\vartheta}_\varphi$, with $\bar{\epsilon} = (\epsilon_1 + \epsilon_2)/2$. For a given opening angle 2φ, Eq. (5.5) returns a discrete set of WGP (or GGP) eigenmodes. Crucially, all the wavevector and frequency dependence stems from the left-hand side of the previous equation. Therefore, the eigenvalues $\vartheta_\varphi \equiv \vartheta(\varphi)$ carry a purely geometric meaning, since they depend uniquely on the configuration of the system (i.e., namely the opening angle and cladding dielectrics).[5]

In particular, using graphene's Drude-like conductivity with negligible damping [cf. Eq. (3.22)], one obtains the following "universal scaling law" for the dispersion relation of graphene plasmons in triangular wedges and grooves:

$$\omega_\varphi(q) = \omega_{\text{flat}}(q)\sqrt{\vartheta_\varphi} \,, \tag{5.6}$$

[4]There is, however, a one-dimensional numerical integration that still needs to be carrier out (whose integral is embedded in the definition of the Green's function; see Supporting Information of Publication N for details).

[5]Recall that, for the planar structures discussed in Sect. 4.2.1, the eigenvalues stemming from the corresponding nonretarded framework where *independent* of the specific value of the neighboring dielectrics. That was because, in the strictly planar case, the substrate and the superstrate contributed on equal footing to the rescaling of the in-plane (nonretarded) Coulomb interaction. In the nonplanar case studied here, that is no longer the case, which is then reflected in eigenvalues that *depend* on the dielectric functions $\epsilon_{1,2}$.

where $\omega_{\text{flat}}(q) = \hbar^{-1}\sqrt{\frac{2\alpha}{\bar{\epsilon}} E_F \hbar c\, q}$ is simply the dispersion relation of GPs in flat graphene [recall Eq. (4.7)]. Importantly, we stress that once the eigenvalues ϑ_φ are obtained, the entire spectrum [i.e., in the whole (q, ω)–space] is fully determined; all of this by performing a *single* computation, once and for all. In fact, one may even plot the dispersion of distinct 2D materials that support SPPs modes from such (single) computation, since the eigenvalues ϑ_φ do not depend on the 2D conductivity that characterizes the specific 2D material under consideration [see Eq. (5.5)].

The computed dispersion relation of graphene plasmons guided along the apex of different triangular wedges and grooves is shown in Fig. 5.3. The spectrum contains a discrete set of well-defined eigenmodes, whose number depends on the opening angle 2φ (indicated in the insets). Moreover, the figure plainly shows that, for the same frequency, smaller angles yield correspondingly larger plasmon wavevectors, which is an indication of stronger field confinement near the apex of sharper wedges (or grooves). In addition, notice the large distance between the dispersion of the WGPs and GGPs eigenmodes and the light line in the dielectric (which is almost vertical and extremely close to the y-axis). This fact reflects the extremely high field localization provided by the aforementioned eigenmodes, which push further graphene's capability of rendering plasmons that are deeply subwavelength. On a side note, this also justifies the high accuracy of our nonretarded framework, as retardations effects are negligible.

Furthermore, note the outstanding agreement between the semi-analytical theory detailed above and the electrodynamic simulations performed using the finite-element method (FEM) technique (whose data is depicted in the figure as colored symbols). This constitutes further evidence of the ability of our semi-analytic method to render accurate results, and at the same time it provides a deeper and more transparent fundamental understanding of the scaling properties of WGPs and GGPs.

Strikingly, the results plotted in Fig. 5.3 also convey a stark contrast between the dispersion curves akin to WGPs and GGPs, demonstrating the superiority of the former in squeezing light below the diffraction limit (because they attain larger wavevectors for the same angle of the structure). We further note that one can transform a wedge into a groove and vice-versa either by swapping the values of ϵ_1 and ϵ_2, or by applying the angular transformation $\varphi \to \pi - \varphi$ (this essentially interchanges the Green's functions G_1 and G_2). For the sake of completeness, in Fig. 5.3f we have portrayed the spectrum of GPs guided along a triangular channel embedded in a homogeneous dielectric medium with the same average relative permittivity as its wedge and groove counterparts shown in the figure. It can be observed that—for the same angle—each of the modes attain increasingly larger wavevectors as we move from a groove, embedded, and wedge configuration (in this order). This observation hints us that the ability to reach deep subwavelength regimes strongly depends on the ϵ_2/ϵ_1 ratio, for a fixed (acute) angle. Such an educated guess is supported by the data shown in Fig. 5.4, where it can be observed that the geometrical scaling factor entering in Eq. (5.6) decreases monotonically with increasing ϵ_2/ϵ_1. In turn, this translates into higher effective indexes, $n_{\text{eff}} \equiv q/k_0$, for larger quotients ϵ_2/ϵ_1. As an example, at a frequency of 20 THz, the fundamental WGP eigenmode in Fig. 5.3b

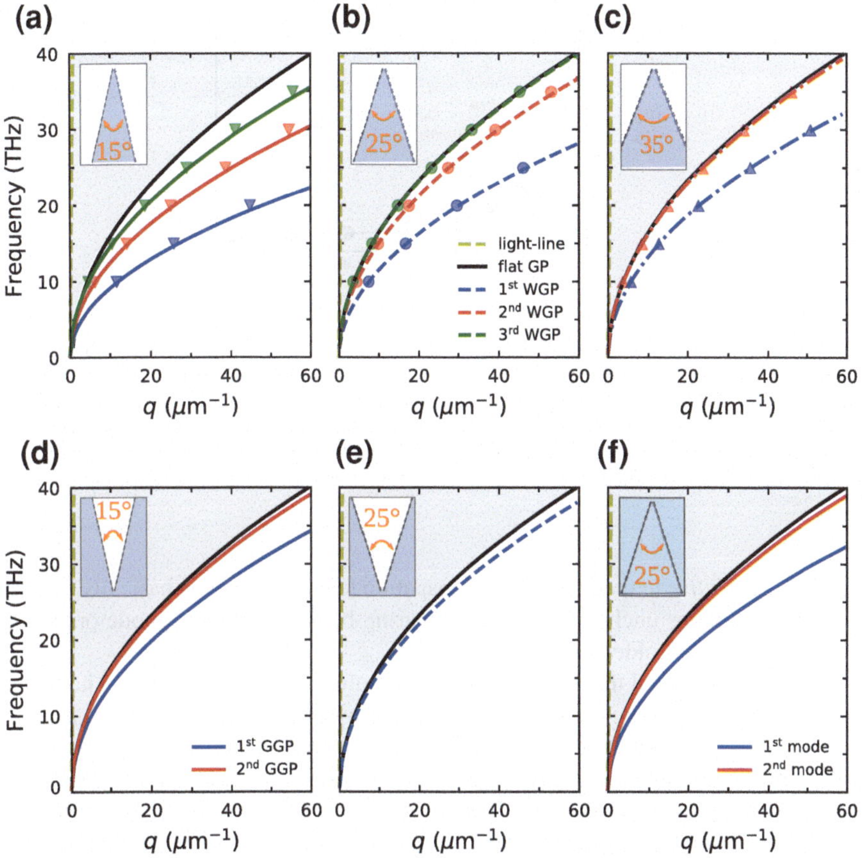

Fig. 5.3 Dispersion relations of graphene plasmons in triangular wedges and grooves (odd-parity modes). Spectrum of WGPs [upper row; **a–c**] and GGPs [first two bottom panels; **d–e**] associated different graphene-covered triangular configurations, with different opening angles, 2φ (indicated in the insets), as given by Eq. (5.6) [we take $E_F = 0.4$ eV]. The solid black line represents the dispersion of GPs in a flat interface and serves as a guide to the eye, while the straight yellowish dashed line indicates the light line. The colored symbols $\triangledown$, $\bigcirc$, and $\triangle$, in the upper row correspond to the results for the WGP dispersion as obtained from full-wave numerical simulations (COMSOL's finite-element method [44]). The insets' shading represents a dielectric with $\epsilon_d = 4$, whereas the white regions denote a dielectric medium with $\epsilon_d = 1$ (e.g., air). **f** Spectrum of GPs guided along a V-shaped graphene channel embedded in a homogeneous medium with $\epsilon_1 = \epsilon_2 = \bar{\epsilon} = 2.5$ (shaded in the inset as)

exhibits an effective index of $n_{\text{eff}} \simeq 72$. This already large effective index increases even further at higher frequencies (for instance, for the CO_2 laser wavelength of $\lambda_0 = 10.6\,\mu\text{m}$ (around 28.3 THz) this value climbs to $n_{\text{eff}} \simeq 100$).

Naturally, the higher amount of field localization promoted by the WGPs or GPPs comes hand in hand with slightly larger *absolute* propagation losses, a trait that is well-known in plasmonics. Nevertheless, we have found that the number of plasmon

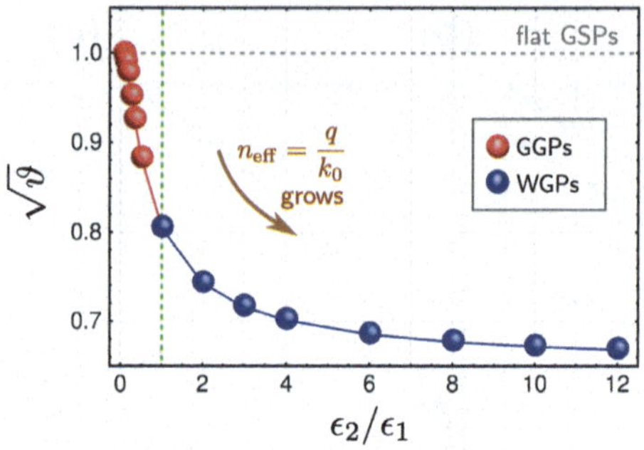

Fig. 5.4 Dependence of the scaling factor $\sqrt{\vartheta}$, for eigenvalues associated with the fundamental mode, as a function of the ratio ϵ_2/ϵ_1. In computing the data in blue we have fixed $\epsilon_1 = 1$ while varying ϵ_2, and vice-versa for the red data points. The vertical green dashed line indicates the point where $\epsilon_1 = \epsilon_2 = 1$, whereas the horizontal gray dashed line sets the upper bound corresponding to the flat GSP dispersion. We further note that the specific value of ϑ is naturally sensitive to the absolute value of the dielectric constant that is kept constant. For all configurations we take $2\varphi = 25°$

oscillations *relative* to the plasmon's propagation length—which is basically a quality factor—remains unchanged when comparing between different mode orders or plasmons in flat, unfolded graphene.[6]

In closing, we recall that once the potential evaluated in the 2D material has been determined—using the expansion coefficients provided in the eigenvectors associated with the eigenvalue problem (5.3)—one can construct the scalar electric potential in the whole coordinate space using Eq. (5.2). From here, the corresponding induced electric field follows directly via $\mathbf{E}(\mathbf{r}) = -\nabla\Phi(\mathbf{r})$. The cross-sectional potential and electric field distributions (computed semi-analytically) associated with the two lowest-frequency WGPs in a representative $2\varphi = 25°$ dielectric wedge are shown Fig. 5.5. The figure exhibits telling evidence of the remarkable confinement that WGP modes are able to deliver near the apex of the wedge (the results for the groove are qualitatively similar, albeit with slightly less localization for the same resonant frequency); this can be observed from either the potential or from the corresponding electric field. The vertical confinement, with respect to the channel's apex, decreases rather rapidly as we move from the fundamental resonance to the higher-order branches of the plasmonic spectrum. For the parameters and configuration portrayed in Fig. 5.5 the lowest energy mode essentially remains localized within the first 200 nm (which is substantial, given that the corresponding free-space

---

[6] This feature is a direct consequence of the "universal scaling" epitomized by Eqs. (5.5)–(5.6), and can be clearly seen from Eq. (5.5) by noting that

$$q(\omega) = \frac{2i\omega\epsilon_0\bar{\epsilon}}{\sigma(\omega)}\frac{1}{\vartheta_\varphi} \quad \Rightarrow \quad \#\,\text{plasmon osc.} = \frac{L_p}{\lambda_{\mathrm{WGP}}} = \frac{1}{4\pi}\frac{\mathrm{Re}\,q}{\mathrm{Im}\,q} \quad \text{(independent of } \vartheta) \tag{5.7}$$

which is clearly independent of the eigenvalue ϑ [and thus the quality factor for WGPs and GGPs is the same as for plasmons in flat graphene at the same frequency (also, note that $\frac{\mathrm{Re}\,q}{\mathrm{Im}\,q} = \frac{\mathrm{Im}\,\sigma}{\mathrm{Re}\,\sigma}$)].

wavelength is $\lambda_0 \simeq 15\,\mu m$), whereas for the second order mode that distance grows to about $3.6\,\mu m$. Moreover, Fig. 5.5 also serves to support our previous claim that, despite the inherent simplifications in our theory, it can still be applied to realistic wedges (grooves) of finite height (depth), provided that height (depth) is larger that the tail of the potential and of the electric field.

The behavior of the spatial distributions displayed in Fig. 5.5 is qualitatively maintained throughout most of the dispersion curve (and similarly for different angles), the only important difference being the degree of confinement in the whereabouts of the apex. Therefore, an infinitely vast number of triangularly-shaped configurations—with different angles, heights or depths—can be engineered depending on the required level of localization and/or frequency window of interest. In this regard, our theoretical model can be readily applied in the design of tailored nanophotonic devices that meet *a priori* the requirements in terms of operating frequency range and physical dimensions. We thus expect that this work will fuel future experimental realizations of WGPs/GGPs, as we believe that such modes may hold interesting implications for future all-photonic circuitries at the nanoscale.

5.2 Effective-Index Description of Channel Plasmons in Two-Dimensional Materials

In the previous section (Sect. 5.1) we have developed a rigorous theoretical model to describe 2D channel plasmons supported by triangular wedges or grooves, and, in particular, to calculate their spectral features and plasmonic properties, focusing on their capabilities for achieving deep subwavelength waveguiding.

Here, we present an alternative (approximate) method for describing 2D channel plasmons guided along the apex of a triangularly-shaped sheet of a 2D plasmon-supporting material (like graphene, a conventional 2DEG, a highly-doped TMDC, ultrathin metal films, etc). As it will become apparent in what follows, our perturbative approach has been designed to provide accurate results for triangular channels with small opening angles, while maintaining an inexpensive theoretical description of the underlying electrodynamics. More specifically, the approach developed here follows in part the effective-index framework previously applied to CPPs occurring in traditional 3D metals[7] [8, 45, 46].

Briefly, the concept behind the effective-index method (EIM) is that, at each height (with respect to the bottom of the channel), one can identify a one-dimensional (1D) dielectric–graphene–dielectric–graphene–dielectric (DGDGD) waveguide in which the core dielectric has a width given by $w(z)$. The combination of such 1D

[7] Interestingly, the effective-index method for plasmonic structures based on customary 3D metals provides closed-form, *analytical* results for the 3D CPPs' dispersion and mode profiles. As we detail below, and to the best of our knowledge, for 2D CPPs this is not the case anymore (with the reason for that essentially being the more complex form of the plasmon dispersion of the corresponding flat interface).

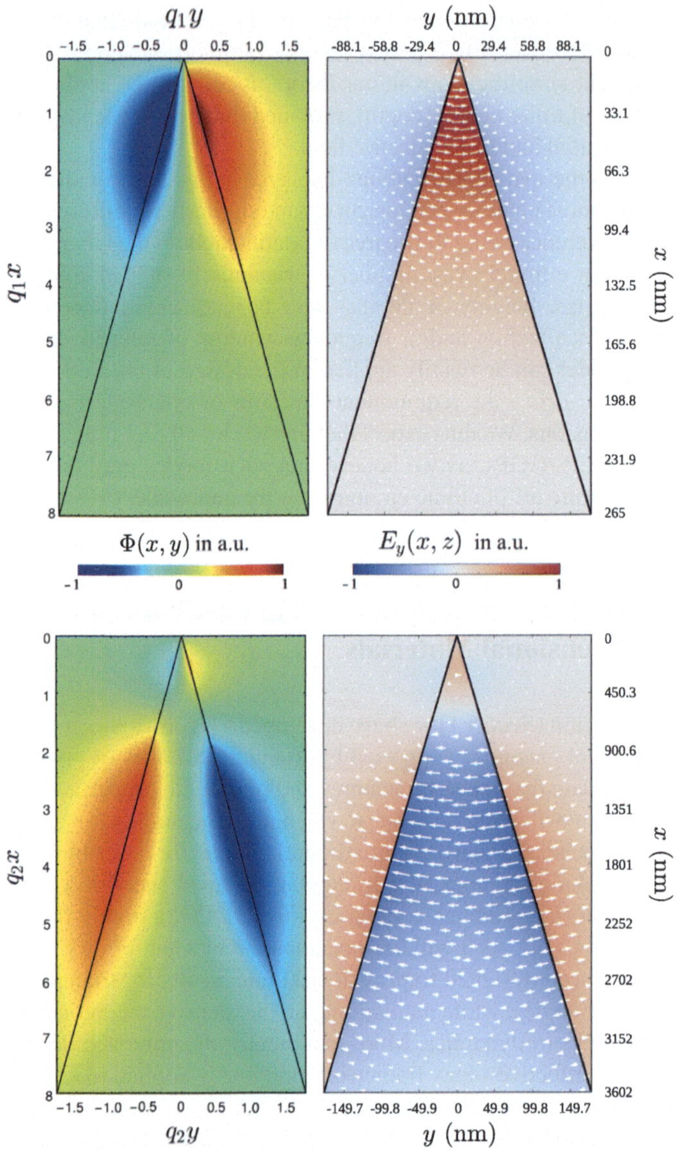

Fig. 5.5 Two-dimensional colormaps of potential, $\Phi(x, y)$, and electric field distributions, $\mathbf{E}(x, y) = -\nabla_{2D}\Phi(x, y)$, ascribed to the fundamental (upper panels) and second order (lower panels) WGPs modes, for a resonant frequency of $f = 20\,\text{THz}$. Note that the electric field in Cartesian coordinates can be fetched from its polar version by applying a rotation matrix, namely $[E_x, E_y]^{\text{T}} = R(\varphi)[E_r, E_\theta]^{\text{T}}$. The plotted region is the same in each pair of 2D plots, where the axes of the panels on the left are written in dimensionless units (normalized to the corresponding propagation constants) and the axes of the panels on the right are given in nanometers. The colormaps in the background of the electric field plots correspond to the (normalized) $E_y(x, y)$, which is dominant in most of the plotted spatial regions. Material setup: $E_F = 0.4\,\text{eV}$, $\epsilon_1 = 1$, and $\epsilon_2 = 4$

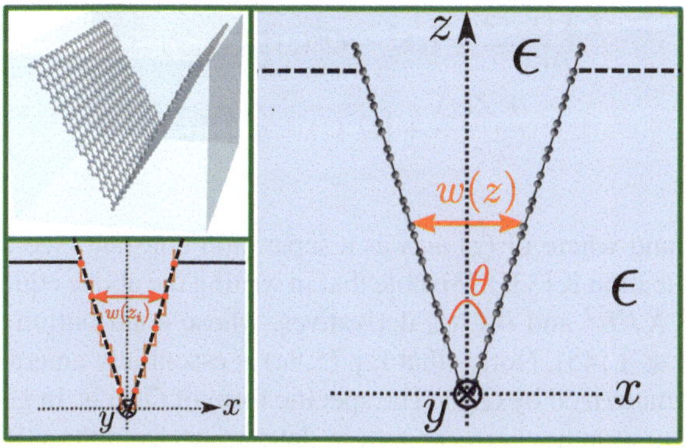

Fig. 5.6 Conceptual illustration of the effective-index description of 2D CPPs. Two-dimensional material (here represented by a graphene monolayer) folded into a V-shaped channel and embedded in a dielectric medium with relative permittivity ϵ. The channel width as a function of the z-coordinate follows $w(z) = 2z \tan(\theta/2)$ for a triangular cross-sections like the one sketched here

waveguide configurations—i.e., DGDGD slabs stacked together—can then be used to construct the whole two-dimensional channel waveguide akin to graphene CPPs; see Fig. 5.6. Below, we demonstrate that this conceptually intuitive picture can be used to determine the dispersion of CPPs and corresponding electric field distributions by solving a Schrödinger-like equation whose eigenvalues depend only on the angle θ of the channel. Finally, the results obtained using this unified description are then benchmarked against the ones calculated using more rigorous theories, like the one described in Sect. 5.1 [32].

We consider an atomically-thin material folded into V-shape forming an angle θ, as illustrated in Fig. 5.6; we assume henceforth that $\theta \ll 1$ in order for the EIM to be valid. Within this limit, one can make use of the transverse magnetic (TM) and electric (TE) representations, and seek TM-like solutions in which the main component of the electric field lies along the x-axis. Recall that the underlying idea behind the EIM applied to the description of CPPs in V-grooves is to model the inner region as two coupled 1D waveguides. This foundational principle does not change when considering a dielectric V-channel covered with graphene (or any other 2D material). However, the corresponding equations will be quite different, asserting the natural differences between the two materials (e.g., 3D metal and 2D graphene). With the above-noted considerations in mind, the (x-component) of the field can be factorized as [8, 45, 46]

$$E_x(\mathbf{r}) = X(x, z)Z(z)e^{iqy}\,, \tag{5.8}$$

where q denotes the propagation constant of the 2D material's channel plasmon. With this ansatz, one readily obtains the coupled waveguide equation [derivable from the Helmholtz equation for the electric field (5.8)]:

$$\frac{\partial^2 X(x,z)}{\partial x^2} + \left[\epsilon k_0^2 - Q^2(z)\right] X(x,z) = 0\,, \tag{5.9a}$$

$$\frac{\partial^2 Z(z)}{\partial z^2} + \left[Q^2(z) - q^2\right] Z(z) = 0\,, \tag{5.9b}$$

with $k_0 = \omega/c$, and where $Q^2(z)$ acts as a separation constant (see Supplementary Material of Publication K [33]). We note that in writing the above equations, we have neglected the $\partial^2 X/\partial z^2$ and $\partial X/\partial z$ derivatives, whose contribution is small in the regime where $\theta \ll 1$ [45]. Notice that Eq. (5.9a) is essentially an equation for a 1D waveguide, parameterized by $Q(z)$. The specific form of $Q(z)$ is, in general, nontrivial. Therefore, our strategy here consists in developing from the GP condition at a planar graphene double-layer with the same dielectric media—DGDGD—, which, for the mode with the sought-after symmetry, is given by[8] [25]

$$1 + \coth\left[\frac{w(z)}{2}\kappa_Q\right] + \frac{i\sigma(\omega)}{\omega\epsilon_0\epsilon}\kappa_Q = 0\,, \tag{5.10}$$

with $\kappa_Q = \sqrt{Q^2(z) - \epsilon k_0^2}$, and where $w(z) = 2z\tan(\theta/2) \approx \theta z$ is the gap-width as a function of the z-coordinate. Here, $\sigma(\omega)$ is the dynamical *surface* conductivity of graphene (or any conducting 2D medium). A closed-form expression for $Q^2(z)$ is obtainable by expanding the previous equation for small widths,[9] which yields [33]

$$Q^2(z) = \epsilon k_0^2 + \frac{\epsilon^2}{2f_\sigma^2}\left[1 + \frac{4f_\sigma}{\epsilon w(z)} + \sqrt{1 + \frac{8f_\sigma}{\epsilon w(z)}}\right]\,, \tag{5.11}$$

where we have introduced the function $f_\sigma(\omega) = \mathrm{Im}\{\sigma_{2D}\}/(\omega\epsilon_0)$ for short-hand notation. In the specific case of graphene, $f_\sigma(\omega) = 4\alpha\hbar c\frac{E_F}{(\hbar\omega)^2}$ materializes the assumption that the conductivity is given by its Drude-like expression with negligible damping [cf. Eq. (3.22)]. In possession of an explicit relation for $Q^2(z)$, a solution for the coupled waveguide equation (5.9) may be fetched. The solution to the first equation is trivial, simply being $X(x,z) = A\cosh([Q^2(z) - \epsilon k_0^2]^{1/2}x)$ in the inner region, where A is a constant. The remaining differential equation (5.9b) is not as straightforward. We can make such equation more affordable by defining the wavevector $q_0 = 2\epsilon/f_\sigma$ associated with plasmons propagating in the planar (i.e., unfolded) 2D material [25], while at the same time introducing the dimensionless variable $\zeta = \theta q_0 z$. Performing these transformations, we arrive at a dimensionless eigenvalue equation[10]

---

[8] The interested reader might want to revisit Sect. 4.1.2, and, in particular, Eq. (4.10b).

[9] In fact, more rigorously, for $w(z)/2[Q^2(z) - \epsilon k_0^2]^{1/2} \ll 1$.

[10] In writing Eq. (5.12a) we have approximated the width (for small angles) as $w(z) \approx \theta z$, for the sake of clarity. The expression using the *exact* form is recoverable upon making the replacement $\theta \to 2\tan(\theta/2)$ (we have used the *exact* expression in all our results).

$$-\theta^2 \frac{\partial^2 Z(\zeta)}{\partial \zeta^2} + V(\zeta) Z(\zeta) = \mathcal{E}_\theta Z(\zeta)\,, \tag{5.12a}$$

which clearly resembles a Schrödinger equation with an "energy potential"

$$V(\zeta) = -\frac{8 + \zeta + \sqrt{\zeta^2 + 16\zeta}}{8\zeta}\,, \tag{5.12b}$$

where the 2D CPP dispersion relation $q^2 = \epsilon k_0^2 - \mathcal{E}_\theta q_0^2$ is given in terms of the dimensionless eigenvalue $\mathcal{E}_\theta$. Interestingly, notice that $\mathcal{E}_\theta$ mimics an "effective dielectric function" that is entirely determined by the geometry—namely the angle θ—of the channel. We stress that the solution of Eq. (5.12) yields the eigenfunctions and propagation constants akin to the respective CPP eigenmodes. Crucially, this economical treatment paints a clear picture of the underlying physics, in which smaller angles leads to concomitantly deeper "energy potentials" and consequently modes with larger wavevectors q (and thus, in turn, stronger field confinement). Furthermore, the "universal scaling" found previously also emerges here in an equally natural fashion, where—as before—the 2D CPP wavevector depends entirely on the geometry of the V-channel.

The differential equation (5.12) can be solved effortlessly using standard numerical methods. In this regard, the results presented in the following were obtained by solving Eq. (5.12) by employing the so-called shooting method [47].

The eigenfunctions $Z(z)$ resulting from the numerical solution of Eq. (5.12) are depicted in Fig. 5.7, and correspond to the vertical profile of 2D CPP modes in a V-shaped channel with $\theta = 15°$. They are clearly reminiscent of bound modes in a potential well like the ones encountered in typical quantum mechanical problems. The fundamental 2D CPP eigenmode is highly confined within the bottom of the groove (i.e., near the apex of the channel), while the successively higher order modes tend to be concomitantly more spread along the vertical direction of the channel [here, corresponding to the (ζ-) z-direction].

The corresponding two-dimensional field distributions inside the channel simply follow from Eq. (5.8). These are shown in Fig. 5.8 for V-shaped graphene channels with different opening angles and excited at different resonant frequencies. The top panels illustrate the modal distributions for the fundamental (Fig. 5.8a) and second-order (Fig. 5.8b) graphene CPP eigenmodes, whose main features naturally echo the ones previously observed in Fig. 5.7. The influence of the groove angle is reflected in the two bottom panels, with smaller opening angles delivering correspondingly higher field confinements near the channel's apex, a behavior that is directly attributable to the stronger "confining potential" for CPPs in sharper V-shaped structures.

We now revisit the dispersion relation of the 2D CPPs, but now calculated within the EIM, that is, using Eq. (5.12). Figure 5.9a–c show the obtained dispersion relations of 2D CPP in representative triangular channels. The dispersion curves akin to graphene CPPs lie to the right of their corresponding flat GP counterpart, thereby

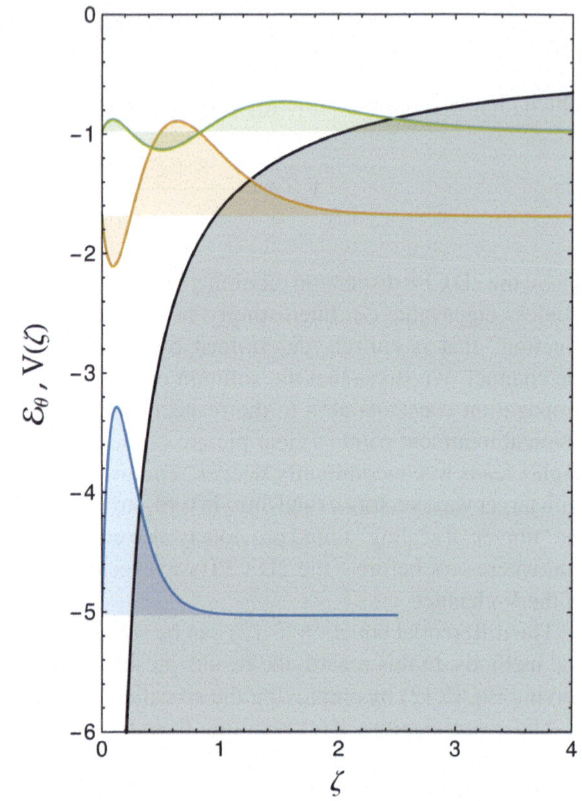

Fig. 5.7 Calculated field distributions along the height of the channel, $Z_n(\zeta)$, associated with the first three 2D CPPs eigenmodes in a triangular channel with $\theta = 15°$. We have plotted the eigenfunctions in an energy diagram along with the potential $V(\zeta)$ (black line), to highlight the similarity with typical quantum mechanical problems. The vertical axis for each Z_n starts at the position of the corresponding eigenvalue $\mathcal{E}_\theta^{(n)}$

indicating the higher amount of field localization attained by the former when compared to the latter. The degree of subwavelength localization of the electromagnetic field is even more dramatic for smaller angles (compare Fig. 5.9b, with c).

In order to gauge the level of fidelity of the perturbative effective-index approach outlined above we have also included the dispersions obtained using the rigorous method described in Sect. 5.1 (dashed lines in Fig. 5.9a–c). The results obtained using both methods exhibit a very good agreement; this illustrates that, despite its inherent simplicity, the EIM framework can still provide consistent results, thus being a very valuable tool in judging the salient features of 2D CPPs. Unfortunately, the EIM does not capture the higher-order modes in the case of $\theta = 25°$ (not shown); the justification for this lies in the breakage of the small-argument expansion performed in the cotangent figuring in Eq. (5.10). Still, we note that for smaller angles (e.g., $\theta = 15°$) the second-order CPP eigenmode is already well described by the EIM owing to the smaller angle.

A more transparent analysis can be performed by exploiting the universal scaling of the 2D CPPs spectrum. In this spirit, Fig. 5.9d shows the geometrical scaling factor, $q/q_0 \simeq \sqrt{-\mathcal{E}_\theta}$ (since $q, q_0 \gg k_0$) as a function of the channel's angle, together with

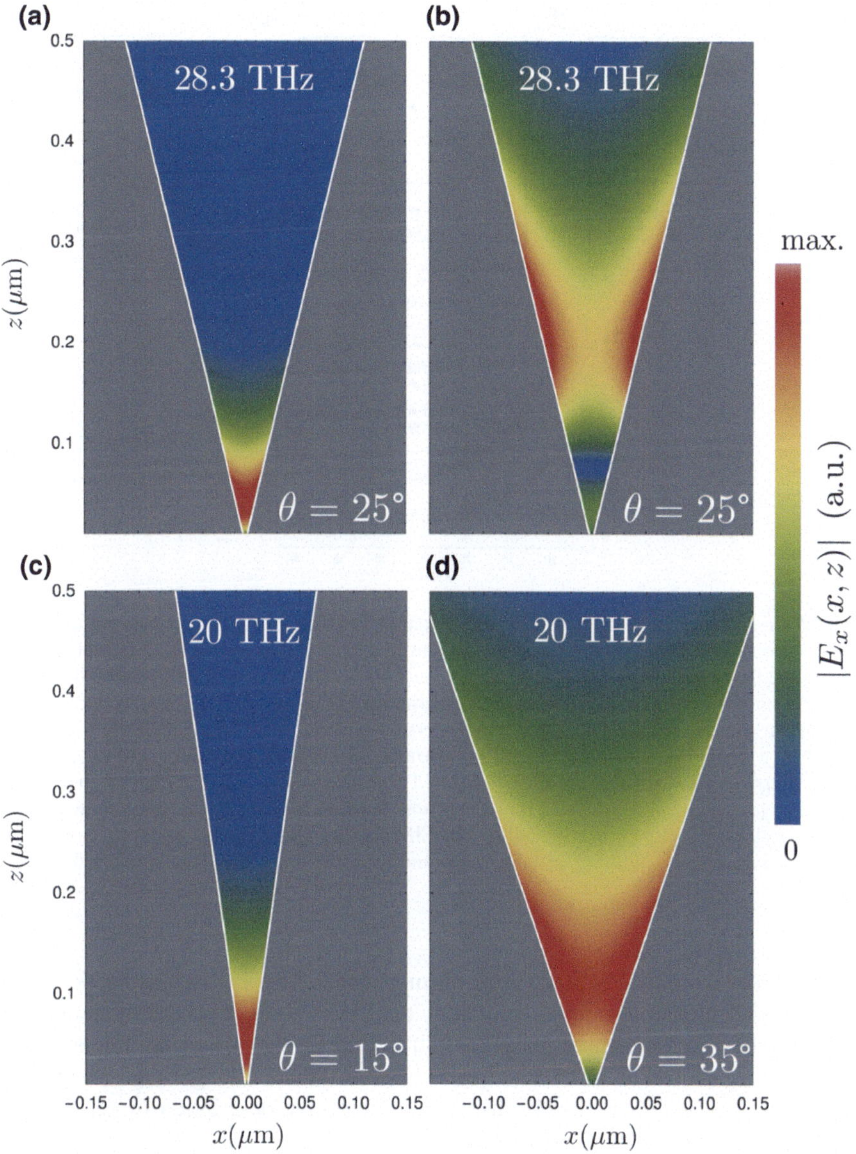

Fig. 5.8 Spatial distributions of the electric field magnitude, $|E_x^{(n)}(x, z)|$, for graphene CPPs in various triangularly-shaped channels: fundamental **a**, **c**, **d**, and second-order **b** guided plasmonic eigenmodes. The resonant frequencies and opening angles are explicitly indicated in each panel. The field is depicted only in the inner region. Material parameters: $E_F = 0.5\,\mathrm{eV}$ and $\epsilon = 2.1$

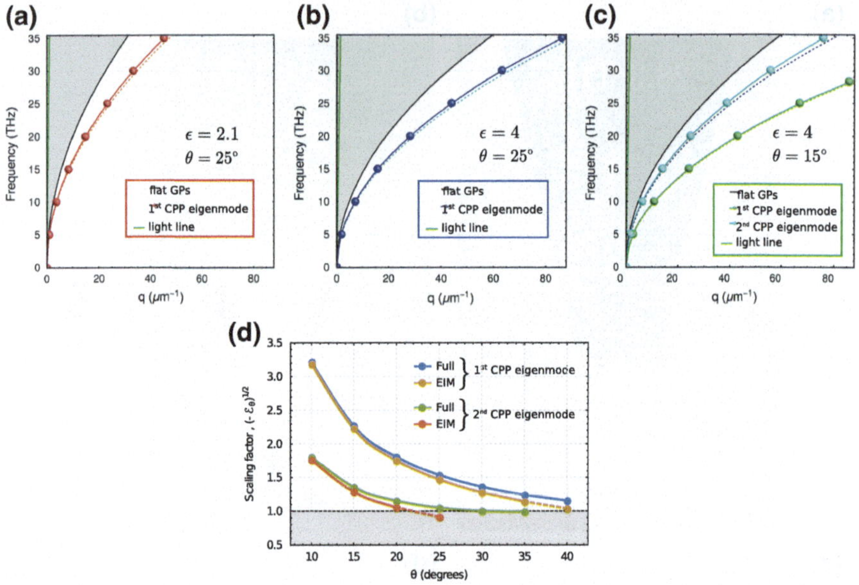

Fig. 5.9 **a–c** Dispersion relationd of guided graphene CPPs in V-shaped channels embedded in homogeneous dielectric environments with different dielectric constants ϵ, for two different opening angles; see figure's labels. As an eye-guide, the solid black lines indicate the dispersion of GPs in the corresponding flat interface. We assume a Fermi energy of $E_F = 0.5\,\text{eV}$ in the calculations. For comparison, the dashed lines show the dispersion relation of CPPs obtained using the non-perturbative theory described in Sect. 5.1 [32]. **d** Universal scaling exhibited by 2D CPPs, namely $\sqrt{-\mathcal{E}_\theta} \simeq q/q_0$, as a function of the angle θ (EIM). The results obtained using the approach outlined in Sect. 5.1 [32] (Full) are also plotted for comparison. The dashed lines indicate the point where the EIM seems to surpass its regime of validity. The EIM curves follow a simple analytical expression of the form $a_n + b_n\theta^{-1}$ with $a_n = \{0.33, 0.36\}$ and $b_n = \{28.7, 14.1\}$ (where n refers the mode order)

the equivalent quantity computed using the more general theory discussed in Sect. 5.1. Figure 5.9d demonstrates quite remarkably the EIM's ability to correctly determine the properties of 2D channel plasmons. Strikingly, in the case of the fundamental mode, the agreement is maintained even beyond the small angle regime. On the other hand—as we have already discussed above—the description of the higher-order modes is restricted to smaller angles.

Figure 5.9d plays the role of an epilogue: it plainly shows the universal scaling of the 2D CPP's propagation constant as a function of the angle—valid independently of the frequency, 2D conductivity, or dielectric constant—while at the same time outlining the regime of validity of the EIM. Finally, we have found that the curves plotted in the figure follow a simple analytical expression, in the form of $a + b\,\theta^{-1}$, to which we have fitted the constants a and b (whose values are stated in the figure's caption). This feature should be convenient for designing such nanostructures, functioning as a "ruler" or "recipe" to architecture 2D CPPs with tailored plasmonic properties.

References

1. Atwater HA (2007) Sci Am 296:56. https://www.scientificamerican.com/article/the-promise-of-plasmonics/, https://doi.org/10.1038/scientificamerican0407-56
2. Ozbay E (2006) Science 311(5758):189. http://science.sciencemag.org/content/311/5758/189, https://doi.org/10.1126/science.1114849
3. Ebbesen TW, Genet C, Bozhevolnyi SI (2008) Phys Today 61(5):44. https://doi.org/10.1063/1.2930735
4. Gramotnev DK, Bozhevolnyi SI (2010) Nat Photonics 4(2):83. https://doi.org/10.1038/nphoton.2010.282
5. Fang Y, Sun M (2015) Light Sci Appl 4(6):e294. https://www.nature.com/articles/lsa201567, https://doi.org/10.1038/lsa.2015.67
6. Bozhevolnyi S (2008) Plasmonic nanoguides and circuits, 1st edn. Pan Stanford
7. Smith CLC, Stenger N, Kristensen A, Mortensen NA, Bozhevolnyi SI (2015) Nanoscale 7:9355. https://doi.org/10.1039/C5NR01282A
8. Bozhevolnyi SI (2006) Opt Exp 14(20):9467. http://www.opticsexpress.org/abstract.cfm?URI=oe-14-20-9467, https://doi.org/10.1364/OE.14.009467
9. Moreno E, García-Vidal FJ, Rodrigo SG, Martín-Moreno L, Bozhevolnyi SI (2006) Opt Lett 31(23):3447. http://ol.osa.org/abstract.cfm?URI=ol-31-23-3447, https://doi.org/10.1364/OL.31.003447
10. Chen Y, Nielsen TR, Gregersen N, Lodahl P, Mørk J (2010) Phys Rev B 81:125431. https://doi.org/10.1103/PhysRevB.81.125431
11. Chen Y, Gregersen N, Nielsen TR, Mørk J, Lodahl P (2010) Opt Exp 18(12):12489. http://www.opticsexpress.org/abstract.cfm?URI=oe-18-12-12489, https://doi.org/10.1364/OE.18.012489
12. Bozhevolnyi SI, Volkov VS, Devaux E, Ebbesen TW (2005) Phys Rev Lett 95:046802. https://doi.org/10.1103/PhysRevLett.95.046802
13. Søndergaard T, Novikov SM, Holmgaard T, Eriksen RL, Beermann J, Han Z, Pedersen K, Bozhevolnyi SI (2012) Nat Commun 3:969. https://doi.org/10.1038/ncomms1976
14. Raza S, Stenger N, Pors A, Holmgaard T, Kadkhodazadeh S, Wagner JB, Pedersen K, Wubs M, Bozhevolnyi SI, Mortensen NA (2014) Nat Commun 5:4125. https://www.nature.com/articles/ncomms5125, https://doi.org/10.1038/ncomms5125
15. Bozhevolnyi SI, Volkov VS, Devaux E, Laluet JY, Ebbesen TW (2006) Nature 440:508. https://doi.org/10.1038/nature04594
16. Haffner C, Heni W, Fedoryshyn Y, Niegemann J, Melikyan A, Elder DL, Baeuerle B, Salamin Y, Josten A, Koch U, Hoessbacher C, Ducry F, Juchli L, Emboras A, Hillerkuss D, Kohl M, Dalton LR, Hafner C, Leuthold J (2015) Nat Photonics 9:525. https://doi.org/10.1038/nphoton.2015.127
17. Nielsen RB, Fernandez-Cuesta I, Boltasseva A, Volkov VS, Bozhevolnyi SI, Klukowska A, Kristensen A (2008) Opt Lett 33(23):2800. https://doi.org/10.1364/OL.33.002800
18. Gramotnev DK, Pile DFP (2004) Appl Phys Lett 85(26):6323. https://doi.org/10.1063/1.1839283
19. Li X, Jiang T, Shen L, Deng X (2013) Appl Phys Lett 102(3):031606. https://doi.org/10.1063/1.4789440
20. Pile DFP, Gramotnev DK (2004) Opt Lett 29(10):1069. http://ol.osa.org/abstract.cfm?URI=ol-29-10-1069, https://doi.org/10.1364/OL.29.001069

21. Lotan O, Smith CLC, Bar-David J, Mortensen NA, Kristensen A, Levy U (2016) ACS Photonics 3(11):2150. https://doi.org/10.1021/acsphotonics.6b00648
22. Volkov VS, Bozhevolnyi SI, Rodrigo SG, Martín-Moreno L, García-Vidal FJ, Devaux E, Ebbesen TW (2009) Nano Lett 9(3):1278. https://doi.org/10.1021/nl900268v
23. Bozhevolnyi SI, Mortensen NA (2017) Nanophotonics 6:1185. https://doi.org/10.1515/nanoph-2016-0179
24. Bermúdez-Ureña E, Gonzalez-Ballestero C, Geiselmann M, Marty R, Radko IP, Holmgaard T, Alaverdyan Y, Moreno E, García-Vidal FJ, Bozhevolnyi SI, Quidant R (2015) Nat Commun 6:7883. https://doi.org/10.1038/ncomms8883
25. Gonçalves PAD, Peres NMR (2016) An introduction to graphene plasmonics, 1st edn. World Scientific, Singapore. https://doi.org/10.1142/9948
26. García de Abajo FJ (2014) ACS Photonics 1(3):135. https://doi.org/10.1021/ph400147y
27. Grigorenko AN, Polini M, Novoselov KS (2012) Nat Photonics 6:749.https://doi.org/10.1038/nphoton.2012.262
28. Low T, Avouris P (2014) ACS Nano 8(2):1086. https://doi.org/10.1021/nn406627u
29. Christensen T, Jauho AP, Wubs M, Mortensen NA (2015) Phys Rev B 91:125414. https://doi.org/10.1103/PhysRevB.91.125414
30. Riso M, Cuevas M, Depine RA (2015) J Opt 17(7):075001. https://doi.org/10.1088/2040-8978/17/7/075001
31. Davoyan AR, Engheta N (2016) ACS Photonics 3(5):737. https://doi.org/10.1021/acsphotonics.5b00503
32. Gonçalves PAD, Dias EJC, Xiao S, Vasilevskiy MI, Mortensen NA, Peres NMR (2016) ACS Photonics 3(11):2176. https://doi.org/10.1021/acsphotonics.6b00674
33. Gonçalves PAD, Bozhevolnyi SI, Mortensen NA, Peres NMR (2017) Optica 4(6):595. http://www.osapublishing.org/optica/abstract.cfm?URI=optica-4-6-595, https://doi.org/10.1364/OPTICA.4.000595
34. Smirnova D, Mousavi SH, Wang Z, Kivshar YS, Khanikaev AB (2016) ACS Photonics 3(5):875. https://doi.org/10.1021/acsphotonics.6b00116
35. Liu P, Zhang X, Ma Z, Cai W, Wang L, Xu J (2013) Opt. Express 21(26):32432. http://www.opticsexpress.org/abstract.cfm?URI=oe-21-26-32432, https://doi.org/10.1364/OE.21.032432
36. Kim K, Lee Z, Malone BD, Chan KT, Alemán B, Regan W, Gannett W, Crommie MF, Cohen ML, Zettl A (2011) Phys Rev B 83:245433. https://doi.org/10.1103/PhysRevB.83.245433
37. Pakhnevich AA, Golod SV, Prinz VY (2015) J Phys D Appl Phys 48(43):435303. https://doi.org/10.1088/0022-3727/48/43/435303
38. Song Q, An M, Chen X, Peng Z, Zang J, Yang N (2016) Nanoscale 8:14943. https://doi.org/10.1039/C6NR01992G
39. Zhu W, Low T, Perebeinos V, Bol AA, Zhu Y, Yan H, Tersoff J, Avouris P (2012) Nano Lett 12(7):3431. https://doi.org/10.1021/nl300563h
40. Yang S, Wang C, Sahin H, Chen H, Li Y, Li SS, Suslu A, Peeters FM, Liu Q, Li J, Tongay S (2015) Nano Lett 15(3):1660. https://doi.org/10.1021/nl504276u
41. Wang B, Huang M, Kim NY, Cunning BV, Huang Y, Qu D, Chen X, Jin S, Biswal M, Zhang X, Lee SH, Lim H, Yoo WJ, Lee Z, Ruoff RS (2017) Nano Lett 17(3):1467. https://doi.org/10.1021/acs.nanolett.6b04459
42. Yan H, Low T, Zhu W, Wu Y, Freitag M, Li X, Guinea F, Avouris P, Xia F (2013) Nat Photonics 7(5):394. https://doi.org/10.1038/nphoton.2013.57
43. Abramowitz M, Stegun IA (1972) Handbook of mathematical functions: with formulas, graphs, and mathematical tables. Dover, New York. http://people.math.sfu.ca/~cbm/aands/
44. COMSOL Multiphysics. www.comsol.com
45. Bozhevolnyi SI, Nerkararyan KV (2009) Opt Lett 34(13):2039. http://ol.osa.org/abstract.cfm?URI=ol-34-13-2039, https://doi.org/10.1364/OL.34.002039
46. Bozhevolnyi SI, Nerkararyan KV (2009) Opt Exp. 17(12):10327. http://www.opticsexpress.org/abstract.cfm?URI=oe-17-12-10327, https://doi.org/10.1364/OE.17.010327
47. Press WH, Teukolsky SA, Vetterling WT, Flannery BP (2007) Numerical recipes: the art of scientific computing. Cambridge University Press, New York

Part II
Quantum Plasmonics

Chapter 6
Electrodynamics of Metals Beyond the Local-Response Approximation: Nonlocal Effects

The well-established theory of classical electrodynamics allows us to describe the optical response of metals, and their subsequent plasmonic properties, in most settings—a cursory account of classical plasmonics in terms of classical electrodynamics has been provided in Chap. 2. Broadly speaking, it relies on the macroscopic formulation of Maxwell's equations and corresponding boundary conditions, together with the premise that the materials' response to external stimuli is local in nature (cf. Sect. 2.1.3). However, this oversimplified picture—albeit undeniably accurate enough in many scenarios—inevitably fails to account for the proper electrodynamics governing the optical response of (truly nanometric) nanostructures, including processes that involve large momentum transfers and/or optical interactions in the extreme near-field regime.

In this chapter, we shall remedy some—though not all—of the inaccuracies related with the classical, local response treatment of plasmonics. Here, we will concern ourselves primarily with the incorporation of nonlocality, which is formally equivalent to consider response functions that include both frequency and wavevector dependencies, and discuss how this augmentation modifies the predictions pertaining to the local-response approximation (LRA). Specifically, we will introduce the hydrodynamical model for plasmonics in order to describe the lowest-order corrections due to nonlocality. We then close the chapter by going one step further and discuss the nonlocal and quantum features of the optical response of a three-dimensional (3D) homogeneous electron gas within the framework of the nonlocal random-phase approximation (RPA)—also known as the Lindhard dielectric function of the free-electron gas. It should be noted that both of the above-noted extensions are nevertheless still based on homogeneous, bulk-like descriptions of the plasmon-supporting electron gas, accompanied by "hard-wall" material boundaries that prevent the electronic density to spill over across dielectric–metal interfaces. We shall return to this point in Chap. 8, were we will present a mesoscopic formalism that remedies that deficiency, while at the same time incorporates nonlocality and other quantum mechanical processes.

P. A. D. Gonçalves, *Plasmonics and Light–Matter Interactions in Two-Dimensional Materials and in Metal Nanostructures*, Springer Theses,
https://doi.org/10.1007/978-3-030-38291-9_6

Shortcomings of the Classical Descriptions of Plasmonics

Before moving on to the actual description of nonlocal effects in plasmonics, it is instructive to summarize a selection [1] of the most prominent idealizations associated with a description of plasmonics rooted in classical theories.

(i) Locality | Conventional considerations of plasmons and plasmon-mediated phenomena rely on the assumption that the materials' electromagnetic response is embodied through local constitutive relations, as pointed out in Sect. 2.1.3. In reality, however, the constitutive relations are nonlocal, as clearly stated by Eqs. (2.5)–(2.6), in which the response at position $\mathbf{r}$ is ultimately dependent on the perturbation at all[1] points $\mathbf{r}'$ at an earlier time. In momentum space, this observation translates into wavevector-dependent response functions, which is often used interchangeably in the literature (and here) with terminologies such as nonlocality or spatial dispersion.

(ii) Incompleteness | Classical treatments of the optical response of metals only account for a restricted, finite subset of all available transitions, namely, electronic intraband processes described by the Drude model, potentially augmented by the *ad hoc* addition of vertical interband transitions (recall Sect. 2.1.4). Importantly, even in the simple case of a single-band jellium metal, the local optics description inevitably fails to include single-particle excitations, which contribute to a particularly important mechanism of plasmon decay, by which high-momentum plasmons experience damping due to the generation of electron-hole pairs. This mechanism is commonly known as Landau damping, and it is intertwined with a nonlocal description of the electron gas.

Finally, the LRA also fails to account for bulk plasmon photoexcitation [1, 5]: even though that, in a homogeneous medium, transverse and longitudinal electromagnetic waves do not couple, this is not longer true in the presence of an interface, through which transverse and longitudinal waves can actually mix. Therefore, electromagnetic waves of frequency $\omega > \omega_{\rm p}$ can in principle excite bulk normal modes [6–8] following $\omega_{\rm BP}^2 \simeq \omega_{\rm p}^2 + \beta^2 q^2$, with $\beta^2 = \frac{3}{5} v_F^2$. In finite-sized metallic nanostructures these longitudinal excitations can then give rise to a manifold of quantized bulk plasmons [4, 9–16].

(iii) Discontinuity | The customary assumption of an abruptly defined, piecewise dielectric function across material boundaries, say, a planar dielectric–jellium interface (at $z = 0$), combined with the boundary condition that ensures the continuity of the normal component of the displacement field, automatically implies that the normal component of the electric field is discontinuous, i.e., $E_z(0^-) = \dfrac{\epsilon_{\rm d}}{\epsilon_{\rm m}} E_z(0^+)$. Imperatively, hand in hand with this discontinuity is the materialization of an induced surface charge density, that for a vacuum–jellium interface takes the form: $\rho_{\rm ind}(\mathbf{r}) = \epsilon_0 \left[E_z(0^+) - E_z(0^-)\right] \delta(z)$ [1]. This divergent behavior is clearly unphys-

[1] Leading-order corrections are sufficiently well captured by considering a finite set of those, since they are typically dominated by the neighboring points $\mathbf{r}$' within the spatial extend around $\mathbf{r}$ defined by the nonlocal length parameter $\xi_{\rm NL} \sim v_F/\omega$ [2–4].

ical (inasmuch as the discontinuity of E_z is unphysical too).[2] However, at the microscopic level there must be a transition region where ρ_{ind} is peaked, but finite (and E_z varies continuously). Intuitively, one expects that the extent of this surface region to be roughly on the order of the screening length or the Fermi wavelength, which are typically on the order of a couple of angstroms ($\sim$Å). This is indeed the case, as shown by various theoretical calculations [1, 5, 17–21]. We shall see this explicitly when revisiting this point in Chap. 8.

In closing, we note that the above-stated division is arguably somewhat ambiguous, in the sense that a strict separation between points i–iii is not always unequivocal because there is a substantial overlap between them. Nevertheless, this conceptual categorization is advantageous and facilitates the identifications of the shortcomings that a given nonclassical formalism remedies (and the issues that potentially remain). Lastly, we comment on the fact the issues listed above are not—by all means—exhaustive. For example, in extremely small metallic clusters [22–24] or nanoparticles [24–27], the manifestation of quantum finite-size effects or features related with the atomic structure eventually lead to the breakdown of the jellium picture [5, 24, 28]. In that regime, atomistic [29, 30] or *ab initio* methods such as DFT [31] or TDDFT [32] need to be employed.

6.1 Hydrodynamic Theory of an Electron Gas

The origin of the hydrodynamic description of the electron gas is commonly attributed to Bloch [33], and subsequently elaborated upon by Jensen [34] and others [6, 35]. Over the last couple of decades, the topic has resurfaced and has seen alternative derivations and further extensions [4, 16, 36–43]. In broad strokes, the concept behind the hydrodynamic model (HDM) relies on the description of the inhomogeneous electron gas through the appropriate equation of motions (obtained, e.g., via Hamilton's equations) coupled with the electromagnetic field [14]. The macroscopic hydrodynamic theory of the electron gas has been formulated in a number of ways, ranging from rigorously applying analytical mechanics to the quantum many-body Schrödinger equation [6, 35, 39, 40], to semiclassical treatments that incorporate both classical and quantum contributions [4, 14, 36, 38, 42, 43]. Herein, we mostly follow the latter approach, and simply highlight the most important steps of the derivation rather than presenting an overly detailed account of it.[3]

Under the HDM, the dynamics of the electrons are governed by the continuity equation jointly with Cauchy's momentum equation, that is,

[2]The discontinuity of E_z and the associated divergent behavior of ρ_{ind} have practical repercussions, e.g., in electromagnetic phenomena driven by strong near-field interactions, where the LRA tends to flagrantly overestimate the strength of processes mediated by large momentum-transfers [1, 17].

[3]In that respect, the interested reader may consult, for instance, Refs. [4, 6, 14, 35, 42] and references therein.

$$\frac{\partial n}{\partial t} + \nabla \cdot (n\mathbf{v}) = 0\,, \tag{6.1a}$$

$$mn\left[\frac{\partial}{\partial t} + (\mathbf{v}\cdot\nabla)\right]\mathbf{v} = -en\left[\mathbf{E} + \mathbf{v}\times\mathbf{B}\right] - mn\gamma\mathbf{v} - n\nabla\frac{\delta G[n]}{\delta n}\,, \tag{6.1b}$$

respectively. In the previous equations, $n \equiv n(\mathbf{r}, t)$ is the particle density, $\mathbf{v} \equiv \mathbf{v}(\mathbf{r}, t)$ denotes the macroscopic particle velocity, γ is a phenomenological relaxation rate, and $G[n]$ is a density-dependent functional through which quantum mechanical effects are included (e.g., internal kinetic energy and/or xc terms). In its simplest version, $G[n]$ is modeled by the Thomas–Fermi functional [44]

$$G[n(\mathbf{r}, t)] \approx T_{\text{TF}}[n(\mathbf{r}, t)] = \frac{3\hbar^2}{10m}\left(3\pi^2\right)^{\frac{2}{3}} \int n^{\frac{5}{3}}(\mathbf{r}, t)\, d\mathbf{r}\,, \tag{6.2}$$

and thus only the internal kinetic energy of the noninteracting electron gas is taken into account. Hence, in this case, the $n\nabla\delta G[n]/\delta n$ term amounts to a generalization of the density-dependent pressure gradient, and thus this term will be responsible to take into account the Fermi degeneracy pressure of the electron gas. Performing the functional derivative, one then obtains

$$\frac{\delta G[n]}{\delta n} = \frac{\hbar^2}{2m}\left(3\pi^2\right)^{\frac{2}{3}} n^{\frac{2}{3}} = E_F\left(\frac{n}{n_0}\right)^{\frac{2}{3}}\,, \tag{6.3}$$

where we have identified $\hbar^2(3\pi^2 n_0)^{\frac{2}{3}}/(2m) = \hbar^2 k_F^2/(2m) = E_F$, with n_0 being the equilibrium (i.e., unperturbed) electron density. As they currently stand, the hydrodynamic equations (6.1) contain nonlinearities; thus, in the spirit of linear response theory, we assume that the external perturbation is small in such a way that the various quantities can be split into a equilibrium part (which is static and homogeneous) and a out-of-equilibrium part (representing dynamic and inhomogeneous quantities). Explicitly, this translates into writing $n = n_0 + n_1$ and $\mathbf{v} = \mathbf{v}_1$ (and where the fields are obviously first-order only). With these considerations in mind, the linearization of Eq. (6.1) produces

$$\frac{\partial \rho_1}{\partial t} + \nabla \cdot \mathbf{J}_1 = 0\,, \tag{6.4a}$$

$$\frac{\partial \mathbf{J}_1}{\partial t} = \frac{e^2 n_0}{m}\mathbf{E}_1 - \gamma\mathbf{J}_1 - \beta^2\nabla\rho_1\,, \tag{6.4b}$$

where we have introduced the induced charge density, $\rho_1 = -en_1$, and the induced current density, $\mathbf{J}_1 = -en_0\mathbf{v}_1$. Importantly, we have also made the identification $\beta^2 = \frac{1}{3}v_F^2$, where β is a velocity. At this point, a remark is in order: this value for β holds in the low-frequency regime where $\omega \ll \gamma$, but in plasmonics we are interested in the opposite, high-frequency regime where $\omega \gg \gamma$. In the latter, the appropriate[4] value for β is given by $\beta^2 = \frac{3}{5}v_F^2$, thereby coinciding with the leading-order nonlocal correction resulting from the expansion of the Lindhard function of a 3DEG. For this reason, we take $\beta = \sqrt{\frac{3}{5}}v_F$ hereafter.

Now, by combining the linearized equations of motion (6.4), one finally arrives at (where the subscripts have been dropped) [4, 48]

$$\boxed{\frac{\beta^2}{\omega(\omega+i\gamma)}\nabla\left[\nabla\cdot\mathbf{J}(\mathbf{r},\omega)\right]+\mathbf{J}(\mathbf{r},\omega)=\sigma(\omega)\mathbf{E}(\mathbf{r},\omega)}\,, \tag{6.5}$$

after converting to the frequency domain. Notice that, in the absence of nonlocality we have $\beta = 0$, and thus the familiar LRA result $\mathbf{J}(\mathbf{r},\omega) = \sigma(\omega)\mathbf{E}(\mathbf{r},\omega)$ is promptly recovered in the $\beta \to 0$ limit. The HDM result (6.5) supplemented with Maxwell's wave equation

$$\nabla\times\nabla\times\mathbf{E}(\mathbf{r},\omega) - k_0^2\epsilon_\infty\mathbf{E}(\mathbf{r},\omega) = i\omega\mu_0\mathbf{J}(\mathbf{r},\omega)\,, \tag{6.6}$$

then form together a closed set of coupled equations describing the electrodynamics of the electron gas within the framework of the hydrodynamic model.

Furthermore, with the hydrodynamic differential equation (6.5) comes the need of an additional boundary condition [4, 14, 49]. For an arbitrary dielectric–metal interface $\partial\Omega$, and assuming that the metal's equilibrium electron density exhibits a step profile (i.e., being constant inside the metal and zero in the dielectric region), the aforementioned additional boundary condition reads [4]

$$\mathbf{J}\cdot\hat{\mathbf{n}}\Big|_{\mathbf{r}\in\partial\Omega} = 0\,, \tag{6.7}$$

at the boundary $\partial\Omega$ with associated unit normal vector $\hat{\mathbf{n}}$. Equation (6.7) therefore enforces that the normal component of the induced current density is zero at the interface, and thus it is continuous across the same. This "hard wall" boundary condition is therefore tantamount to the assumption of an infinite work function, and hence electronic spill-out effects are inherently overlooked in the HDM.

---

[4] The *proper* value of β—the "*speed of propagation of [the] hydrodynamic disturbance in the electron gas*" [6]—has been discussed extensively in the literature—see, for instance, Refs. [14, 36, 42, 45, 46]. Among these, a particularly elegant result has been presented by Halevi in Ref. [36], where he derived the result $\beta_{\text{Halevi}}(\omega) = \sqrt{\frac{\frac{3}{5}\omega+\frac{1}{3}i\gamma}{\omega+i\gamma}}\,v_F$. This frequency-dependent, complex-valued β stems from the exploitation of the Mermin-corrected Lindhard function [44, 47] in the limit of weak spatial dispersion.

Box 6.1-Longitudinal and transverse fields in the HDM

The two coupled Eqs. (6.5) and (6.6) may be combined to yield [4, 41, 48]

$$\nabla \times \nabla \times \mathbf{E}(\mathbf{r}, \omega) - k_0^2 \epsilon_{\rm m}(\omega) \mathbf{E}(\mathbf{r}, \omega) = k_0^2 \, \xi_{\rm NL}^2 \nabla \left[\nabla \cdot \mathbf{E}(\mathbf{r}, \omega) \right] , \tag{6.8}$$

where $\xi_{\rm NL}$ is a nonlocal length defined via

$$\xi_{\rm NL}^2 = \frac{\epsilon_\infty}{\omega(\omega + i\gamma)} \beta^2 \, . \tag{6.9}$$

Equation (6.8) can be thought as a generalized nonlocal wave equation within the HDM [48].

Next, it is instructive to consider longitudinal and transverse fields separately. Starting from the latter, and noting that a transverse electric field is divergence-free ($\nabla \cdot \mathbf{E}_{\rm T} = 0$), then from Eq. (6.8) we see that

$$\nabla \times \nabla \times \mathbf{E}_{\rm T} - k_0^2 \epsilon_{\rm m}(\omega) \mathbf{E}_{\rm T} = 0 \, , \tag{6.10}$$

and thus, assuming plane-wave solutions, one finds the bulk wavevector

$$k_{\rm T}^2 = k_0^2 \epsilon_{\rm m}(\omega) \, , \tag{6.11}$$

which is the familiar dispersion relation of light propagating in homogeneous medium (here a metal) with dielectric function $\epsilon_{\rm m}(\omega)$. On the other hand, a purely longitudinal electric field is curl-free ($\nabla \times \mathbf{E}_{\rm L} = 0$), and thus Eq. (6.8) yields

$$\xi_{\rm NL}^2 \nabla^2 \mathbf{E}_{\rm L} + \epsilon_{\rm m}(\omega) \mathbf{E}_{\rm L} = 0 \, , \tag{6.12}$$

where the identity $\nabla (\nabla \cdot \mathbf{F}) = \nabla \times \nabla \times \mathbf{F} + \nabla^2 \mathbf{F}$ has been used. Hence, the bulk wavevector associated with the longitudinal electric field is

$$k_{\rm L}^2 = \frac{\epsilon_{\rm m}(\omega)}{\xi_{\rm NL}^2} \, . \tag{6.13}$$

Crucially, these considerations reveal that the nonlocal correction in the framework of the HDM only affects longitudinal waves, and leaves transverse waves unchanged. Therefore, all nonlocal effects under the hydrodynamic approach can only enter via the presence of a longitudinal part of the electric field [4, 14, 48, 50].

6.2 Nonlocal Optical Response of Metals

Having established the underpinnings of the hydrodynamic theory of the electron fluid in Sect. 6.1, we are now in position to apply the HDM and investigate nonlocal effects in plasmonic systems. Below, we provide an overview of the key modifications to LRA-predicted plasmonic response of a planar dielectric–metal interface and of metallic spheres.

6.2.1 Nonlocal Effects in the SPP Dispersion at a Planar Dielectric–Metal Interface

We consider a prototypical planar dielectric–metal interface (Fig. 6.1), but where now the metal occupying the $z > 0$ half-space exhibits a nonlocal response in accordance with the premises of the HDM.

In order to determine the nonlocal SPP dispersion relation, we shall look for p-polarized surface electromagnetic waves that satisfy Maxwell's equations both in the dielectric and in the metal [and, in the latter, that further satisfy Eq. (6.5)]. The electromagnetic field in the dielectric half-space ($z < 0$) is purely transverse, and may thus be written as

$$\mathbf{E}_\mathrm{d}(\mathbf{r}, \omega) = \left[E_{x,\mathrm{d}}, 0, E_{z,\mathrm{d}}\right] e^{\kappa_\mathrm{d} z} e^{i(qx-\omega t)}\,, \tag{6.14a}$$

$$\mathbf{H}_\mathrm{d}(\mathbf{r}, \omega) = \left[0, H_{y,\mathrm{d}}, 0\right] e^{\kappa_\mathrm{d} z} e^{i(qx-\omega t)}\,, \tag{6.14b}$$

where $\kappa_\mathrm{d} = \sqrt{q^2 - k_\mathrm{d}^2}$, with $k_\mathrm{d} = k_0\sqrt{\epsilon_\mathrm{d}}$, and together with the relations[5]

$$E_{z,\mathrm{d}} = -i\frac{q}{\kappa_\mathrm{d}} E_{x,\mathrm{d}}\,, \tag{6.15a}$$

$$E_{x,\mathrm{d}} = \frac{\kappa_\mathrm{d}}{i\omega\epsilon_0\epsilon_\mathrm{d}} H_{y,\mathrm{d}}\,, \tag{6.15b}$$

$$E_{z,\mathrm{d}} = -\frac{q}{\omega\epsilon_0\epsilon_\mathrm{d}} H_{y,\mathrm{d}}\,. \tag{6.15c}$$

In the metal region ($z > 0$), both transverse and longitudinal waves (i.e., divergence-free and curl-free waves, respectively) are allowed; hence, we have

$$\mathbf{E}^\mathrm{T/L}_\mathrm{m}(\mathbf{r}, \omega) = \left[E^\mathrm{T/L}_{x,\mathrm{m}}, 0, E^\mathrm{T/L}_{z,\mathrm{m}}\right] e^{-\kappa_{\mathrm{m/nl}} z} e^{i(qx-\omega t)}\,, \tag{6.16a}$$

$$\mathbf{H}_\mathrm{m}(\mathbf{r}, \omega) = \left[0, H_{y,\mathrm{m}}, 0\right] e^{-\kappa_\mathrm{m} z} e^{i(qx-\omega t)}\,, \tag{6.16b}$$

[5] Stemming from Maxwell's equations $\nabla \times \mathbf{E}_\mathrm{d}(\mathbf{r}, \omega) = i\omega\mu_0\mathbf{H}_\mathrm{d}(\mathbf{r}, \omega)$ and $\nabla \cdot \mathbf{E}_\mathrm{d}(\mathbf{r}, \omega) = 0$.

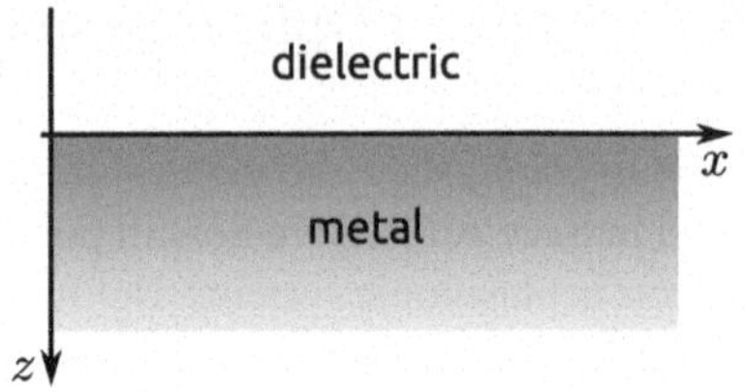

Fig. 6.1 Schematic representation of a planar dielectric–metal interface, where the boundary between the two media is defined by the $z = 0$ plane. The system is assumed to be uniform along the y-direction

where[6]

$$\kappa_{\rm m} = \sqrt{q^2 - k_{\rm m}^2} \quad \text{with} \quad k_{\rm m} = k_0\sqrt{\epsilon_{\rm m}}\,, \tag{6.17a}$$

$$\kappa_{\rm nl} = \sqrt{q^2 - k_{\rm nl}^2} \quad \text{with} \quad k_{\rm nl} = \beta^{-1}\sqrt{\omega(\omega + i\gamma) - \omega_p^2/\epsilon_\infty}\,. \tag{6.17b}$$

Notice that the magnetic field remains purely transverse and thus we do not need the superscripts to distinguish the two types of waves. The relations between the fields' amplitudes can be found using Maxwell's equations. Explicitly, these are:

$$E^{\rm T}_{z,\rm m} = i\frac{q}{\kappa_{\rm m}} E^{\rm T}_{x,\rm m}\,, \tag{6.18a}$$

$$E^{\rm T}_{x,\rm m} = \frac{i\kappa_{\rm m}}{\omega\epsilon_0\epsilon_{\rm m}} H_{y,\rm m}\,, \tag{6.18b}$$

$$E^{\rm T}_{z,\rm m} = -\frac{q}{\omega\epsilon_0\epsilon_{\rm m}} H_{y,\rm m}\,, \tag{6.18c}$$

$$E^{\rm L}_{z,\rm m} = i\frac{\kappa_{\rm nl}}{q} E^{\rm L}_{x,\rm m}\,. \tag{6.18d}$$

We now proceed by invoking the boundary conditions, which are now three: the two ubiquitous Maxwell's boundary conditions, $\hat{\mathbf{z}} \times (\mathbf{E}_{\rm m} - \mathbf{E}_{\rm d})\,|_{z=0} = 0$ and $\hat{\mathbf{z}} \times (\mathbf{H}_{\rm m} - \mathbf{H}_{\rm d})\,|_{z=0} = 0$, plus a third one stated by Eq. (6.7). The latter can be equivalently expressed as

$$\left(\epsilon_\infty \mathbf{E}_{\rm m} \cdot \hat{\mathbf{z}} - \epsilon_{\rm d}\mathbf{E}_{\rm d} \cdot \hat{\mathbf{z}}\right)|_{z=0} = 0\,. \tag{6.19}$$

Putting together Eqs. (6.14)–(6.18) jointly with the corresponding boundaries conditions, leads to the following determinantal equation (where, for convenience, we have defined $\bar{E}^{\rm L}_{x,\rm m} = i\omega\epsilon_0 E^{\rm L}_{x,\rm m}$)

[6]Recall Box 6.1, with the identifications between both notations being $k_{\rm T} \equiv k_{\rm m}$ and $k_{\rm L} \equiv k_{\rm nl}$.

$$\begin{pmatrix} \frac{\kappa_d}{\epsilon_d} & \frac{\kappa_m}{\epsilon_m} & -1 \\ 1 & -1 & 0 \\ -\frac{1}{\epsilon_\infty} & \frac{1}{\epsilon_m} & -\frac{\kappa_{nl}}{q^2} \end{pmatrix} \begin{pmatrix} H_{y,d} \\ H_{y,m} \\ \bar{E}^{L}_{x,m} \end{pmatrix} = 0 . \tag{6.20}$$

Finally, the implicit condition for the nonlocal SPP's dispersion then follows from equating the determinant of the previous matrix to zero, which yields [14, 50]

$$1 + \frac{\epsilon_m \kappa_d}{\epsilon_d \kappa_m} + \delta_{nl} = 0 , \tag{6.21a}$$

where the nonlocal correction δ_{nl} amounts to

$$\delta_{nl} = \frac{q^2}{\kappa_{nl}\kappa_m} \frac{\epsilon_m - \epsilon_\infty}{\epsilon_\infty} . \tag{6.21b}$$

Equation (6.21a) determines the dispersion relation of SPPs at an interface between a dielectric and a metal, taking into account nonlocal effects within the framework of the HDM. The LRA SPP condition (2.23) is naturally recovered upon taking the limit $\delta_{nl} \to 0$.

In the nonretarded limit, Eqs. (6.21) become

$$1 + \frac{\epsilon_m}{\epsilon_d} + \delta^{nr}_{nl} = 0 \qquad \text{with} \qquad \delta^{nr}_{nl} = \frac{q}{\kappa_{nl}} \frac{\epsilon_m - \epsilon_\infty}{\epsilon_\infty} . \tag{6.22}$$

Moreover, in the specific case of a vacuum–jellium interface with negligible loss (i.e., $\epsilon_d = \epsilon_\infty = 1$ and $\gamma = 0$), the solutions of Eq. (6.22) take the form

$$\omega(q) = \frac{1}{\sqrt{2}} \sqrt{\omega_p^2 + \beta^2 q^2 + \beta q \sqrt{2\omega_p^2 + \beta^2 q^2}} . \tag{6.23}$$

The solutions of Eqs. (6.21)–(6.22) are shown in Fig. 6.2. The corresponding LRA-solutions are also plotted for comparison. It is apparent from the figure that for large wavevectors there is a clear deviation from the LRA prediction towards higher frequencies; this blueshift of the SPP resonance—for a fixed SPP wavevector q—becomes particularly evident upon inspection of the figure's inset. Notably, notice that in the presence of nonlocality the SPP dispersion even crosses the dispersionless nonretarded surface plasmon resonance associated with the LRA, i.e., $\omega_{SP} = \omega_p/\sqrt{2}$. Lastly, it also transpires that in the nonlocal case the SPP group velocity remains finite, which is in stark contrast with the LRA prediction.

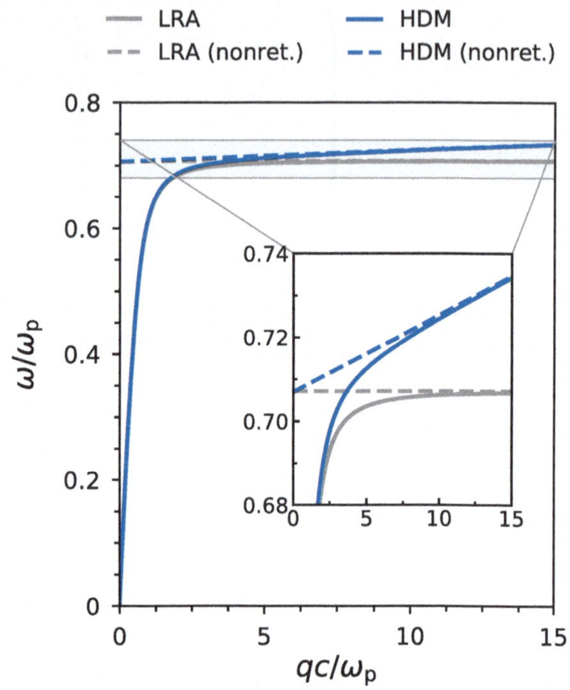

Fig. 6.2 Surface plasmon polariton dispersion in a flat dielectric–metal interface. The dielectric is assumed to be air ($\epsilon_d = 1$) and the metal is modeled within the HDM as a lossless Drude-like metal with $\epsilon_\infty = 1$ and $\gamma = 0$. Here, $\beta = \sqrt{3/5}\, v_F$ with $v_F = 1.4 \times 10^6$ m s^{-1}, which is adequate for silver and gold [51]. The inset shows in detail the region marked by the shaded area in the main plot

6.2.2 *Nonlocal Plasmonic Response of Metallic Nanospheres*

Like the planar dielectric–metal interface, the spherical metal particle corresponds to another archetypal plasmonic structure. In the following, we give a concise description of nonlocality in plasmonic nanospheres within the HDM description.[7]

As we have seen in Sect. 2.2.2, the gold standard for theoretically modeling the electromagnetic response of a spherical particle is through the application of Mie theory. The first attempt to generalize Mie's theory by incorporating nonlocal effects was set forth by Ruppin [11, 12] in the 1970s. As we have just seen in the case of the planar interface, the hydrodynamic description for a metal sphere also involves the incorporating of a curl-free (or irrotational) wave [16, 53].

In particular, the Mie scattering coefficients within the HDM are given by [16]

$$a_l^{\text{TM}} = \frac{\epsilon_{\text{m}} j_l(x_{\text{m}}) \Psi_l'(x_{\text{d}}) - \epsilon_{\text{d}} j_l(x_{\text{d}}) \Psi_l'(x_{\text{m}}) \left[1 + \delta_l^{\text{HDM}}\right]}{\epsilon_{\text{m}} j_l(x_{\text{m}}) \xi_l'(x_{\text{d}}) - \epsilon_{\text{d}} h_l^{(1)}(x_{\text{d}}) \Psi_l'(x_{\text{m}}) \left[1 + \delta_l^{\text{HDM}}\right]}, \tag{6.24a}$$

$$b_l^{\text{TE}} = \frac{j_l(x_{\text{m}}) \Psi_l'(x_{\text{d}}) - j_l(x_{\text{d}}) \Psi_l'(x_{\text{m}})}{j_l(x_{\text{m}}) \xi_l'(x_{\text{d}}) - h_l^{(1)}(x_{\text{d}}) \Psi_l'(x_{\text{m}})}, \tag{6.24b}$$

[7]A detailed account is given, for instance, in Refs. [16, 52].

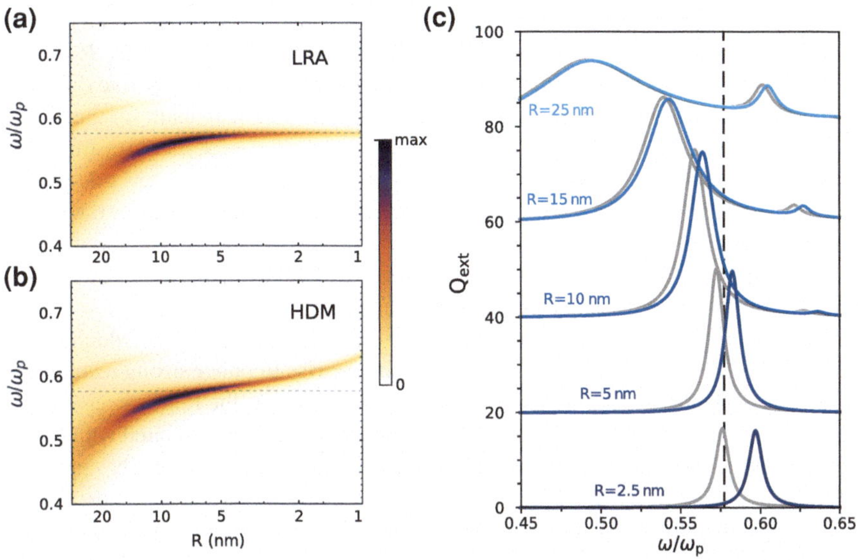

Fig. 6.3 Colormap of the normalized extinction cross-section of plasmonic nanopheres, $Q_{\text{ext}} \equiv \sigma_{\text{ext}}/(\pi R^2)$. **a** Local Mie theory and **b** nonlocal hydrodynamic Mie theory for metallic spheres of varying radii. **c** Local (gray lines) and nonlocal (blue lines) extinction spectra for metal nanoparticles with five different radii. Parameters: $\epsilon_{\text{d}} = \epsilon_\infty = 1$, $\hbar\omega_{\text{p}} = 10\,\text{eV}$, $\hbar\gamma = 0.1\,\text{eV}$, and $\beta = \sqrt{3/5}\,v_F$ with $v_F = 1.4 \times 10^6\,\text{m s}^{-1}$

where the nonlocal correction in the case of a metal sphere takes the form [16]

$$\delta_l^{\text{HDM}} = l(l+1)\frac{j_l(x_{\text{m}})}{\Psi_l'(x_{\text{m}})}\frac{j_l(x_{\text{nl}})}{x_{\text{nl}}\, j_l'(x_{\text{nl}})}\frac{\epsilon_{\text{m}} - \epsilon_\infty}{\epsilon_\infty}, \tag{6.25}$$

with $x_j = k_j R$ for $j \in \{\text{d, m, nl}\}$ (and where the wavevectors are defined in the same way as in Sect. 6.2.1). The definitions of Riccati–Bessel functions Ψ_l and ξ_l are the same as in Sect. 2.2.2. Notice that the TE Mie coefficient b_l^{TE} does not exhibit any nonlocal correction; this should not be surprising, since TE-polarized waves do not couple to the longitudinal component of the electric field that is responsible for the introduction of nonlocality in the HDM. Again, in the $\beta \to 0$ limit the nonlocal correction vanishes and the conventional LRA Mie coefficients (2.31) are reinstated.

In possession of the hydrodynamic Mie coefficients (6.24), the optical response of metal spheres to an impinging plane-wave can be unambiguously calculated using the formulas for the extinction, scattering, and absorption cross-sections given by Eq. (2.32).

The outcome of the application of the nonlocal Mie theory—i.e., in accordance with Eq. (6.24)—is depicted in Fig. 6.3. In particular, Fig. 6.3a, b show, respectively, the local and nonlocal normalized extinction cross-sections, $Q_{\text{ext}} \equiv \sigma_{\text{ext}}/(\pi R^2)$, for metal spheres of varying radii. Its main features are clearly evocative of the ones exhibited by the SPP dispersion in the planar case, cf. Fig. 6.2, but where the role of the in-plane wavevector is now played by the inverse of the sphere's radius, R^{-1}. The observed blueshift of the LSP resonances is reiterated in Fig. 6.3c, for five different radii. For small, deeply subwavelength nanospheres the classical result for the dipole LSP approaches the nonretarded limit of $\omega_{l=1} = \omega_p/\sqrt{\epsilon_\infty + 2\epsilon_d}$; contrasting this, the HDM prediction crosses this value and continues to blueshift towards higher frequencies (again, mimicking the nonlocal correction previously seen for the planar dielectric–metal interface).

Furthermore, owing to the intrinsically finite size of a sphere, there are also higher-order (multipolar) LSP resonances at higher frequencies, one of which ($l = 2$) can readily seen in Fig. 6.3 for spheres of moderate size. In larger spheres—which we do not consider here—such higher-order LSPs can give rise to rather strong resonances (recall Fig. 2.8), which are then progressively suppressed as the radius decreases (a signature of the so-called dipole limit). Nevertheless, the LSPs associated with $l \geq 2$, like their dipole sibling, also undergo blueshifts in the hydrodynamically-corrected setting.

Before concluding, we briefly comment on the fact that the HDM also predicts the existence of "confined" bulk plasmons for frequencies above the screened plasma frequency $\omega_p/\sqrt{\epsilon_\infty}$, which have no parallel in the LRA [16]. They are, however, extremely challenging to observe experimentally due to their inherently weak response.

6.3 Nonlocality at a Planar Interface between a Dielectric and Lindhard Metal

One of the most famous properties of SPPs is their ability to confine the electromagnetic field below the diffraction limit of conventional optics [54–56], and thus plasmonics constitutes a route towards extreme photonics and plasmon–emitter interactions [17, 57–61]. However, in the extremely deep subwavelength regime, and, in particular, when the surface plasmon's wavevector approaches the intrinsic length scales associated with the electron's dynamics, a description of the underlying physics in terms of classical electrodynamics inevitably breaks down [5, 21, 62] (as we have discussed in the beginning of the present chapter). In the previous section, we have studied nonlocal effects within the simplest version of the hydrodynamic model, which is able to capture the leading-order nonlocal correction impacting electrodynamics of the electron gas. Nevertheless, for $q \gtrsim k_F$ (or, perhaps more accurately, for $q \gtrsim \frac{\omega}{v_F}$), the small-momentum expansion that is at the heart of the HDM invariably becomes insufficient. Going further, it should be also mentioned that, despite its merits [4, 63], the HDM is still a crude approximation, in particular in what Landau damping is concerned. Indeed, this rather important mechanism of nonlocal damping—that accounts for the possibility of plasmon decay into excited electron-hole pairs—is utterly overlooked[8] in the HDM.

To address the above-mentioned shortcomings, we shall describe the electrodynamic problem using the full dielectric function of the electron gas. Concretely, we treat the metal's response at the level of the nonlocal RPA [44, 65] which is equivalent to the use of the Lindhard dielectric function of the homogeneous 3DEG [66] (its explicit form is indicated in Appendix A). To that end, in the following we employ the so-called specular reflection model (SRM) [8, 67] together with the concept of surface impedance to construct a generalized reflection coefficient for p-polarized waves that incorporates the nonlocal effects through the Lindhard dielectric function [44, 66]. The mathematical details of this formalism are described in Appendix B and thus we shall avoid repeating them here; instead, we will simply state the relevant expressions, commenting along the way. Finally, we end this section with an explicit calculation of the surface plasmon dispersion obtained using the SRM.

The optical response of the system can be uncovered from the structure's scattering coefficients. In the case of a planar dielectric–metal interface, the poles of the reflection coefficient for p-polarization indicate the existence of surface plasmons. A slightly less conventional way to write r_p is to exploit the notion of surface impedance,[9] defined as $Z_p = \mathbf{E}_\parallel / \left(\mathbf{H}_\parallel \times \hat{\mathbf{n}}_{\text{in}}\right)$ [67, 71, 72] where the subscript in the fields denotes their component tangent to the interface and $\hat{\mathbf{n}}_{\text{in}}$ is a unit normal vector pointing inwards (i.e., into the metal). In particular—as shown in Appendix B—the reflection coefficient for p-polarized waves can be written in terms of the corresponding surface impedance via [67]

$$r_p = \frac{\dfrac{k_{z,\text{d}}}{\epsilon_\text{d} k_0} - Z_p/Z_0}{\dfrac{k_{z,\text{d}}}{\epsilon_\text{d} k_0} + Z_p/Z_0}, \tag{6.26}$$

where the quantity $Z_0 = \sqrt{\mu_0/\epsilon_0}$ is the impedance of free-space and $k_{z,\text{d}}^2 = \epsilon_\text{d} k_0^2 - q^2$. Inserting the expression for the surface impedance, given by Eq. (B.18), one finds [67]

---

[8] In fairness, some attempts to include nonlocal damping *phenomenologically* in the hydrodynamic theory of the electron gas have been put forth, e.g., the Kreibig model [64] for size-dependent *bulk* damping and the semiclassical generalized nonlocal optical response (GNOR) [41] framework that includes electron diffusion. The former phenomenologically modifies the Drude-type bulk damping according to $\gamma \to \gamma + A\frac{v_F}{R}$ for small spheres (and where A is a fitting constant), whereas the latter can be obatined from the HDM simply by performing the replacement $\beta^2 \to \beta^2 + D(\gamma - i\omega)$ (where D is too a fitting constant). Nevertheless, an accurate account of Landau damping can only be administered by using quantum mechanics, where the *actual electronic transitions* are properly included.

[9] The introduction of the concept of surface impedance is generally attributed to Leontovich [68], although both Shchukin [69] and Rytov [70] have also made significant contributions [71]. It has been widely applied to describe the so-called skin effect in metals at low frequencies [71], the penetration depth of the magnetic field in superconductors [72], and metasurfaces [73, 74].

$$r_p = \frac{k_{z,\mathrm{d}} - i\dfrac{2\epsilon_\mathrm{d}}{\pi}\displaystyle\int_0^\infty \frac{dk_z}{q^2+k_z^2}\left[\frac{q^2}{\epsilon_\mathrm{L}(k,\omega)} + \frac{k_z^2}{\epsilon_\mathrm{T}(k,\omega) - (q^2+k_z^2)/k_0^2}\right]}{k_{z,\mathrm{d}} + i\dfrac{2\epsilon_\mathrm{d}}{\pi}\displaystyle\int_0^\infty \frac{dk_z}{q^2+k_z^2}\left[\frac{q^2}{\epsilon_\mathrm{L}(k,\omega)} + \frac{k_z^2}{\epsilon_\mathrm{T}(k,\omega) - (q^2+k_z^2)/k_0^2}\right]}, \quad (6.27)$$

where $k = |\mathbf{k}| = \sqrt{q^2 + k_z^2}$.

In the electrostatic limit—which typically coincides with the regime where nonlocal effects are more pronounced—the previous expression simplifies considerably to [8, 67]

$$r_p = \frac{1 - 2\pi^{-1}\epsilon_\mathrm{d}\, I(q,\omega)}{1 + 2\pi^{-1}\epsilon_\mathrm{d}\, I(q,\omega)}, \quad (6.28a)$$

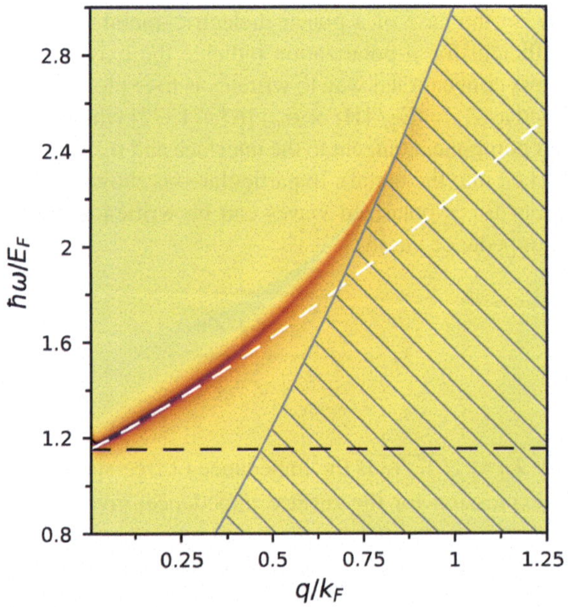

Fig. 6.4 Loss function $\mathrm{Im}\, r_p$ showing the nonretarded surface plasmon dispersion at a planar dielectric–metal interface calculated according to the specular reflection model with the Lindhard dielectric function (including Mermin's RTA correction). The colormap is plotted in logarithmic scale. The white dashed curve represents the surface plasmon dispersion predicted by the HDM, and thus contains only the lowest-order nonlocal correction to the LRA. The horizontal black dashed line indicates the nonretarded LRA surface plasmon frequency $\omega_\mathrm{p}/\sqrt{\epsilon_\infty + \epsilon_\mathrm{d}}$. The hatched region marks the electron-hole continuum of a bulk 3DEG. Material parameters: jellium metal with $\epsilon_\infty = 1$ and a Wigner–Seitz radius of $r_s = 3$ (in units of the Bohr radius), which corresponds to a plasma frequency of $\hbar\omega_\mathrm{p} = 9.07\,\mathrm{eV}$. We further take a Drude-type damping of $\hbar\gamma = 20\,\mathrm{meV}$ and assume that the dielectric medium is air ($\epsilon_\mathrm{d} = 1$)

with

$$I(q, \omega) = q \int_0^\infty \frac{dk_z}{q^2 + k_z^2} \frac{1}{\epsilon_{\rm L}\left(\sqrt{q^2 + k_z^2}, \omega\right)}. \tag{6.28b}$$

The nonlocal dispersion relation of surface plasmons at a planar dielectric–metal interface, computed using the nonretarded SRM result (6.28a), is shown in Fig. 6.4. Specifically, the figure's colormap depicts the loss function via Im r_p; the surface plasmon dispersion is therefore registered by the continuous peak in the loss function (indicated by the darker band starting at $\omega_{\rm p}/\sqrt{\epsilon_\infty + \epsilon_{\rm d}}$ and subsequently blueshifting towards higher frequencies). In addition, we have plotted the dispersion curve calculated within the simpler HDM (dashed white line) for comparison. Markedly, the HDM prediction provides a good approximation to the surface plasmon dispersion up to about $q/k_F \lesssim 0.4$. Beyond this point it progressively deteriorates, particularly failing to a great extent when it approaches the region of *bulk* intraband Landau damping—starting at $\hbar^2 q^2/(2m) + \hbar v_F q$—after which it practically ceases to exist due to the presence of this nonlocal damping mechanism. In fact, in the case of a semi-infinite metal, and as opposed to the case of an infinite, bulk metal, the region of finite Landau damping (even in the collisionless case!) extends down to low wavevectors due to the lack of translational invariance along the direction normal to the metal's surface [67], as can be readily recognized upon inspecting Eq. (6.28).

References

1. Feibelman PJ (1982) Prog Surf Sci 12(4):287. https://doi.org/10.1016/0079-6816(82)90001-6
2. Ginzburg P, Zayats AV (2013) ACS Nano 7(5):4334. https://doi.org/10.1021/nn400842m
3. Mortensen NA (2013) Photonics nanostructures. Fundam Appl 11(4):303. http://www.sciencedirect.com/science/article/pii/S1569441013000370, https://doi.org/10.1016/j.photonics.2013.06.002
4. Raza S, Bozhevolnyi SI, Wubs M, Mortensen NA (2015) J Phys: Condens Matter 27(18):183204. https://doi.org/10.1088/0953-8984/27/18/183204
5. Liebsch A (1997) Electronic excitations at metal surfaces. Springer, New York
6. Ritchie RH, Wilems RE (1969) Phys Rev 178:372. https://doi.org/10.1103/PhysRev.178.372
7. Barton G (1979) Rep Prog Phys 42(6):963. https://doi.org/10.1088/0034-4885/42/6/001.
8. Pitarke JM, Silkin VM, Chulkov EV, Echenique PM (2006) Rep Prog Phys 70(1):1. https://doi.org/10.1088/0034-4885/70/1/R01
9. Lindau I, Nilsson PO (1971) Phys Scr 3(2):87. https://doi.org/10.1088/0031-8949/3/2/007
10. Anderegg M, Feuerbacher B, Fitton B (1971) Phys Rev Lett 27:1565. https://doi.org/10.1103/PhysRevLett.27.1565
11. Ruppin R (1973) Phys Rev Lett 31:1434. https://doi.org/10.1103/PhysRevLett.31.1434
12. Ruppin R (1975) Phys Rev B 11:2871. https://doi.org/10.1103/PhysRevB.11.2871
13. Boardman AD, Paranjape BV (1977) J Phys F: Met Phys 7(9):1935. https://doi.org/10.1088/0305-4608/7/9/036
14. Boardman AD (1982) Electromagnetic surface modes. Wiley
15. Özer MM, Moon EJ, Eguiluz AG, Weitering HH (2011) Phys Rev Lett 106:197601. https://doi.org/10.1103/PhysRevLett.106.197601

16. Christensen T, Yan W, Raza S, Jauho AP, Mortensen NA, Wubs M (2014) ACS Nano 8(2):1745. https://doi.org/10.1021/nn406153k
17. Gonçalves PAD, Christensen T, Rivera N, Jauho AP, Mortensen NA, Soljačić M (2020) Plasmon-emitter interactions at the nanoscale. Nat Commun 11:366. https://doi.org/10.1038/s41467-019-13820-z
18. Lang ND, Kohn W (1970) Phys Rev B 1:4555. https://doi.org/10.1103/PhysRevB.1.4555
19. Lang ND, Kohn W (1971) Phys Rev B 3:1215. https://doi.org/10.1103/PhysRevB.3.1215
20. Kenner VE, Allen RE, Saslow WM (1972) Phys Lett A 38(4):255. http://www.sciencedirect.com/science/article/pii/0375960172900680, https://doi.org/10.1016/0375-9601(72)90068-0
21. Christensen T, Yan W, Jauho AP, Soljačić M, Mortensen NA (2017) Phys Rev Lett 118:157402. https://doi.org/10.1103/PhysRevLett.118.157402
22. Reiners T, Ellert C, Schmidt M, Haberland H (1995) Phys Rev Lett 74:1558. https://doi.org/10.1103/PhysRevLett.74.1558
23. Mandal S, Wang J, Winans RE, Jensen L, Sen A (2013) J Phys Chem C 117(13):6741. https://doi.org/10.1021/jp310514z
24. Varas A, García-González P, Feist J, García-Vidal FJ, Rubio A (2016) Nanophotonics 5(3):409. https://doi.org/10.1515/nanoph-2015-0141
25. He Y, Zeng T (2010) J Phys Chem C 114(42):18023. https://doi.org/10.1021/jp101598j
26. Scholl JA, Koh AL, Dionne JA (2012) Nature 483(7390):421. https://doi.org/10.1038/nature10904
27. Campos A, Troc N, Cottancin E, Pellarin M, Weissker HC, Lermé J, Kociakand M, Hillenkamp M (2018) Nat Phys. https://doi.org/10.1038/s41567-018-0345-z
28. Zhang P, Feist J, Rubio A, García-González P, García-Vidal FJ (2014) Phys Rev B 90:161407. https://doi.org/10.1103/PhysRevB.90.161407
29. Thongrattanasiri S, Manjavacas A, García de Abajo FJ (2012) ACS Nano 6(2):1766. https://doi.org/10.1021/nn204780e
30. Christensen T, Wang W, Jauho AP, Wubs M, Mortensen NA (2014) Phys Rev B 90:241414. https://doi.org/10.1103/PhysRevB.90.241414
31. Fiolhais C, Marques MA, Ullrich CA, Nogueira F (eds) (2003) A primer in density functional theory. Lecture notes in physics. Springer, New York
32. Marques MA, Ullrich CA, Nogueira F, Rubio A, Burke K, Gross EKU (2006) Time-dependent density functional theory. Lecture notes in physics. Springer, New York
33. Bloch F (1933) Z Phys 81(5):363. https://doi.org/10.1007/BF01344553
34. Jensen H (1937) Z Phys 106(9):620. https://doi.org/10.1007/BF01339901
35. Eguiluz A, Quinn JJ (1976) Phys Rev B 14:1347. https://doi.org/10.1103/PhysRevB.14.1347
36. Halevi P (1995) Phys Rev B 51:7497. https://doi.org/10.1103/PhysRevB.51.7497
37. Tokatly I, Pankratov O (1999) Phys Rev B 60:15550. https://doi.org/10.1103/PhysRevB.60.15550
38. Marklund M, Brodin G, Stenflo L, Liu CS (2008) EPL 84(1):17006. https://doi.org/10.1209/0295-5075/84/17006
39. Tao J, Gao X, Vignale G, Tokatly IV (2009) Phys Rev Lett 103:086401. https://doi.org/10.1103/PhysRevLett
40. Gao X, Tao J, Vignale G, Tokatly IV (2010) Phys Rev B 81:195106. https://doi.org/10.1103/PhysRevB.81.195106
41. Mortensen NA, Raza S, Wubs M, Søndergaard T, Bozhevolnyi SI (2014) Nat Commun 5:3809. https://doi.org/10.1038/ncomms4809
42. Yan W (2015) Phys Rev B 91:115416. https://doi.org/10.1103/PhysRevB.91.115416
43. Moldabekov ZA, Bonitz M, Ramazanov TS (2018) Phys Plasmas 25(3):031903. https://doi.org/10.1063/1.5003910
44. Giuliani G, Vignale G (2005) Quantum theory of the electron liquid. Cambridge University Press. https://doi.org/10.1017/CBO9780511619915
45. Fetter AL (1973) Ann Phys 81(2):367. http://www.sciencedirect.com/science/article/pii/0003491673901619, https://doi.org/10.1016/0003-4916(73)90161-9
46. Kleinman L (1973) Phys Rev B 7:2288. https://doi.org/10.1103/PhysRevB.7.2288

47. Mermin ND (1970) Phys Rev B 1:2362. https://doi.org/10.1103/PhysRevB.1.2362
48. Toscano G, Raza S, Yan W, Jeppesen C, Xiao S, Wubs M, Jauho AP, Bozhevolnyi SI, Mortensen NA (2013) Nanophotonics 2(3):161. https://doi.org/10.1515/nanoph-2013-0014
49. Melnyk AR, Harrison MJ (1970) Phys Rev B 2:835. https://doi.org/10.1103/PhysRevB.2.835
50. Raza S, Christensen T, Wubs M, Bozhevolnyi SI, Mortensen NA (2013) Phys Rev B 88:115401. https://doi.org/10.1103/PhysRevB.88.115401
51. Ashcroft NW, Mermin ND (1976) Solid state physics. Harcourt College Publishers, New York
52. David C, García de Abajo FJ (2011) J Phys Chem C 115(40):19470. https://doi.org/10.1021/jp204261u
53. Stratton JA (1941) Electromagnetic theory. McGraw-Hill, New York
54. Gonçalves PAD, Peres NMR (2016) An introduction to graphene plasmonics, 1st edn. World Scientific, Singapore. https://doi.org/10.1142/9948
55. Maradudin AA, Barnes WL, Sambles JR (eds) (2014) Modern plasmonics, 1st edn. Elsevier
56. Gramotnev DK, Bozhevolnyi SI (2010) Nat Photonics 4(2):83. https://doi.org/10.1038/nphoton.2010.282
57. Flick J, Rivera N, Narang P (2018) Nanophotonics 7(9):1479. https://doi.org/10.1515/nanoph-2018-0067
58. Akselrod GM, Argyropoulos C, Hoang TB, Ciracì C, Fang C, Huang J, Smith DR, Mikkelsen MH (2014) Nat Photonics 8:835. https://doi.org/10.1038/nphoton.2014.228
59. Pelton M (2015) Nat Photonics 9:427. https://doi.org/10.1038/nphoton.2015.103
60. Chikkaraddy R, de Nijs B, Benz F, Barrow SJ, Scherman OA, Rosta E, Demetriadou A, Fox P, Hess O, Baumberg JJ (2016) Nature 535(7610):127. https://doi.org/10.1038/nature17974
61. Liu R, Zhou ZK, Yu YC, Zhang T, Wang H, Liu G, Wei Y, Chen H, Wang XH (2017) Phys Rev Lett 118:237401. https://doi.org/10.1103/PhysRevLett.118.237401
62. Bozhevolnyi SI, Martin-Moreno L, García-Vidal F (2017) Quantum plasmonics. Springer
63. Raza S, Kadkhodazadeh S, Christensen T, Di Vece M, Wubs M, Mortensen NA, Stenger N (2015) Nat Commun 6:8788. https://doi.org/10.1038/ncomms9788
64. Kreibig U, Fragstein CV (1969) Z Phys 224(4):307. https://doi.org/10.1007/BF01393059
65. Sólyom J (2010) Fundamentals of the physics of solids. Springer
66. Lindhard J (1954) Kgl. Danske Videnskab. Selskab Mat.-Fys. Medd. 28
67. Ford GW, Weber WH (1984) Phys Rep 113(4):195. https://doi.org/10.1016/0370-1573(84)90098-X
68. Leontovich MA (1948) On the approximate boundary conditions for electromagnetic field on the surface of highly conducting bodies. Propagation of electromagnetic waves, Part II, Moscow
69. Shchukin AN (1940) Propagation of radio waves. Svyazizdatt, Moscow
70. Rytov SM, Eksp Zh (1940) Teor Fiz 10:180
71. Yuferev SV, Ida N (2009) Surface impedance boundary conditions: a comprehensive approach. CRC Press
72. Landau LD, Lifshitz EM (1960) Electrodynamics of continuous media. Pergamon Press, New York
73. Tretyakov S (2003) Analytical modeling in applied electromagnetics. Artech House
74. Tretyakov SA (2015) Phil Trans R Soc A 373(2049):20140362. https://doi.org/10.1098/rsta.2014.0362

Chapter 7
Quantum Nonlocal Effects Probed by Ultraconfined Graphene Plasmons

The remarkable degree of field confinement brought about by graphene plasmons makes them particularly well-suited for studying and enhancing light–matter interactions, and, from a practical viewpoint, can be regarded as an appealing platform for delivering miniaturized nanophotonic devices. We have seen in Chap. 4 that plasmons in extended graphene can attain wavelengths about $\sim\alpha^{-1}\sim 10^2$ times smaller than that of a photon of the same frequency, and that this already quite extraordinary confinement can be pushed even further by patterning graphene into plasmonic nanoresonator structures. Yet another route is to exploit the lower frequency acoustic plasmon of double-layer graphene formed by plasmon hybridization (cf. Sect. 4.1.2), since this mode reaches increasingly larger wavevectors as the separation between the graphene layers is decreased. Unfortunately, this comes hand in hand with the difficulty of exciting such graphene acoustic plasmons from the far-field owing to (i) the deep subwavelength nature of the mode itself, and (ii) to the symmetry of its field distribution (induced charges in opposite graphene layers are in anti-phase [1]). Recently, this adversity has been circumvented by using a van der Waals heterostructure [2]—composed of graphene encapsulated in hexagonal boron nitride (hBN)—placed onto a metal gate [3–5]. Crucially, the latter, besides enabling active control over the Fermi level of graphene, also screens the collective charge oscillations (plasmons) in the graphene monolayer that lies just above the metal, thereby yielding graphene plasmons whose properties are reminiscent of acoustic plasmons in a graphene double-layer [6].

In the present chapter we provide a comprehensive overview of the electrodynamics governing these ultraconfined acoustic-like plasmons supported by a graphene monolayer lying at a small distance d from a metal substrate. Most notably, their deeply subwavelength nature makes them highly susceptible to nonlocal effects, and thus the aforementioned modes may be used to uncover the nonlocal response of the system and to probe the quantum mechanical features of the graphene electron liquid [4] (and of the metal [6, 7]). We begin our considerations first by analyzing

P. A. D. Gonçalves, *Plasmonics and Light–Matter Interactions in Two-Dimensional Materials and in Metal Nanostructures*, Springer Theses,
https://doi.org/10.1007/978-3-030-38291-9_7

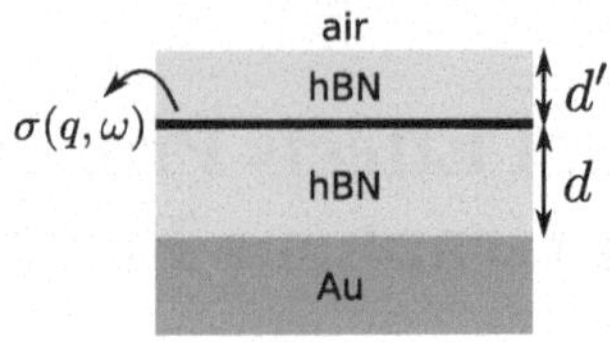

Fig. 7.1 Illustration of the configuration in which acoustic-like graphene plasmons can be realized. The heterostructure consists in hBN-encapsulated graphene where the latter is separated from the bottom metal gate by a distance d. A thin (to be accessible by the SNOM) hBN slab of thickness d' then protects the entire structure

quantum nonlocal effects ascribed to the electrodynamics of graphene alone, and then move on towards the general case where both the graphene and the metal are treated beyond the local-response approximation.

7.1 Acoustic-Like Graphene Plasmons in a Nutshell

Before invoking quantum mechanical effects beyond the framework of the nonlocal[1] RPA it is instructive to analyze the salient features of the acoustic-like graphene plasmons (AGPs) using the theoretical foundations already established in Chap. 3 and also with the benefit of hindsight on the fundamental properties of graphene plasmons studied in Chap. 4.

The pioneering experimental works [3, 4] that have reported the excitation and corresponding observation of acoustic-like graphene plasmons used a setup like the one schematically illustrated in Fig. 7.1; it consists in hBN-encapsulated graphene deposited onto an underlying metal gate (typically made of gold with a thin titanium adhesion layer). For graphene–metal separations d below the penetration length of the graphene plasmon's electric field, the screening exerted by the nearby metallic gate gives rise to the above-noted AGPs. The spectral properties of the latter are then measured using a scanning near-field optical microscope (SNOM), either in scattering or in photocurrent mode.[2]

From a theoretical standpoint, the spectral properties of AGPs can be determined by solving the associated electrodynamic problem or, alternatively, solving the quasi-static Poisson equation since the nonretarded regime is a very good approximation here. We have performed both[3] calculations and the results were indistinguishable, and thus we will focus on the simpler case of electrostatics in what follows. Furthermore, our calculations have shown that in the THz regime considered here the

[1]Throughout this chapter, whenever we mention the random-phase approximation (RPA) we implicitly mean its *nonlocal* version (cf. Sect. 3.2.2).

[2]See Refs. [3, 4] and references therein for details on the experiments.

[3]For the sake of explicitness, we have solved the full electrodynamic problem using the transfer-matrix method (TMM) [1] and analyzed the poles of the heterostructure's reflection coefficient for p-polarized waves.

metal gate is well approximated by a semi-infinite perfect electric conductor. With these assumptions in mind, the dispersion of these screened graphene plasmons can be fetched from the zeros of the RPA dielectric function

$$\epsilon^{\mathrm{RPA}}(q,\omega) = 1 - V(q,\omega)\chi_0^{\tau}(q,\omega), \tag{7.1}$$

where, as before, $\chi_0^{\tau}(q,\omega)$ is the Mermin-corrected graphene's polarization function [cf. Eq. (3.42)], and $V(q,\omega)$ describes the dressed Coulomb interaction akin to the layered geometry depicted in Fig. 7.1, in particular [3]

$$V(q,\omega) = v_q \frac{\tanh\Delta_+ + \frac{\sinh\Delta_-}{\cosh\Delta_+} + \frac{1}{\sqrt{\epsilon_x\epsilon_z}}\left(1 - \frac{\cosh\Delta_-}{\cosh\Delta_+}\right)}{\sqrt{\epsilon_x\epsilon_z} + \tanh\Delta_+}, \tag{7.2}$$

where $v_q = \frac{e^2}{2\epsilon_0 q}$ is the bare Coulomb interaction, and we have defined the auxiliary functions $\Delta_\pm \equiv \Delta_\pm(q,\omega) = q(d \pm d')\sqrt{\epsilon_x/\epsilon_z}$ with d and d' being the thicknesses of hBN slabs below and above the graphene sheet, respectively. Moreover, in the previous expression, $\epsilon_x \equiv \epsilon_x(\omega)$ and $\epsilon_z \equiv \epsilon_z(\omega)$ correspond to the

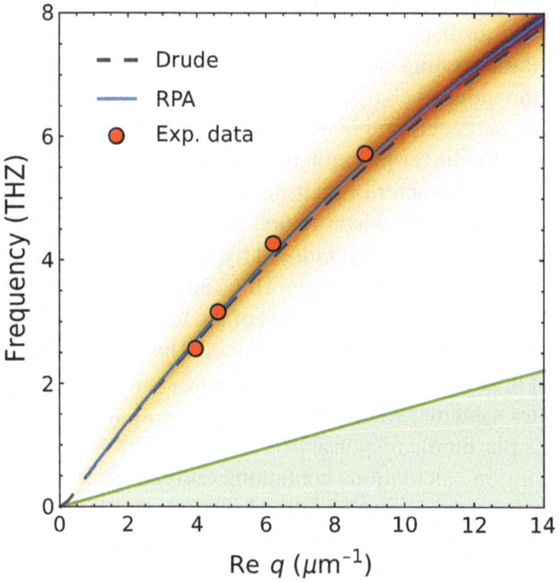

Fig. 7.2 Dispersion relation of acoustic-like graphene plasmons. The blue solid line represents the nonlocal RPA calculation and the corresponding local limit under the Drude model is represented by the gray dashed line. The experimental data is taken from Alonso-González et al. [3]. The colormap in the background shows the loss function via $-\mathrm{Im}\left\{\frac{1}{\epsilon^{\mathrm{RPA}}(q,\omega)}\right\}$ [cf. Eq. (3.41)]. The green shaded region denotes the intraband electron-hole continuum. Parameters [3]: $d = 42\,\mathrm{nm}$, $d' = 13\,\mathrm{nm}$, $E_F = 0.123\,\mathrm{eV}$, $\hbar\gamma = 2.07\,\mathrm{meV}$; the hBN dielectric function is taken from Ref. [8]

in-plane and perpendicular components of the hBN dielectric tensor, i.e., $\overleftrightarrow{\epsilon}_{\text{hBN}}(\omega) = \text{diag}\left[\epsilon_x(\omega), \epsilon_x(\omega), \epsilon_z(\omega)\right]$. Notice that the frequency dependence of $V(q, \omega)$ comes solely from the hBN's optical phonons.

The dispersion relation of AGPs obtained from the zeros of the corresponding RPA dielectric function (7.1) is shown in Fig. 7.2, and it is represented by the blue solid curve; the colormap in the background indicates the loss function via $-\text{Im}\left\{\frac{1}{\epsilon^{\text{RPA}}(q,\omega)}\right\}$. The LRA, Drude-type result is also represented as a gray dashed line for comparison purposes. Notice that for the parameters considered here—chosen in accordance with the experimental setup described in Ref. [3]—nonlocal effects are relatively weak, although the nonlocal RPA result seems to correlate better with the experimental data (depicted in the figure in the form of red circles). Lastly, notice that AGP's dispersion exhibited in the figure is clearly more linear-like and contrasts with the familiar $\propto \sqrt{q}$ behavior of conventional graphene plasmons.

7.2 Nonlocal and Quantum Graphene Plasmonics

In the previous section we have seen that by placing a graphene sheet in the vicinity of a metallic substrate it is possible to excite acoustic-like plasmons (i.e., with an approximately linear dispersion) in graphene. As previously mentioned, this is achievable due to the screening exercised by the metal, in which image charges build up in response to the induced charge oscillations associated with the graphene plasmons: the result are screened graphene plasmons exhibiting a linear dispersion, in close resemblance to the lower plasmon branch in double-layer graphene [1].

Crucially, due to the character of their dispersion relation, AGPs have the potential to achieve extremely large wavevectors. At the same time, the corresponding plasmon (group) velocity, v_{p}, can be tailored by controlling the distance d between the graphene layer and the metallic gate. This can be done with *atomic precision* since the distance d amounts to the thickness of the lower hBN slab, and thus may in principle be chosen with single-atom definiteness. This feature therefore makes the system considered here a unique playground to study nonlocal and quantum effects in condensed matter systems, where they can be probed and subsequently inferred from the material's plasmonic response [4].

Before embarking in calculations containing corrections to the RPA formalism, let us first demonstrate the concept outlined in the preceding paragraph with the theoretical tools that we have already learned so far. In Fig. 7.3 we explicitly show how the dispersion of graphene plasmons can be tailored by tuning the graphene–metal separation d (the setup under consideration is portrayed in Fig. 7.3a). Notably, the plasmon velocity becomes increasingly slower as the graphene–metal spacing is reduced (Fig. 7.3b), a feature that is accompanied by a significant change of the plasmon dispersion towards larger wavevectors (Fig. 7.3c). Furthermore, although the classical result qualitatively captures the above-noted trends, it incurs in significant quantitative deviations from the proper nonlocal plasmon dispersion. Moreover—

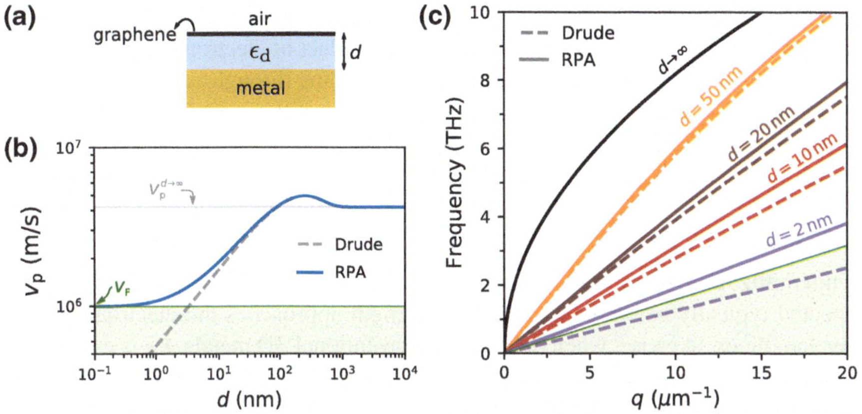

Fig. 7.3 Transition from the local to the nonlocal response regime by controlling the thickness d of the dielectric separating a graphene layer from a metal substrate. **a** Pictorial representation of the graphene–dielectric–metal heterostructure under consideration. **b** Plasmon velocity $v_p \equiv \partial\omega/\partial q$ as a function of the graphene–metal separation d (for a resonant frequency of $\omega/(2\pi) = 5\,\text{THz}$). **c** Dispersion curves of graphene plasmons supported by the graphene–dielectric–metal system for four different graphene–metal separations (indicated in the plot). The dispersion of conventional ("unscreened") graphene plasmons is shown by the solid black line to facilitate the comparison and the interpretation of the results. Setup parameters: $\epsilon_d = \sqrt{\epsilon_x^{\text{hBN}}(0)\epsilon_z^{\text{hBN}}(0)} \simeq 4.884$, $n_e = 1 \times 10^{12}\,\text{cm}^{-2}$ (which roughly amounts to $E_F = 0.1166\,\text{eV}$, for $v_F = 1 \times 10^6\,\text{m s}^{-1}$); the metal is assumed to be well approximated by a perfect electric conductor in the THz regime

and perhaps more troubling—is the fact that, when the graphene–metal separation is on the order of a few nanometers, the classical theory predicts plasmon dispersion diagrams falling within the electron-hole continuum and corresponding plasmon velocities that fall below the Fermi velocity of electrons (or holes) in graphene. Clearly, such behavior is prohibited by the full nonlocal theory, the reason lying in the divergent feature of the density-density correlation function $\chi_0(q, \omega)$ as $q \to \omega/v_F$.

The calculations of the AGPs spectral properties presented in Fig. 7.3 underscore the importance of employing an adequate nonlocal description of plasmonic excitations in such a graphene–metal configuration, particularly when the separation between the metallic substrate and the graphene sheet enters the few-nanometer regime. In what follows, we exploit the tunability and flexibility of this heterostructure—provided by the control over the separation d, but also over the carrier density n_e—to investigate electron-electron interactions in graphene.

Signatures of many-body interactions revealed by acoustic graphene plasmons

The RPA is arguably the most widely used model to describe the optical response of graphene and it provides a quite accurate account of the electrodynamics of the material in nearly all scenarios that are relevant for plasmonics. In spite of this, one should be aware that the RPA framework is still an approximation of the full quantum mechanical picture and therefore has also limitations [9–11]. One of which is the complete lack of exchange terms, as well as the neglect of electron-electron correlations (the only "correlation" that is included is screening through an effective average potential). A detailed, all-encompassing description of the many-body interactions in an electron liquid goes beyond the scope of this thesis—indeed, in the words of Giuliani and Vignale, "*it quickly becomes very difficult to go beyond the simple RPA*" [9]—and thus, following Ref. [4], below we simply give a cursory overview of the many-body corrections experimentally reported by Lundeberg et al. [4].

Renormalization of v_F. The manifestation of nonlocal effects in plasmonics is expected typically when the plasmon wavelength approaches the electronic Fermi wavelength, or, likewise, when $q \gtrsim k_F$. In conventional 3D metals, k_F is essentially fixed and the only way is to increase q (e.g., by exploiting high diffraction orders using gratings [12] or reducing the system's dimensions [13–15]). However, in graphene, the Fermi wavevector $k_F = \sqrt{\pi n_e}$ can be tuned easily by controlling the carrier density n_e using a gate. Thus, by exploiting graphene's acoustic plasmons, which attain large wavevectors q, in combination with relatively low carrier densities, one may straightforwardly increase the q/k_F ratio across a relatively broad parameter space. The net effect is the slow down of the plasmon velocity close to the Fermi velocity of electrons, together with the exacerbation of nonlocal and quantum effects. For graphene at low carrier densities, one of the most important contributions for the latter is the renormalization of v_F owing to many-body interactions[4] [16–23], which results in values for the Fermi velocity larger than the conventional value of $v_F \approx 1 \times 10^6\ \mathrm{m\,s^{-1}}$ [24]. In particular, according to first-order perturbation theory the (renormalized) Fermi velocity in graphene varies logarithmically with the carrier density [16–23]

$$v_F(n_e) = v_F^0 \left[1 + A \ln\left(\frac{\Lambda}{n_e}\right)\right], \tag{7.3}$$

where Λ is a cut-off and A is proportional to the effective coupling constant [16, 17] which also depends on the screening of the Coulomb interaction. Following Ref. [16], here we take $v_F^0 = 0.85 \times 10^6$ m/s as the "bare" Fermi velocity for graphene structures with weak electron-electron interactions, whereas the remaining quantities are obtained by fitting Eq. (7.3) to the results of the microscopic calculations provided in Ref. [4]. We emphasize that the n_e-dependence of the Fermi velocity arises from interaction-induced deviations from the conic energy-momentum dispersion of graphene's massless Dirac fermions, and hence the RPA is not actually "changed" *per se* in this case (in contrast to the compressibility correction below); instead, it solely changes the value of v_F that enters in the RPA.

Compressibility correction. As we have remarked earlier, in many-body condensed matter physics going beyond the RPA quickly becomes rather demanding. Fortunately, the concept of many-body *local field factors* enables the introduction of

[4]This is consequence of the divergent character of the screening length at the charge neutrality point, and thus electron-electron correlations become important. At moderate to high doping levels, these correlations become screened and the picture of a gas of noninteracting fermions is recovered.

exchange and correlation corrections to the RPA while maintaining the simple mathematical structure of the latter [9]. The main concept is to augment the effective field of the RPA with a local effective potential accounting for such effects. Concretely, the generalization of the density-density response function including a local field factor, $G(q, \omega)$, follows from[5] [9]

$$\frac{1}{\chi(q, \omega)} = \frac{1}{\chi_0(q, \omega)} + V(q, \omega)G(q, \omega). \tag{7.4}$$

Naturally, the usefulness of this approach depends on whether or not one can provide an approximate expression for $G(q, \omega)$ with physical meaning. In what follows, following Ref. [4], we will introduce a local field factor in order to correct the compressibility of the electron liquid. In doing so, we further assume that the dynamical field factor can be well approximated by its frequency-independent version, i.e., $G(q, \omega) \simeq G(q, 0)$ (this is the case where one expects the spectra of $\chi(q, \omega)$ and $\chi_0(q, \omega)$ to be qualitatively similar [9]), which can be written as [4]

$$G(q, 0) = -\frac{1 - \kappa_0/\kappa}{V(q, 0)\chi_0(q, 0)}, \tag{7.5}$$

so that the relation $\chi(q, 0)/\chi_0(q, 0) = \kappa/\kappa_0$ holds. The ratio κ/κ_0 has to be calculated using a suitable microscopic theory [18, 19, 25, 26]; alternatively, it can also be measured experimentally [26–28]. Here, we will use the data from the results reported in Ref. [4]. Inserting the previous expression into Eq. (7.4), one finally obtains

$$\chi(q, \omega) = \left[\chi_0^{-1}(q, \omega) - \frac{V(q, \omega)}{V(q, 0)} \frac{1 - \kappa_0/\kappa}{\chi_0(q, 0)}\right]^{-1}, \tag{7.6}$$

where the n_e-dependence of $\chi_0(q, \omega)$ (thereby including the renormalization of v_F) and κ_0/κ is implicitly assumed in the above expression. From here, the "interaction corrected" plasmon dispersion then stems from the zeros of $\epsilon(q, \omega) = 1 - V(q, \omega)\chi(q, \omega)$.

Results and discussion. Having outlined the specific many-body corrections we have set out to address, we have now all the ingredients necessary to theoretically model the experimental work performed by Lundeberg et al. [4]. In that experiment, the configuration of the setup is essentially the same as the one illustrated in Fig. 7.1, that is, an air–hBN–graphene–hBN–metal heterostructure. Three samples, with different graphene–metal separations (controlled by the thickness of the bottom hBN slab) were studied, namely with $d = 27$ nm, $d = 14.5$ nm, and $d = 5.5$ nm. Making use of the theoretical framework detailed above, we have calculated the dispersion—

---

[5] Note that, in terms of a dielectric function, it transpires from the result (7.4) that $\epsilon(q, \omega) = 1 - \dfrac{V(q, \omega)\chi_0(q, \omega)}{1 + V(q, \omega)G(q, \omega)\chi_0(q, \omega)}$. Clearly, the RPA dielectric function $\epsilon^{\text{RPA}}(q, \omega) = 1 - V(q, \omega)\chi_0(q, \omega)$ is appropriately recovered upon neglecting the many-body local field factor.

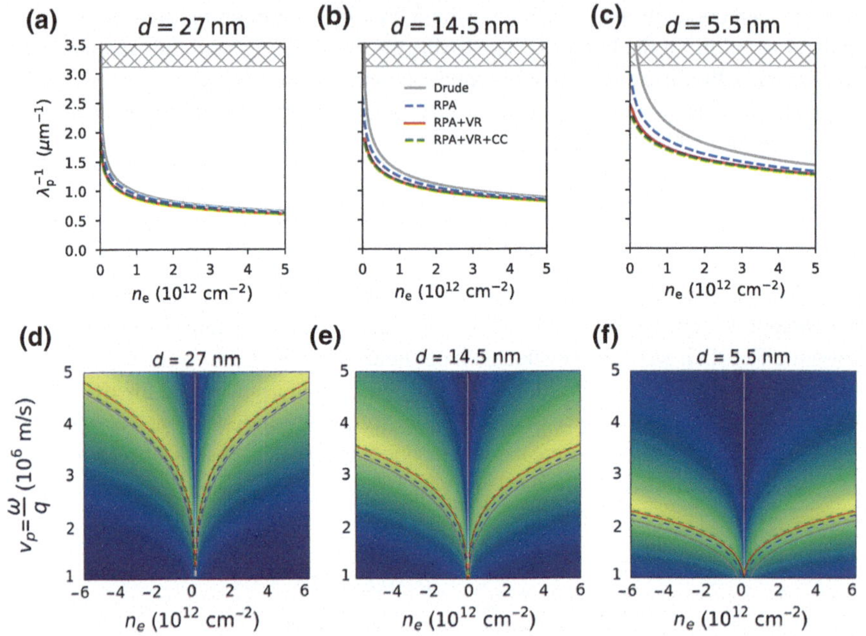

Fig. 7.4 Quantum nonlocal effects in the spectral properties of acoustic-like graphene plasmons. **a–c** Plasmon wavenumber as a function of graphene's carrier density (i.e., $\mathrm{Re}\, q/(2\pi)$ versus n_e) for various device geometries (see parameters below). The hatched regions indicate wavenumbers beyond the intraband electron-hole continuum. **d–f** Plasmon velocity $v_\mathrm{p} \approx \omega/\,\mathrm{Re}\, q$ versus electronic density for the same devices; the colormap in the background represents the calculated loss function $-\,\mathrm{Im}\{\epsilon^{-1}_{\mathrm{RPA+VR+CC}}(q, \omega, n_e)\}$. The sample configuration is an air–hBN–graphene–hBN–metal heterostructure like the one portrayed in Fig. 7.1. In accordance with the experiments carried out by Lundeberg et al. [4], we take $\omega/(2\pi) = 3.11$ THz together with the following parameters for the three different samples: (i) $d = 27$ nm and $d' = 9$ nm; (ii) $d = 14.5$ nm and $d' = 15.5$ nm; and (iii) $d = 5.5$ nm and $d' = 5.5$ nm. The metal gate is assumed to be a perfect conductor (we have also calculated the full electrodynamic theory with the experimental optical constants of gold in the THz [29] using the transfer-matrix method, but we have found negligible differences when compared with a the perfect conductor approximation), and the components of the hBN dielectric tensor are $\epsilon_x^{\mathrm{hBN}} = \epsilon_y^{\mathrm{hBN}} = 6.7$ and $\epsilon_z^{\mathrm{hBN}} = 3.56$ [4]. The density-density response function of graphene (under different approximations) is computed within the RTA with $\hbar\gamma = 8.271$ meV (corresponding to a $\tau = 500$ fs) [4]

and related properties—of the acoustic-like graphene plasmons in those structures, where the parameters were taken in conformity to the experiment. The results of such undertaking are summarized in Fig. 7.4.

We have calculated the AGPs n_e-dependent spectral properties starting from the simplest LRA using the Drude-type conductivity of graphene up to the nonlocal RPA (with and without interaction corrections). In the latter, we have firstly considered the nonlocal RPA without interactions, and subsequently added the many-body corrections in succession, i.e., the Fermi velocity renormalization (VR) and the compressibility correction (CC). The results exhibited in Fig. 7.4 further endorse our previous conclusions (cf. Fig. 7.3) that a nonlocal response formulation is needed in order to satisfactorily determine the AGPs' dispersion in structures where the graphene–metal separation is small. However, notice how for low carrier densities the nonlocal RPA has to be further modified in order to include many-body corrections, especially in the devices with shorter graphene–metal distances. Finally, it should be noted that the effect of the so-called compressibility correction term seems to be negligible,[6] the main effect being the renormalization of the Fermi velocity of graphene. Strikingly, this hints that the RPA (with a n_e-corrected v_F) remains a fairly accurate formalism to describe plasmons in graphene, since the influence of the many-body local field factor is seemingly unimportant. Naturally, for moderate and high carrier densities, the RPA with the usual nominal (and constant) value of $v_F = 1 \times 10^6\ \mathrm{m\,s^{-1}}$ provides a good account of the electrodynamics of graphene.

As a minor point of criticism we would like point out that, in principle, the optimal approach would be to use the actual, experimentally-measured $v_F(n_e)$, rather than the one obtained through microscopic calculations, in order to eliminate the possibility of introducing artifacts. Ideally, one would also ought to experimentally measure the dielectric function of that particular hBN sample that encapsulates the graphene sheet, since it too can affect the calculated AGPs' dispersion as well. However, the latter is likely to be nontrivial due to the geometry of the sample.

Here, our considerations regarding nonlocal and quantum effects are limited to the graphene layer itself, while the metal gate is treated as being as a perfect conductor or as exhibiting a local response [as in our TMM calculations (not shown)]. This assumption will be relaxed in the next section, where we shall model the optical response of the metal under the nonlocal framework provided by the hydrodynamic model described in Sect. 6.1.

7.3 Probing the Nonlocal Response of Metals with Graphene Plasmons

In our investigations so far we have treated the metal at the bottom of the dielectric–graphene–dielectric–metal heterostructure within the LRA. The justification for doing so has been that the dynamics affecting the graphene plasmons in the considered system are dominated by the electrodynamics of the graphene itself. Nevertheless, given the high momenta attained by the AGPs studied here, at some point the nonlocal response of the metal will need to be accounted for. In what follows, we shall depart the LRA picture for the metal and consider its nonlocal response within the hydrodynamic model previously introduced in Sect. 6.1. In particular, we will derive the dispersion relation of graphene plasmons in the aforementioned configuration taking into account the nonlocal local response of both the graphene and the metal.

[6] In the data provided by Lundeberg et al. [4], the deviation due to the compressibility correction is even within the experimental error-bar.

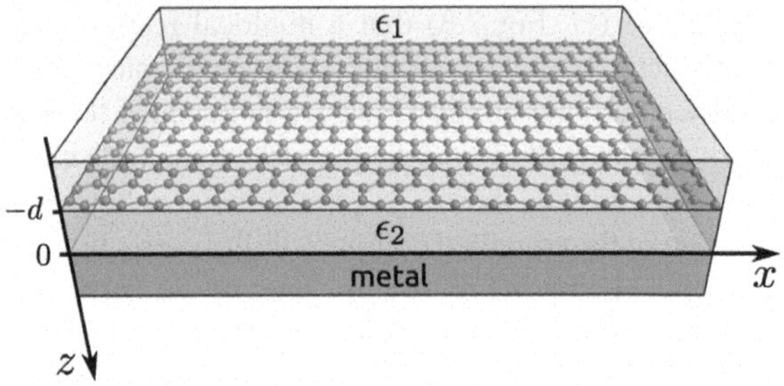

Fig. 7.5 Illustration of the considered dielectric–graphene–dielectric–metal planar heterostructure. The graphene monolayer is located in the $z = -d$ plane, and it is sandwiched between two dielectric media with relative permittivities ϵ_1 and ϵ_2 as indicated in the figure

Hereafter, we consider a planar, layered dielectric–graphene–dielectric–metal heterostructure as depicted in Fig. 7.5.

Dispersion relation. The derivation of the graphene plasmon dispersion in the layered heterostructure under consideration is reminiscent of the steps detailed in Sect. 6.2.1. For this reason, we ask the reader to indulge our concise derivation below, where we only highlight the main steps leading to the determination of the implicit condition for the dispersion of AGPs sustained by the system.

We seek TM surface wave solutions, which in the half-space spanned by $z < -d$ take the form

$$\mathbf{E}_1(\mathbf{r}, \omega) = \left[E_{x,1}, 0, E_{z,1}\right] e^{\kappa_1(z+d)} e^{i(qx-\omega t)}, \tag{7.7a}$$

$$\mathbf{H}_1(\mathbf{r}, \omega) = \left[0, H_{y,1}, 0\right] e^{\kappa_1(z+d)} e^{i(qx-\omega t)}, \tag{7.7b}$$

where $\kappa_1 = \sqrt{q^2 - \epsilon_1 k_0^2}$, and the fields' amplitudes are connected via the interrelations

$$E_{x,1} = \frac{\kappa_1}{i\omega\epsilon_0\epsilon_1} H_{y,1} \quad \text{and} \quad E_{z,1} = -\frac{q}{\omega\epsilon_0\epsilon_1} H_{y,1}. \tag{7.8}$$

On the other hand, in the metal region ($z > 0$), we write the electromagnetic field as

$$\mathbf{E}_\mathrm{m}^{\mathrm{T/L}}(\mathbf{r}, \omega) = \left[E_{x,\mathrm{m}}^{\mathrm{T/L}}, 0, E_{z,\mathrm{m}}^{\mathrm{T/L}}\right] e^{-\kappa_{\mathrm{m/nl}} z} e^{i(qx-\omega t)}, \tag{7.9a}$$

$$\mathbf{H}_\mathrm{m}(\mathbf{r}, \omega) = \left[0, H_{y,\mathrm{m}}, 0\right] e^{-\kappa_\mathrm{m} z} e^{i(qx-\omega t)}, \tag{7.9b}$$

where, as in Sect. 6.2.1, $\kappa_\mathrm{m} = \sqrt{q^2 - \epsilon_\mathrm{m} k_0^2}$ and $\kappa_\mathrm{nl} = \sqrt{q^2 - \beta^{-2}[\omega(\omega + i\gamma_\mathrm{m}) - \omega_p^2/\epsilon_\infty]}$, and where

$$E_{x,\mathrm{m}}^{\mathrm{T}} = \frac{i\kappa_\mathrm{m}}{\omega\epsilon_0\epsilon_\mathrm{m}} H_{y,\mathrm{m}} \quad \text{and} \quad E_{z,\mathrm{m}}^{\mathrm{T}} = -\frac{q}{\omega\epsilon_0\epsilon_\mathrm{m}} H_{y,\mathrm{m}}, \tag{7.10a}$$

$$E_{z,\mathrm{m}}^{\mathrm{L}} = i\frac{\kappa_\mathrm{nl}}{q} E_{x,\mathrm{m}}^{\mathrm{L}}. \tag{7.10b}$$

Lastly, in the dielectric region defined by $-d < z < 0$, both growing and decaying exponentials are allowed, and thus one may write the electromagnetic field in the dielectric slab characterized by the relative permittivity ϵ_2 as

$$\mathbf{E}_2(\mathbf{r}, \omega) = \left\{\left[E^+_{x,2}, 0, E^+_{z,2}\right] e^{\kappa_2 z} + \left[E^-_{x,2}, 0, E^-_{z,2}\right] e^{-\kappa_2 z}\right\} e^{i(qx-\omega t)}, \tag{7.11a}$$

$$\mathbf{H}_2(\mathbf{r}, \omega) = \left\{H^+_{y,2} e^{\kappa_2 z} + H^-_{y,2} e^{-\kappa_2 z}\right\} \hat{\mathbf{y}} e^{i(qx-\omega t)}, \tag{7.11b}$$

with $\kappa_2 = \sqrt{q^2 - \epsilon_2 k_0^2}$, and where the following relations hold

$$E^\pm_{x,2} = \pm \frac{\kappa_2}{i\omega\epsilon_0\epsilon_2} H^\pm_{y,2} \quad \text{and} \quad E^\pm_{z,2} = -\frac{q}{\omega\epsilon_0\epsilon_2} H^\pm_{y,2}. \tag{7.12}$$

In the present case, we now have two sets of boundary conditions: one for at $z = -d$, and a second set at $z = 0$ (recall that both the dielectric with ϵ_1 and the metal are assumed to be semi-infinite). Explicitly, we have

$$E_{x,1} = E^+_{x,2} e^{-\kappa_2 d} + E^-_{x,2} e^{\kappa_2 d}, \tag{7.13a}$$

$$H_{y,1} = H^+_{y,2} e^{-\kappa_2 d} + H^-_{y,2} e^{\kappa_2 d} + \sigma E_{x,1}, \tag{7.13b}$$

at $z = -d$, together with

$$E^+_{x,2} + E^-_{x,2} = E^{\mathrm{T}}_{x,\mathrm{m}} + E^{\mathrm{L}}_{x,\mathrm{m}}, \tag{7.14a}$$

$$H^+_{y,2} + H^-_{y,2} = H_{y,\mathrm{m}}, \tag{7.14b}$$

$$E^+_{z,2} + E^-_{z,2} = \frac{\epsilon_\infty}{\epsilon_2}\left(E^{\mathrm{T}}_{z,\mathrm{m}} + E^{\mathrm{L}}_{z,\mathrm{m}}\right), \tag{7.14c}$$

at $z = 0$. Using the relations Eqs. (7.8), (7.10) and (7.12) to rewrite Eqs. (7.13)–(7.14) in terms of the amplitudes $H_{y,1}$, $H^\pm_{y,2}$, $H_{y,\mathrm{m}}$, and $\bar{E}^{\mathrm{L}}_{x,\mathrm{m}} = i\omega\epsilon_0 E^{\mathrm{L}}_{x,\mathrm{m}}$, one arrives at the following equation for the determinant akin to the aforementioned linear system:

$$\begin{vmatrix} -\frac{\kappa_1}{\epsilon_1} & \frac{\kappa_2}{\epsilon_2} e^{-\kappa_2 d} & -\frac{\kappa_2}{\epsilon_2} e^{\kappa_2 d} & 0 & 0 \\ 1 + \frac{\kappa_1}{\epsilon_1}\frac{i\sigma}{\omega\epsilon_0} & -e^{-\kappa_2 d} & -e^{\kappa_2 d} & 0 & 0 \\ 0 & \frac{\kappa_2}{\epsilon_2} & -\frac{\kappa_2}{\epsilon_2} & \frac{\kappa_\mathrm{m}}{\epsilon_\mathrm{m}} & -1 \\ 0 & 1 & 1 & -1 & 0 \\ 0 & \frac{1}{\epsilon_\infty} & \frac{1}{\epsilon_\infty} & -\frac{1}{\epsilon_\mathrm{m}} & \frac{\kappa_{\mathrm{nl}}}{q^2} \end{vmatrix} = 0, \tag{7.15}$$

whose solution yields the implicit condition for the dispersion relation of acoustic-like graphene plasmons supported by the heterostructure, that is [6]

$$\left(\frac{\epsilon_1}{\kappa_1}+\frac{\epsilon_2}{\kappa_2}+\frac{i\sigma}{\omega\epsilon_0}\right)\left(1+\frac{\epsilon_m\kappa_2}{\epsilon_2\kappa_m}+\delta_{nl}\right)=\left(\frac{\epsilon_1}{\kappa_1}-\frac{\epsilon_2}{\kappa_2}+\frac{i\sigma}{\omega\epsilon_0}\right)\left(-1+\frac{\epsilon_m\kappa_2}{\epsilon_2\kappa_m}-\delta_{nl}\right)e^{-2\kappa_2 d}, \tag{7.16a}$$

where, as before,

$$\delta_{nl}=\frac{q^2}{\kappa_{nl}\kappa_m}\frac{\epsilon_m-\epsilon_\infty}{\epsilon_\infty}. \tag{7.16b}$$

Crucially, the transcendental equation posed by the condition (7.16a) determines the dispersion relation of AGPs in planar dielectric–graphene–dielectric–metal structures[7] (see Fig. 7.5) including nonlocal effects in graphene as well as in the metal. In the former, these are accounted for via a suitable expression for the material's surface conductivity, i.e., $\sigma \equiv \sigma(q,\omega)$, whereas in the latter nonlocality enters within the scope of the HDM through the nonlocal correction parameter δ_{nl}.

Before proceeding further, it is instructive to take a closer look at Eq. (7.16). We note that the term on the left-hand side is simply the multiplication of the "bare" dispersion relations of the uncoupled interfaces: that of graphene cladded by two dielectric media with relative permittivities ϵ_1 and ϵ_2 times that of a dielectric–metal interface (recall Eqs. (4.2) and (6.21), respectively). On the other hand, the term figuring on the right-hand side of Eq. (7.16a) mediates the interaction between the two interfaces. Clearly, in the limit of large separations between the graphene and the metal, or, more precisely, when $\kappa_2 d \to \infty$ (corresponding to $qd \gg 1$ in the nonretarded regime), the two interfaces effectively decouple.

Lastly, we conclude by bringing to the reader's attention that the condition (7.16) that determines the spectrum of AGPs simplifies considerably in the nonretarded regime, where it reduces to

$$\left(\epsilon_1+\epsilon_2+\frac{i\sigma}{\omega\epsilon_0}q\right)\left(1+\frac{\epsilon_m}{\epsilon_2}+\delta_{nl}^{nr}\right)=\left(\epsilon_1-\epsilon_2+\frac{i\sigma}{\omega\epsilon_0}q\right)\left(-1+\frac{\epsilon_m}{\epsilon_2}-\delta_{nl}^{nr}\right)e^{-2qd}, \tag{7.17}$$

with $\delta_{nl}^{nr}=q\,\kappa_{nl}^{-1}(\epsilon_m-\epsilon_\infty)/\epsilon_\infty$.

Results and discussion. In possession of the implicit condition that yields the dispersion relation of graphene plasmons screened by a nearby metal [i.e., Eq. (7.16)], we can now investigate the impact of the nonlocal response on the plasmonic spectrum associated with a planar dielectric–graphene-dielectric–metal system (recall Fig. 7.5).

Figure 7.6 shows the AGPs' spectrum computed using three different approaches: (i) both the graphene and the metal are modeled within the LRA (this plays the role of a "base-line"), (ii) the graphene electrodynamics are described within the nonlocal RPA, whereas the metal's optical response is still treated within the LRA, and,

[7] If the dielectric media that encapsulate the graphene sheet are uniaxial $\overleftrightarrow{\epsilon}_j=\text{diag}\left[\epsilon_j^x,\epsilon_j^x,\epsilon_j^z\right]$ (like, for instance, hBN) the corresponding AGPs' dispersion relation is nevertheless readily obtainable from Eq. (7.16) upon making the following replacements: $\epsilon_j \to \epsilon_j^x$ and $\kappa_j \to \sqrt{(\epsilon_j^x/\epsilon_j^z)q^2-\epsilon_j^x k_0^2}$ for $j\in\{1,2\}$. This setting is considered in Publication F, together with the hybridization of the acoustic-like graphene plasmons with the optical phonons in hBN [6].

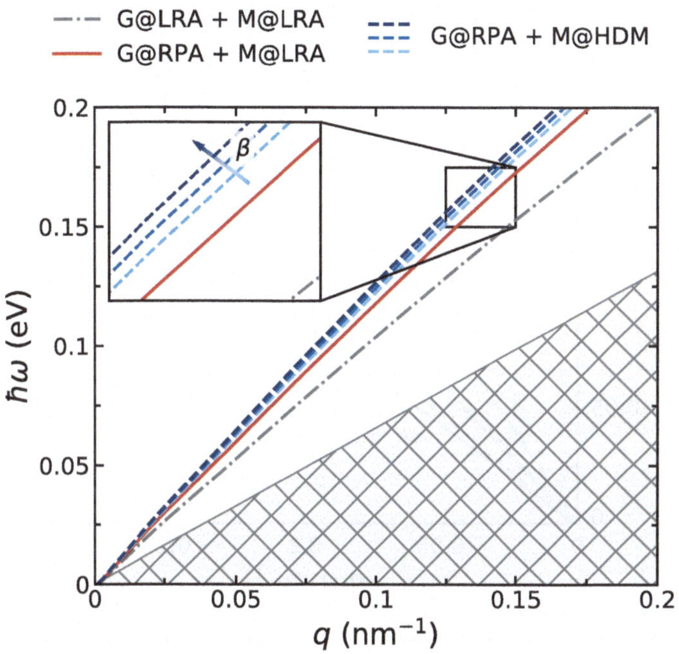

Fig. 7.6 Nonlocal plasmonic response exhibited by acoustic-like graphene plasmons sustained at a dielectric–graphene–dielectric–metal heterostructure (cf. Fig. 7.5). The red solid line shows the calculation of the AGPs' spectrum in the scenario where nonlocal effects are taken into account in the optical response of graphene alone (via the nonlocal RPA with Mermin's RTA) and the metal is still treated within LRA. The blue dashed lines shows the AGPs' dispersion computed using full nonlocal electrodynamics, that is, where the optical response of the graphene and the metal are described using the nonlocal RPA and the nonlocal HDM, respectively. In the latter, we take $\beta = \sqrt{3/5}\, v_{F,\mathrm{m}}$ with $v_{F,\mathrm{m}} = \{1, 1.5, 2\} \times 10^6$ m/s (where darker shades of blue correspond to larger values of the Fermi velocity associated with the metal's conduction electrons). The opposite case, where both materials are treated within the simplest LRA is represented by the gray dot-dashed line (for the purpose of comparison). The hatched region indicates the domain associated with intraband Landau damping in the graphene. Setup parameters: $\epsilon_1 = 1, \epsilon_2 = 4$, and $d = 2$ nm; the graphene is modeled with the parameters $E_F = 0.4$ eV and $\hbar\gamma = 8$ meV; the metal is modeled in accordance with $\epsilon_\mathrm{m} = 1 - \omega_\mathrm{p}^2/(\omega^2 + i\omega\gamma_\mathrm{m})$ with $\hbar\omega_\mathrm{p} = 9$ eV, $\hbar\gamma_\mathrm{m} = 20$ meV (and $\beta = \sqrt{3/5}\, v_{F,\mathrm{m}}$ for the HDM)

finally, (iii) the latter assumption is relaxed further and nonlocality in the metal is taken into account under the framework of the HDM. The results unambiguously demonstrate that the incorporation of nonlocality in the metal's optical response imparts an additional blueshift to the AGPs' dispersion. In particular, the magnitude of the blueshift is proportional to the velocity of the hydrodynamic plasma pressure wave $\beta = \sqrt{3/5}\, v_{F,\mathrm{m}}$ and inversely proportional to the metal's plasma frequency[8]

[8]We note that in the semiclassical Sommerfeld theory of the free-electron gas [30] the Fermi velocity $v_{F,\mathrm{m}}$ and the plasma frequency ω_p are not independent, namely $v_{F,\mathrm{m}} = \frac{\hbar}{m}\left(\frac{9\pi}{4}\right)^{\frac{1}{3}} \frac{1}{r_s}$ and

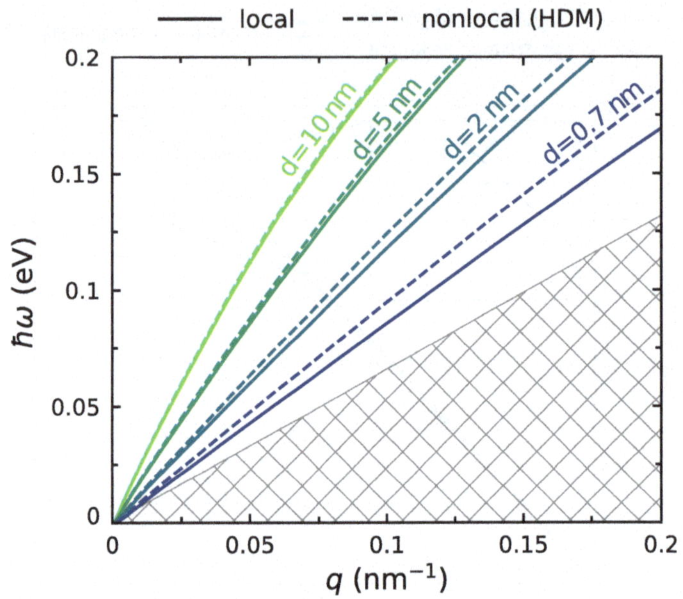

Fig. 7.7 Dependence of the nonlocal plasmonic response of acoustic-like graphene plasmons for varying graphene–metal separations d (indicated in the plot with matching colors). The response of graphene is evaluated at the level of the nonlocal RPA in all cases shown here. As for the metal's optical response, we consider two distinct models: the local response approximation (solid lines) and nonlocal hydrodynamic model (dashed lines). Setup parameters: same as in Fig. 7.6 but with fixed $\beta = \sqrt{3/5}\, v_{F,\mathrm{m}}$, where $v_{F,\mathrm{m}} = 1.4 \times 10^6\ \mathrm{m\,s^{-1}}$ (as in gold or silver)

(not shown; see Publication F). Hence—and provided that all other parameters are well-known experimentally—we envision that one could use the theoretical framework outlined above in order to shed light about the metal's nonlocal response, e.g., deducing the value of β by comparing the experimentally-measured and the theoretical spectra [5, 6, 31]. A perhaps more appropriate reasoning, that squarely rests on the microscopic formalism of the Feibelman d-parameters [32–34] (more on this in Chap. 8), is to establish a connection between the value of β and the position of the centroid of the induced charge density with respect to the dielectric–metal interface (given by $\mathrm{Re}\, d_\perp$, and where $d_\perp$ is the main Feibelman d-parameter); we will return to this point in more depth in Sect. 8.4 and thus at the time being we shall refrain ourselves from pursuing this premise any further.

$\omega_\mathrm{p}^2 = \frac{3e^2}{4\pi m \epsilon_0 r_s^3}$, respectively, where r_s denotes the Wigner–Seitz radius. Nevertheless, the approach taken in Fig. 7.6—varying $v_{F,\mathrm{m}}$ while keeping ω_p fixed—is still legit, and simply corresponds to metals with different carrier densities (controlled by r_s) and effective masses $m \to m^*$. For the sake of explicitness, the metal parameters adopted in the making of the aforementioned figure correspond to $\{(m^*, r_s/a_\mathrm{B})\} \simeq \{(1.644 m_0, 2.555), (0.8951 m_0, 3.129), (0.5814 m_0, 3.613)\}$, with m_0 referring to the electron's rest mass and the pairwise list is written in increasing order of $v_{F,\mathrm{m}}$.

Next, in Fig. 7.7 we examine the dependence of the nonlocal hydrodynamic blueshift on the graphene–metal separation[9] d. For the parameters considered in the figure (well within reach of current experimental capabilities[10] [4, 5]), the nonlocal response of the metal can lead to quite significant blueshifts of the AGPs' dispersion for $d \lesssim 5$ nm. This can be understood by realizing that AGPs attain increasingly larger wavevectors—and therefore are more sensitive to nonlocality in general—upon decreasing d. Furthermore, for graphene–metal separations of only a few-nanometers, d becomes comparable to the penetration length of the electric field into the metal promoted by the smearing of the electronic density and concomitant "inward spill" of the screening charges due to nonlocality [14, 35–40]. Hence, in this regime the sizable manifestation of nonlocal effects should not constitute a surprise. In particular, this notion has been exploited to theoretically describe plasmonic dimers by replacing the actual dimer separation d_0 (which plays a similar role to the graphene–metal separation d in our geometry) by an "effective separation"[11] $d_{\text{eff}} \neq d_0$ in order to capture (albeit rather coarsely) the above-noted nonlocal shift of the induced surface charges [5, 6, 36, 37].

One of the main virtues of the simple HDM is that it enables the incorporation of nonlocal electrodynamics in a tractable fashion. In many ways, it may be viewed as a semiclassical first approximation to the quantum description of the uniform electron gas (e.g., via the more sophisticated RPA or Lindhard dielectric function of a 3DEG). However, as pointed out before (check Sect. 6.3), the HDM flagrantly neglects Landau damping despite its contribution for nonlocal plasmon damping, which becomes increasingly important as q approaches the lower boundary of the intraband (bulk) electron-hole continuum[12] (notice that the latter is typically reached prior to the electron-hole continuum of graphene, as $v_{F,\text{m}} > v_F$ for most metals). In order to address the aforementioned neglect, one possibility is to use the reflection coefficient that we have derived in Sect. 6.3 [cf. Eqs. (6.27)–(6.28a)] and use it as a building block to construct the reflection coefficient of the overall dielectric–graphene–dielectric–metal heterostructure, from which the plasmonic response of the system can be computed. This approach has recently been pursued theoretically [7]. In the next chapter—more concretely in Sect. 8.4—we will present an alternative theoretical model for describing the plasmonic excitations supported in the same dielectric–graphene–dielectric–metal system that is capable of incorporating nonlocality, surface-enhanced Landau damping, as well as the shift of the centroid of induced charge density.

---

[9] Do not confuse it with the d-parameter mentioned in the previous paragraph.

[10] Indeed, even the extreme case of $d = 0.7$ nm [the thickness of a single-layer of hBN (chosen as twice the value of its bulk interlayer separation)] has been achieved experimentally [5].

[11] Notice that d_{eff} is both model- and material-dependent. In the HDM description, owing to its inherently "hard-wall" nature, the model always predicts $d_{\text{eff}} \geq d_0$ (which is at times conflicts with experiment).

[12] This is actually the case in Fig. 7.7 for the $d = 0.7$ nm, where the AGPs' dispersion lies *just before* the boundary of the (bulk) metal's electron-hole continuum.

References

1. Gonçalves PAD, Peres NMR (2016) An introduction to graphene plasmonics, 1st edn. World Scientific, Singapore. https://doi.org/10.1142/9948
2. Geim AK, Grigorieva IV (2013) Nature 499:419. https://doi.org/10.1038/nature12385
3. Alonso-González P, Nikitin AY, Gao Y, Woessner A, Lundeberg MB, Principi A, Forcellini N, Yan W, Vélez S, Huber AJ, Watanabe K, Taniguchi T, Casanova F, Hueso LE, Polini M, Hone J, Koppens FHL, Hillenbrand R (2017) Nat Nanotechnol 12:31. https://doi.org/10.1038/nnano.2016.185
4. Lundeberg MB, Gao Y, Asgari R, Tan C, Van Duppen B, Autore M, Alonso-González P, Woessner A, Watanabe K, Taniguchi T, Hillenbrand R, Hone J, Polini M, Koppens FHL (2017) Science 357(6347):187. https://doi.org/10.1126/science.aan2735
5. Alcaraz Iranzo D, Nanot S, Dias EJC, Epstein I, Peng C, Efetov DK, Lundeberg MB, Parret R, Osmond J, Hong JY, Kong J, Englund DR, Peres NMR, Koppens FHL (2018) Science 360(6386):291. https://doi.org/10.1126/science.aar8438
6. Dias EJC, Iranzo DA, Gonçalves PAD, Hajati Y, Bludov YV, Jauho AP, Mortensen NA, Koppens FHL, Peres NMR (2018) Phys Rev B 97:245405. https://doi.org/10.1103/PhysRevB.97.245405
7. Echarri AR, Cox JD, García de Abajo FJ (2019) Optica 6(5):630. https://doi.org/10.1364/OPTICA.6.000630
8. Woessner A, Lundeberg MB, Gao Y, Principi A, Alonso-González P, Carrega M, Watanabe K, Taniguchi T, Vignale G, Polini M, Hone J, Hillenbrand R, Koppens FHL (2015) Nat Mater 14:421. https://doi.org/10.1038/nmat4169
9. Giuliani G, Vignale G (2005) Quantum theory of the electron liquid. Cambridge University Press, Cambridge. https://doi.org/10.1017/CBO9780511619915
10. Mahan GD (2000) Many-particle physics, 3rd edn. Springer, New York
11. Bruus H, Flensberg K (2004) Many-body quantum theory in condensed matter physics: an introduction. Oxford Graduate Texts, Oxford University Press
12. David C, Christensen J, Mortensen NA (2016) Phys Rev B 94:165410. https://doi.org/10.1103/PhysRevB.94.165410
13. Bozhevolnyi SI, Mortensen NA (2017) Nanophotonics 6:1185. https://doi.org/10.1515/nanoph-2016-0179
14. Mortensen NA (2013) Photonics Nanostructures Fundam Appl 11(4):303. https://doi.org/10.1016/j.photonics.2013.06.002
15. Raza S, Bozhevolnyi SI, Wubs M, Mortensen NA (2015) J Phys Condens Matter 27(18):183204. https://doi.org/10.1088/0953-8984/27/18/183204
16. Elias DC, Gorbachev RV, Mayorov AS, Morozov SV, Zhukov AA, Blake P, Ponomarenko LA, Grigorieva IV, Novoselov KS, Guinea F, Geim AK (2011) Nat Phys 7:701. https://doi.org/10.1038/nphys2049
17. Peres NMR (2010) Rev Mod Phys 82:2673. https://doi.org/10.1103/RevModPhys.82.2673
18. Barlas Y, Pereg-Barnea T, Polini M, Asgari R, MacDonald AH (2007) Phys Rev Lett 98:236601. https://doi.org/10.1103/PhysRevLett.98.236601
19. Polini M, Asgari R, Barlas Y, Pereg-Barnea T, MacDonald A (2007) Exploring graphene. Solid State Commun 143(1):58. https://doi.org/10.1016/j.ssc.2007.04.035
20. Abedinpour SH, Vignale G, Principi A, Polini M, Tse WK, MacDonald AH (2011) Phys Rev B 84:045429. https://doi.org/10.1103/PhysRevB.84.045429
21. Das Sarma S, Hwang EH (2013) Phys Rev B 87:045425. https://doi.org/10.1103/PhysRevB.87.045425
22. Astrakhantsev NY, Braguta VV, Katsnelson MI (2015) Phys Rev B 92:245105. https://doi.org/10.1103/PhysRevB.92.245105
23. Tang HK, Leaw JN, Rodrigues JNB, Herbut IF, Sengupta P, Assaad FF, Adam S (2018) Science 361(6402):570. https://doi.org/10.1126/science.aao2934
24. Castro Neto AH, Guinea F, Peres NMR, Novoselov KS, Geim AK (2009) Rev Mod Phys 81:109. https://doi.org/10.1103/RevModPhys.81.109

25. Hwang EH, Hu BYK, Das Sarma S (2007) Phys Rev Lett 99:226801. https://doi.org/10.1103/PhysRevLett.99.226801
26. Asgari R, Katsnelson MI, Polini M (2014) Ann Phys 526(9–10):359. https://doi.org/10.1002/andp.201400167
27. Yu GL, Jalil R, Belle B, Mayorov AS, Blake P, Schedin F, Morozov SV, Ponomarenko LA, Chiappini F, Wiedmann S, Zeitler U, Katsnelson MI, Geim AK, Novoselov KS, Elias DC (2013) Proc Natl Acad Sci U. S. A. 110(9):3282. https://doi.org/10.1073/pnas.1300599110
28. Tomarken SL, Cao Y, Demir A, Watanabe K, Taniguchi T, Jarillo-Herrero P, Ashoori RC (2019) Phys Rev Lett 123:046601. https://doi.org/10.1103/PhysRevLett.123.046601
29. Ordal MA, Bell RJ, Alexander RW, Long LL, Querry MR (1987) Appl Opt 26(4):744. https://doi.org/10.1364/AO.26.000744
30. Ashcroft NW, Mermin ND (1976) Solid state physics. Harcourt College Publishers, New York
31. Ciracì C, Hill RT, Mock JJ, Urzhumov Y, Fernández-Domínguez AI, Maier SA, Pendry JB, Chilkoti A, Smith DR (2012) Science 337(6098):1072. https://doi.org/10.1126/science.1224823
32. Feibelman PJ (1982) Prog Surf Sci 12(4):287. https://doi.org/10.1016/0079-6816(82)90001-6
33. Christensen T, Yan W, Jauho AP, Soljačić M, Mortensen NA (2017) Phys Rev Lett 118:157402. https://doi.org/10.1103/PhysRevLett.118.157402
34. Gonçalves PAD, Christensen T, Rivera N, Jauho AP, Mortensen NA, Soljačić M (2020) Plasmon-emitter interactions at the nanoscale. Nat Commun 11:366. https://doi.org/10.1038/s41467-019-13820-z
35. Ginzburg P, Zayats AV (2013) ACS Nano 7(5):4334. https://doi.org/10.1021/nn400842m
36. Teperik TV, Nordlander P, Aizpurua J, Borisov AG (2013) Phys Rev Lett 110:263901. https://doi.org/10.1103/PhysRevLett.110.263901
37. Fernández-Domínguez AI, Zhang P, Luo Y, Maier SA, García-Vidal FJ, Pendry JB (2012) Phys Rev B 86:241110. https://doi.org/10.1103/PhysRevB.86.241110
38. Ciracì C, Pendry JB, Smith DR (2013) ChemPhysChem 14(6):1109. https://doi.org/10.1002/cphc.201200992
39. Zhu W, Esteban R, Borisov AG, Baumberg JJ, Nordlander P, Lezec HJ, Aizpurua J, Crozier KB (2016) Nat Commun 7:11495. https://doi.org/10.1038/ncomms11495
40. Toscano G, Straubel J, Kwiatkowski A, Rockstuhl C, Evers F, Xu H, Mortensen NA, Wubs M (2015) Nat Commun 6:7132. https://doi.org/10.1038/ncomms8132

Chapter 8
Quantum Corrections in Plasmonics and Plasmon–Emitter Interactions

Classical descriptions of photonic and plasmonic systems rely on the knowledge of the materials' local, bulk dielectric functions (or conductivities) [1, 2]. These, together with the specification of the structure's geometry—i.e., size and shape—constitute the sole ingredients required for computing the electromagnetic response of the system and its collective excitations [1–3]. Moreover, in the deeply subwavelength regime, the scale-invariance associated with electrostatics renders the aspect of size expendable (as we have seen in Sect. 4.2.1), and thus only the materials' shape[1] and bulk response are required for describing plasmon resonances within a nonretarded local-response framework. Notwithstanding, as the characteristic dimension of metallic nanostructures reaches the few-nanometer regime (which is empirically found to be somewhere below $\lesssim 10 - 20\,\text{nm}$ [3]), the accurateness of traditional descriptions in terms of classical electrodynamics rapidly declines owing to the emergence of nonlocal and quantum mechanical effects [3, 6–15].

In general, the *exact* computation of the optical response of a quantum nanoplasmonic system is an extremely challenging task due to the complex nonlocal and many-body dynamics that govern the plasmon-supporting electron gas. One approach relies on the use of *ab initio* time-dependent density functional theory (TDDFT) [16] in an attempt to describe systems' response in a quantum mechanical setting (where the effects that are included naturally depend on the approximation(s) employed when solving the many-body problem). Nonetheless, techniques fully based on TDDFT are in practice limited to very small systems like metal clusters, whose radii typically fall below $\lesssim 3\,\text{nm}$ (for instance, a Na cluster with around 1000 atoms would span a radius of only about 2 nm [8]). However, the vast majority of nanoplasmonic structures of interest possess characteristic length scales that are significantly larger than that, yet not sufficiently macroscopic in order to be satisfactorily described in terms of

[1] Note that the shape remains of paramount importance in order to properly describe the Coulomb interaction, e.g., instated through the Green's function associated with the corresponding Poisson's equation [1, 4, 5] (recall, also, Sect. 4.2.1).

P. A. D. Gonçalves, *Plasmonics and Light–Matter Interactions in Two-Dimensional Materials and in Metal Nanostructures*, Springer Theses,
https://doi.org/10.1007/978-3-030-38291-9_8

classical electrodynamics alone. Hence, it becomes clear that there is a high-demand for the development of suitable theoretical methods for quantum nanoplasmonics that are capable to treat quantum and nonlocal effects rigorously while at the same time being relatively simple and versatile to implement.

In this chapter, we introduce and extend a mesoscopic formalism for quantum nanoplasmonics that addresses the above-noted predicament. A significant part of the material covered in this chapter derives from Publication A. Specifically, we extend and further develop the formalism of Feibelman d-parameters [6, 7] in order to bridge the "classical to quantum boundary"; see Fig. 8.1. This framework is rooted on the introduction of quantum surface-response functions—here obtained using TDDFT—dubbed as Feibelman $d_\perp$- and $d_\parallel$-parameters. These account for the dynamics of the surface region where the electron gas is inhomogeneous, whereas the bulk is still described in terms of classical (local) response functions.

The formulation in terms of the d-parameters can therefore take into account, to leading-order, intrinsically quantum mechanical effects such as nonlocality, electronic spill-out (or "spill-in"), and surface-enabled Landau damping (i.e., plasmon decay into electron-hole pairs), while preserving an amenable theoretical treatment that is general and that can be practically implemented in a plethora of plasmonic nanostructures [3, 17, 18].

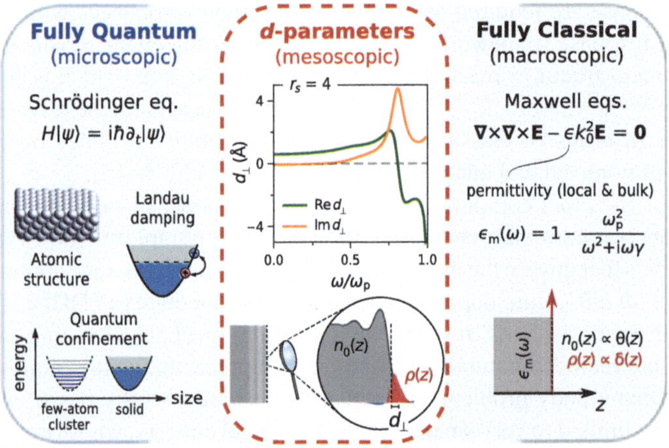

Fig. 8.1 Nonclassical mesoscopic electrodynamics via d-parameters. The nonclassical surface response functions—the Feibelman d-parameters—rigorously incorporate quantum mechanical effects in mesoscopic electrodynamics, bridging the gap between the purely quantum (microscopic) and the purely classical (macroscopic) domains. Inset: $d_\perp$-parameter of an $r_s = 4$ jellium metal computed from TDDFT [3]; in this case the corresponding $d_\parallel$-parameter vanishes (owing to charge-neutrality [7])

Box 8.1—Feibelman d-parameters

The complex-valued Feibelman d-parameters, $d_\perp$ and $d_\parallel$, can be defined from the quantum mechanical induced charge density, $\rho(\mathbf{r}) \equiv \rho(z)e^{iqx}$, and associated induced current density, $\mathbf{J}(\mathbf{r}) \equiv \mathbf{J}(z)e^{iqx}$ (frequency-dependence suppressed, but implicit) [3, 6, 7]:

$$d_\perp = \frac{\int_{-\infty}^{\infty} z\,\rho(z)\,\mathrm{d}z}{\int_{-\infty}^{\infty} \rho(z)\,\mathrm{d}z} \quad \text{and} \quad d_\parallel = \frac{\int_{-\infty}^{\infty} z\,\partial_z J_x(z)\,\mathrm{d}z}{\int_{-\infty}^{\infty} \partial_z J_x(z)\,\mathrm{d}z},$$

here, for an interface spanning the xy-plane at $z = 0$. It is apparent from the previous expressions that $d_\perp$ corresponds to the centroid of the induced charge density (cf. Fig. 8.1), while $d_\parallel$ corresponds to the centroid of the normal derivative of the tangential current (which is identically zero for charge-neutral interfaces) [7]. Unlike the bulk permittivity that characterizes a single material, the d-parameters are surface-response functions that depend on the *two* materials that make up the interface (including, in principle, their surface terminations). In short, the essential appeal of the d-parameters is their facilitation of a practical introduction of the important electronic length scales associated with the dynamics of the electron gas at an interface.

Figure 8.2 shows the Feibelman d-parameters obtained previously in Ref. [3] through TDDFT calculations within the jellium model for the interacting electron gas. The figure depicts the d-parameters for simple metals with different electronic densities, defined by the Wigner–Seitz radius $r_s = \{2, 4\}$ (in units of the Bohr radius, a_B), via $r_s a_\mathrm{B} = (3/4\pi n_e)^{1/3}$. The corresponding plasma frequency is thus $\omega_\mathrm{p} = \sqrt{\frac{3e^2}{4\pi\epsilon_0 m}\frac{1}{(r_s a_\mathrm{B})^3}}$. It should be noted that the d-parameters are microscopic surface-response functions that obey Kramers–Kronig relations and specific sum-rules [19]. The TDDFT data presented here is for air–metal (or vacuum–metal) interfaces (i.e., for $\epsilon_\mathrm{d} = 1$); we note that, as mentioned above, the specific values of the d-parameters depend on the dielectric medium next to the metal [7, 20], since it changes the screened interaction. Lastly, we emphasize that although in Fig. 8.2 we have $d_\parallel \neq 0$ for silver (last panel in Fig. 8.2) in this case silver is of course still charge-neutral. The finiteness of $d_\parallel$ here is merely a consequence of the methodology used to incorporate valence-band screening in the TDDFT calculation [3] (see Refs. [3, 7] for additional details).

In what follows, we begin by reviewing the concepts underpinning the d-parameter formalism in the nonretarded regime and use it to derive the mesoscopic nonretarded reflection and transmission coefficients of a planar dielectric–metal interface [3, 6, 7]. Using this knowledge, we then extend our considerations to the retarded regime by introducing a set of d-parameter-corrected mesoscopic boundary con-

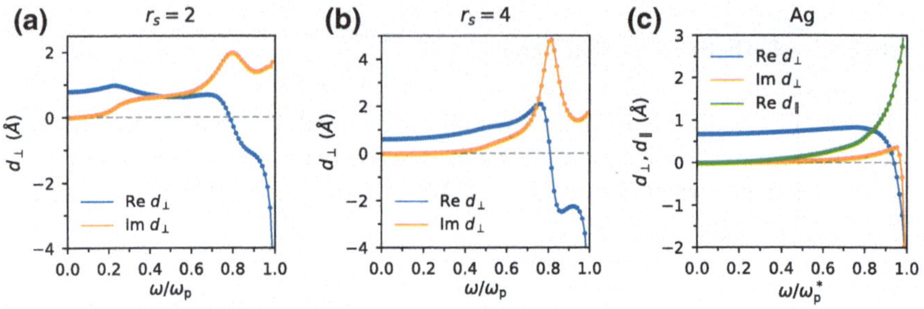

Fig. 8.2 Spectral dependence of the Feibelman d-parameters for jellium metals with different Wigner-Seitz radius, $r_s = \{2, 4\}$ (in units of the Bohr radius), and for silver (Ag). We note that a jellium with $r_s = \{2, 4\}$ is resemblant to aluminum (Al) and sodium (Na), respectively. For silver (Ag), we normalize the frequencies to the *screened* plasma frequency, $\hbar\omega_p^* \simeq 3.81$ eV. The circles represent data obtained from TDDFT calculations by Christensen et al. [3] and the solid lines are the corresponding interpolations. The dielectric medium interfacing the metal is assumed to have $\epsilon_d = 1$

ditions [17, 18, 21]. Next, we discuss the spherically symmetric case, i.e., that of a metal sphere, and derive the corresponding mesoscopic Mie coefficients [17]. In both configurations—the planar interface and the sphere—we calculate the quantum corrections to the corresponding plasmon dispersion, epitomized by emergence of nonclassical spectral shifts and surface-enabled Landau damping [3, 17]. Finally, in possession of the *analytical* mesoscopic scattering coefficients that describe the electromagnetic response of those systems, we extend and apply our formalism for mesoscopic electrodynamics to investigate plasmon–emitter interactions at the extreme nanoscale [17]. In particular, we study the impact of quantum surface corrections incorporated via the d-parameters in the context of plasmon-empowered light–matter interactions such as the modification of the dynamics of emitters in the vicinity of plasmonic nanostructures—encompassing both dipole-allowed and dipole-forbidden transitions—and plasmon-mediated energy transfer between two emitters.

8.1 Quantum Plasmonics at Planar Interfaces: Feibelman d-Parameters

8.1.1 Nonretarded Reflection and Transmission Coefficients and Surface Plasmon Dispersion

We consider that a planar dielectric–metal interface—portrayed in Fig. 8.3—is perturbed by an external potential of the form $\phi_{\text{ext}}(\mathbf{r}) = e^{iqx}e^{qz}$, representing an excitation (a harmonic time-dependence of the form $e^{-i\omega t}$ is assumed throughout) impinging on the planar interface from the dielectric side and parameterized

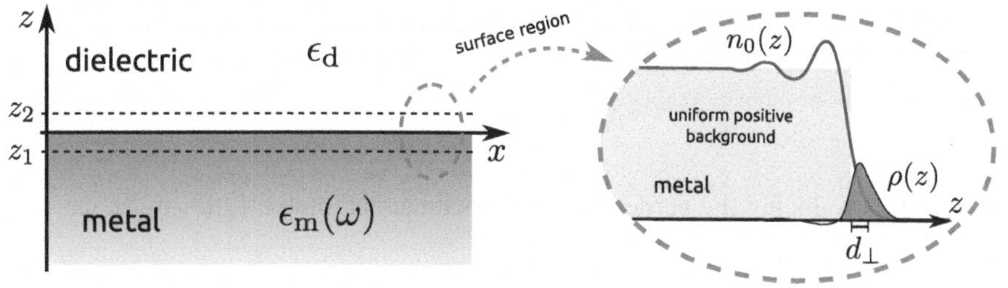

Fig. 8.3 Illustration of a planar dielectric–metal interface, here defined by the $z = 0$ plane. In the asymptotic regions, i.e., for $|z| \geq |z_{1,2}|$, the cladding dielectric is characterized by a dielectric constant ϵ_d whereas the electromagnetic properties of the metal are accounted for by its (local) bulk dielectric function $\epsilon_m(\omega)$. In the surface region, quantum mechanical features lead to a nonuniform equilibrium electron density, $n_0(z)$, and an induced charge density, $\rho(z)$ (that arises due to the system's response to an external perturbation). As indicated schematically in the figure, the Feibelman $d_\perp$-parameter can be regarded as the position of the centroid of the induced charge density with respect to the positive background edge (also dubbed as jellium edge)

by a wavevector q. This perturbation drives the system out of equilibrium, which responds to the perturbation by setting up a (surface-confined) charged density deviation $\rho(\mathbf{r}) = \rho(z)e^{iqx}$. Therefore, the associated induced electrostatic potential then follows from Coulomb's law[2] [4]

$$\phi_{\text{ind}}(\mathbf{r}) = \frac{1}{4\pi\epsilon_0}\int_{-\infty}^{\infty}\frac{\rho(\mathbf{r}')}{|\mathbf{r}-\mathbf{r}'|}d\mathbf{r}' = \frac{e^{iqx}}{2q\epsilon_0}\int_{-\infty}^{\infty}\rho(z')e^{-q|z-z'|}dz', \tag{8.1}$$

and thus, noting that $\phi_{\text{ind}}(\mathbf{r}) = \phi_{\text{ind}}(z)e^{iqx}$, one finds

$$\phi_{\text{ind}}(z) = \frac{1}{2q\epsilon_0}\int_{z_1}^{z_2}\rho(z')e^{-q|z-z'|}dz'. \tag{8.2}$$

Notice that in the previous expression we have replaced the limits of integration by $[z_1, z_2]$, that is, the z-coordinates that enclose the induced charge density and induced potential. This is permitted because those quantities effectively vanish for $|z| > |z_{1,2}|$, by definition [7]. Performing an expansion in Eq. (8.2) around $qz' = 0$, which is justifiable since $\phi_{\text{ind}}(z)$ is surface-peaked, we obtain

$$\begin{aligned}\phi_{\text{ind}}(z) &= \frac{1}{2q\epsilon_0}\int_{z_1}^{z_2}dz'\rho(z')\left\{1+\text{sgn}(z)qz'+\mathcal{O}[(qz')^2]\right\}e^{-q|z|}\\ &\equiv \frac{e^{-q|z|}}{2q\epsilon_0}\rho_s\left\{1+\text{sgn}(z)qd_\perp+\mathcal{O}[(qZ)^2]\right\},\end{aligned} \tag{8.3}$$

[2]We note that in writing Eq. (8.1) the following results $\int_{-\infty}^{\infty}d\bar{x}\frac{e^{-iq\bar{x}}}{\sqrt{\bar{x}^2+\bar{y}^2+\bar{z}^2}} = 2K_0\left(q\sqrt{\bar{y}^2+\bar{z}^2}\right)$ and $\int_{-\infty}^{\infty}d\bar{y}\,K_0\left(q\sqrt{\bar{y}^2+\bar{z}^2}\right) = \frac{\pi}{q}e^{-q|\bar{z}|}$ have been used [22].

where Z is a small (by assumption) length scale, and we have defined

$$\rho_s = \int_{z_1}^{z_2} \rho(z)\, dz\,, \tag{8.4}$$

while also introducing the perpendicular Feibelman parameter $d_\perp$ [3, 6, 7]:

$$d_\perp = \frac{1}{\rho_s} \int_{z_1}^{z_2} z\, \rho(z)\, dz\,. \tag{8.5}$$

The quantities denoted by ρ_s and $d_\perp$ correspond, respectively, to the effective quantum mechanical surface charge density and its first moment. The latter can also be interpreted as the position of the centroid of the induced charge density (with respect to the jellium edge). In addition, we note that $d_\perp$ has the form of a dipole moment.[3] In fact, this resemblance was exploited in a recent computational work by Yan et al. [18].

Nonretarded scattering coefficients of a planar dielectric–metal interface. In the following, our goal is to derive the leading-order quantum surface corrections to the reflection and transmission coefficients for the planar dielectric–metal geometry. To that end, we now introduce the asymptotic potentials [3, 7]

$$\phi_+^\infty(z > 0) \equiv e^{qz} - r e^{-qz}\,, \tag{8.6a}$$
$$\phi_-^\infty(z < 0) \equiv t e^{qz}\,, \tag{8.6b}$$

which are valid away from the interface, that is, in the regions defined by $|z| \geq |z_{1,2}|$. Naturally, these must agree with the actual potential $\phi_{\text{ext}}(z) + \phi_{\text{ind}}(z)$ beyond the surface region, thereby establishing the following relations:

$$r = -\frac{\rho_s}{2q\epsilon_0}\,(1 + q d_\perp)\,, \tag{8.7a}$$
$$t = 1 + \frac{\rho_s}{2q\epsilon_0}\,(1 - q d_\perp)\,. \tag{8.7b}$$

The sought-after coefficients r and t can be determined by imposing the adequate boundary conditions. However, the usual boundary conditions of classical optics that relate the asymptotic (classical) potentials are of no use here due to the inclusion of microscopic (quantum mechanical) surface features. In particular, subtracting the components of the classical potential on either side of the interface produces

[3] Indeed, ρ_s and $d_\perp$ may be regarded as a monopole and a dipole moment, respectively. In particular, the reader may recognize that the expansion that led to Eq. (8.3) has effectively introduced a multipole expansion. Generically, one could in principle include an arbitrary number of higher-order multipoles, albeit such endeavor would necessarily make the ensuing mathematical analysis burdensome.

$$\begin{aligned}\phi_+^\infty(z=0^+) - \phi_-^\infty(z=0^-) = 1 - r - t = \frac{\rho_s}{\epsilon_0} d_\perp\,, \\ \Rightarrow \quad 1 - r - t = \frac{\rho_s}{\epsilon_0} d_\perp\,,\end{aligned} \tag{8.8}$$

where in the last step we have made use of Eqs. (8.7). Similarly, considering the derivatives of the potential, one finds

$$\begin{aligned}\left.\frac{\partial \phi_+^\infty}{\partial z}\right|_{z=0^+} - \left.\frac{\partial \phi_-^\infty}{\partial z}\right|_{z=0^-} = q(1+r) - qt = -\frac{\rho_s}{\epsilon_0}\,, \\ \Rightarrow \quad 1 + r - t = -\frac{\rho_s}{q\epsilon_0}\,.\end{aligned} \tag{8.9}$$

Combining Eqs. (8.8) and (8.9) yields our *first effective boundary condition* [3, 7]:

$$\boxed{(1+qd_\perp) - r(1-qd_\perp) - t(1+qd_\perp) = 0}\,, \tag{8.10}$$

where any explicit reference to ρ_s is now absent.

In order to arrive at the second modified boundary condition, it is useful to introduce an auxiliary charge density and a corresponding auxiliary current density, $\rho^\infty(\mathbf{r})$ and $\mathbf{J}^\infty(\mathbf{r})$, respectively, which naturally obey the continuity equation, $\nabla \cdot \mathbf{J}^\infty = i\omega\rho^\infty$. Moreover, as in the case of $\phi_\pm^\infty$, these classical charge- and current-densities must match the proper quantum mechanical charge- and current-densities, $\rho(\mathbf{r})$ and $\mathbf{J}(\mathbf{r})$, in the asymptotic regions $|z| \geq |z_{1,2}|$. Noting that $\mathbf{J}^\infty(x,z) = [J_x^\infty(z)\hat{\mathbf{x}} + J_z^\infty(z)\hat{\mathbf{z}}]e^{iqx}$ and $\rho^\infty(x,z) = \rho_s^\infty\delta(z)e^{iqx}$, it follows from the continuity equation that

$$i\omega\rho_s^\infty\delta(z) = iqJ_x^\infty(z) + \frac{\partial}{\partial z}J_z^\infty(z)\,,$$

which upon integrating from $z = 0^-$ to $z = 0^+$ leads to

$$\begin{aligned} i\omega\rho_s^\infty = \underbrace{iq\int_{0^-}^{0^+} dz J_x^\infty(z)}_{=0} + J_z^\infty(0^+) - J_z^\infty(0^-)\,, \\ \Rightarrow i\omega\rho_s^\infty = J_z^\infty(0^+) - J_z^\infty(0^-)\,,\end{aligned} \tag{8.11}$$

where the integral over $J_x^\infty(z)$ vanishes since the classical current is nonsingular (i.e., weakly-varying) across the infinitesimal domain [3]. Furthermore, since by construction the asymptotic currents are generated by the asymptotic fields via the corresponding classical bulk conductivities, that is, $J_z^\infty(0^\pm) = \sigma E_z^\infty(0^\pm) = -\sigma\partial_z\phi^\infty|_{z=0^\pm}$, and noting that $\sigma_{d,m} = -i\omega\epsilon_0(\epsilon_{d,m} - 1)$, one can cast the previous equation as

$$\frac{\rho_s^\infty}{q\epsilon_0} = (\epsilon_d - 1)(1+r) - (\epsilon_m - 1)t\,. \tag{8.12}$$

This expression interrelates ρ_s^∞ with the scattering amplitudes r and t and with the (local) bulk dielectric functions (which characterize the media in the asymptotic regions defined by $|z| \geq |z_{1,2}|$). The usefulness of Eq. (8.12) depends on our ability to establish a connection between the charge- and current-densities in the far-field and the actual microscopic charge- and current-densities. To that end, we proceed in a similar fashion to the one that led to Eq. (8.11), namely, integrating the continuity equation but now over an extended domain around the interface up to the asymptotic regions

$$i\omega\rho_s = iq \int_{z_1}^{z_2} dz\, J_x(z) + J_z(z_2) - J_z(z_1)\,, \tag{8.13a}$$

$$i\omega\rho_s^\infty = iq \int_{z_1}^{z_2} dz\, J_x^\infty(z) + J_z^\infty(z_2) - J_z^\infty(z_1)\,. \tag{8.13b}$$

Subtracting the above equations while at the same time realizing that, by definition, $J_z(z_{1,2}) = J_z^\infty(z_{1,2})$, renders

$$\begin{aligned} \rho_s - \rho_s^\infty &= \frac{q}{\omega} \int_{z_1}^{z_2} dz \left[J_x(z) - J_x^\infty(z)\right]\,, \\ &= \frac{q}{\omega} d_\parallel \left[J_x^\infty(0^-) - J_x^\infty(0^+)\right]\,. \end{aligned} \tag{8.14}$$

where we have introduced the parallel Feibelman d-parameter [3, 6, 7]:

$$d_\parallel \equiv \frac{\int_{z_1}^{z_2} dz \left[J_x(z) - J_x^\infty(z)\right]}{J_x^\infty(0^-) - J_x^\infty(0^+)}\,. \tag{8.15}$$

Utilizing a similar argument to the one that prompted us from Eq. (8.11) to the result Eq. (8.12), we can rewrite Eq. (8.14) as

$$\frac{\rho_s - \rho_s^\infty}{q\epsilon_0} = q d_\parallel \left[(\epsilon_d - 1)(1 - r) - (\epsilon_m - 1)t\right]\,. \tag{8.16}$$

On the other hand, the combinations of Eqs. (8.9) and (8.12) enables us to obtain the relation

$$\frac{\rho_s - \rho_s^\infty}{q\epsilon_0} = t\epsilon_m - (1 + r)\epsilon_d\,. \tag{8.17}$$

Finally, equating the previous two expressions grants us with the *second effective boundary condition*, which reads [3, 7]:

$$\boxed{\epsilon_d(1 + qd_\parallel) - qd_\parallel + r\left[\epsilon_d(1 - qd_\parallel) + qd_\parallel\right] - t\left[\epsilon_m(1 + qd_\parallel) - qd_\parallel\right] = 0}\,. \tag{8.18}$$

In possession of the two *effective* boundary conditions—epitomized by Eqs. (8.10) and (8.18)—we can now solve the coupled system constituted by those two expressions, thereby determining the nonretarded *mesoscopic* reflection and transmission coefficients within the d-parameter formalism [3, 6, 7]:

$$r = \frac{\epsilon_\mathrm{m} - \epsilon_\mathrm{d} + (\epsilon_\mathrm{m} - \epsilon_\mathrm{d}) q (d_\perp + d_\parallel)}{\epsilon_\mathrm{m} + \epsilon_\mathrm{d} - (\epsilon_\mathrm{m} - \epsilon_\mathrm{d}) q (d_\perp - d_\parallel)}, \tag{8.19a}$$

and

$$t = \frac{2\epsilon_\mathrm{d}}{\epsilon_\mathrm{m} + \epsilon_\mathrm{d} - (\epsilon_\mathrm{m} - \epsilon_\mathrm{d}) q (d_\perp - d_\parallel)}, \tag{8.19b}$$

where for consistency we have only kept terms which are linear in $qd_{\perp,\parallel}$. These two equations therefore include the lowest-order quantum surface corrections to the classical reflection and transmission amplitudes. Notice that the latter are recovered by taking the limit of vanishing Feibelman d-parameters, i.e., $d_{\perp,\parallel} \to 0$, yielding the well-know classical results [1]

$$r^\mathrm{cl} = \frac{\epsilon_\mathrm{m} - \epsilon_\mathrm{d}}{\epsilon_\mathrm{m} + \epsilon_\mathrm{d}} \quad \text{and} \quad t^\mathrm{cl} = \frac{2\epsilon_\mathrm{d}}{\epsilon_\mathrm{m} + \epsilon_\mathrm{d}}, \tag{8.20}$$

for a planar dielectric–metal interface.

The determination of the nonretarded Feibelman scattering coefficients r and t is pivotal for describing nonlocal quantum surface corrections to, for instance, the surface plasmon dispersion [3, 6, 7, 17, 23], the enhancement of the optical local density of states (LDOS) experienced by an emitter in the vicinity of a metal surface [17, 23, 24], or their role in the van der Waals interaction [23–26]. We will return to some of these points in Sect. 8.3.

Nonretarded surface plasmon dispersion. Equipped with the nonclassical reflection coefficient—cf. Eq. (8.19a)—, the corresponding surface plasmon dispersion for a planar interface is determined from the poles of $r \equiv r(q, \omega)$. Hence, the nonclassical surface plasmon spectrum stems from

$$\epsilon_\mathrm{m} + \epsilon_\mathrm{d} - (\epsilon_\mathrm{m} - \epsilon_\mathrm{d}) q (d_\perp - d_\parallel) = 0. \tag{8.21}$$

Clearly, a closed-form expression for the dispersion in the form of q versus ω can be found; explicitly, it reads

$$q_\mathrm{SP} = \frac{\epsilon_\mathrm{m} + \epsilon_\mathrm{d}}{\epsilon_\mathrm{m} - \epsilon_\mathrm{d}} \frac{1}{d_\perp - d_\parallel}. \tag{8.22}$$

On the other hand, writing an explicit solution in closed-form for the surface plasmon frequency as a function of the wavevector, i.e., ω versus q, is not possible[4] due to the implicit frequency-dependence of the d-parameters. Nevertheless, starting from Eq. (8.21) one may work towards a perturbative solution, herein denoted by $\omega_{\text{SP}}^{(1)}$. In particular, for the exemplary case of a homogeneous 3DEG—that is, a jellium metal so that $\epsilon_\infty = 1$ (in this case one would have $d_\parallel = 0$, but we shall retain it here for the sake of generality and because it can be used to model, e.g., overlayers, adsorption, or surface roughness [3, 7, 27])—interfacing vacuum $\epsilon_{\text{d}} = 1$, we find to first-order in $qd_{\perp,\parallel}$ (and for $\omega_{\text{p}} \gg \gamma$) [3, 6, 7]

$$\text{Re}\ \omega_{\text{SP}}^{(1)} \simeq \text{Re}\ \omega_{\text{SP}}^{(0)} \Big[1 - \underbrace{\frac{1}{2} q\, \text{Re} \left(d_\perp^{(0)} - d_\parallel^{(0)} \right)}_{\substack{\text{electronic} \\ \text{spill-out/spill-in}}} \Big], \tag{8.24a}$$

$$-\text{Im}\ \omega_{\text{SP}}^{(1)} \simeq \underbrace{\frac{\gamma}{2}}_{\substack{\text{classical} \\ \text{bulk} \\ \text{damping}}} + \underbrace{\frac{1}{2}\text{Re}\ \omega_{\text{SP}}^{(0)}\, q\, \text{Im} \left(d_\perp^{(0)} - d_\parallel^{(0)} \right)}_{\substack{\text{surface-assisted} \\ \text{Landau damping}}}, \tag{8.24b}$$

where $\text{Re}\ \omega_{\text{SP}}^{(0)} = \omega_{\text{p}}/\sqrt{2}$ is the classical nonretarded surface plasmon frequency and $d_\alpha^{(0)} \equiv d_\alpha(\text{Re}\ \omega_{\text{SP}}^{(0)})$ for $\alpha = \{\perp, \parallel\}$.

At this point, the reader may appreciate that, despite being only approximate, Eq. (8.24a) delivers a clear message: to lowest-order, the direction of the frequency shift due to nonlocal quantum surface phenomena ultimately depends on the sign of $\text{Re}\left(d_\perp - d_\parallel\right)$. Specifically, these nonclassical corrections lead to red (blue) shift of the surface plasmon resonance—relative to its classical value—if $\text{Re}\left(d_\perp - d_\parallel\right) > 0$ (< 0). Furthermore, for charge-neutral surfaces one has $d_\parallel = 0$ [7]. Recalling the interpretation of $d_\perp$ as the centroid of the induced charge density, then one may associate the positiveness or negativeness of $d_\perp$, respectively, to a spill-out or a spill-in of the electron density with respect to the fixed ionic background of the metal [3, 6, 7, 17]. Concomitantly, the aforementioned nonclassical resonant frequency shift is accompanied by increased broadening of the surface plasmon resonance; the latter is embodied in the imaginary part of the d-parameters and takes into account surface-assisted Landau damping [3, 6, 7, 17].

---

[4] At any rate, assuming a Drude-type dielectric function $\epsilon_{\text{m}}(\omega) = \epsilon_\infty - \omega_{\text{p}}^2/(\omega^2 + i\omega\gamma)$ for the metal, one obtains

$$\omega = -i\frac{\gamma}{2} + \frac{1}{2}\sqrt{4\omega_{\text{p}}^2 \frac{1 - q(d_\perp - d_\parallel)}{\epsilon^+ + \epsilon^- q(d_\perp - d_\parallel)} - \gamma^2}, \tag{8.23}$$

where $\epsilon_\pm \equiv \epsilon_{\text{d}} \pm \epsilon_\infty$. Note, however, that this is still an *implicit* equation due to the frequency-dependence of the d-parameters (and, possibly, of ϵ_∞ and/or ϵ_{d}).

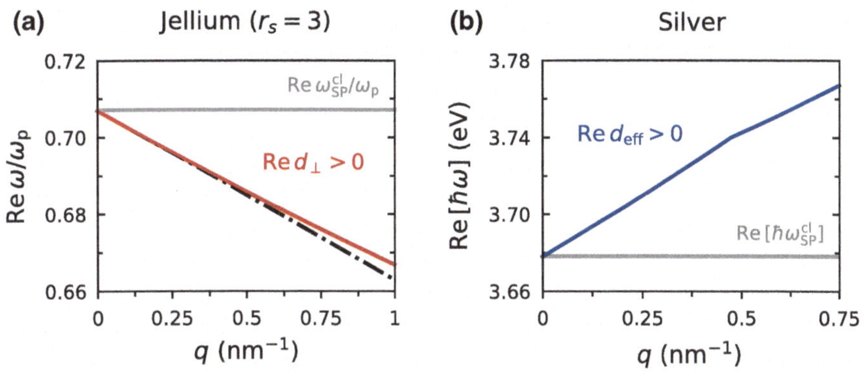

Fig. 8.4 Nonretarded surface plasmon dispersion for vacuum–metal interfaces calculated using the formalism of the Feibelman d-parameters [by solving Eq. (8.21) numerically]. The d-parameters' data are taken from TDDFT calculations performed by Christensen et al. [3, 28]. **a** Nonclassical surface plasmon dispersion for a vacuum–jellium interface [with Wigner–Seitz radius $r_s = 3$ (in units of the Bohr radius); corresponding to $\hbar\omega = 9.07$ eV, $\epsilon_\infty = 1$, and $d_\parallel = 0$] where we have assumed a Drude-type damping of $\gamma = \omega_p/100$. The dispersionless gray line indicates the classical result, whereas the black dot-dashed line represents the perturbative solution Eq. (8.24a). **b** Same as in **a** but now for a vacuum–silver interface. Here, we have defined $d_{eff} = d_\perp - d_\parallel$ [the fact that $d_\parallel \neq 0$ is simply due to the semiclassical account of silver's nontrivial d-band screening, and not a statement about charge-neutrality; see Ref. [3] (and Ref. [7]) for details]. We describe silver using Johnson and Christy's experimental data [29] for $\epsilon_m(\omega)$, parameterized according to the prescription described in the supplemental material of Ref. [3] (in order to be consistent with their TDDFT implementation)

Figure 8.4 shows the nonretarded surface plasmon dispersion relation for a homogeneous 3DEG and for silver. The figure clearly underscores the statements set forth in the previous paragraph, i.e., that the surface plasmon dispersion undergoes either a redshift, for Re $d_{eff} > 0$, or a blueshift, for Re $d_{eff} < 0$; here, $d_{eff} \equiv d_\perp - d_\parallel$. We emphasize that for jellium metals (e.g., simple metals, such as the ones belonging to the alkali group) Re d_{eff} is typically positive around the surface plasmon resonance [7, 10, 30–36]. However, silver (and other noble metals) typically exhibits negative values for Re d_{eff} in experiments [7, 12, 13, 37–39]; this behavior is attributed to the influence of the (substantial) d-band screening (by virtue of the core electron's contribution to the background polarizability) [7, 40, 41].

Finally, we also mention here in passing that for metal structures beyond the planar interface, e.g., metallic nanoparticles with different shapes, the predicted direction of the nonclassical frequency shift (amount of nonclassical broadening) depends not only on the sign of Re d_{eff} (magnitude of Im d_{eff}) but also on their interplay with a set of geometry-dependent shape factors [3].

8.1.2 Surface Plasmon Polaritons: Quantum Surface Corrections Incorporated via d-Parameters

After having determined the mesoscopic nonretarded reflection and transmission coefficients—and ensuing nonretarded surface plasmon dispersion—for the planar interface, we now revisit the same dielectric–metal configuration taking retardation effects into account. To that end, we introduce a set of modified, mesoscopic boundary conditions [17, 18, 21] that differ from the classical, macroscopic ones by a term proportional to the d-parameters. From these, we obtain the corresponding nonclassical retarded reflection and transmission coefficients and investigate the influence of quantum surface corrections on the surface plasmon polariton (SPP) dispersion.

Mesoscopic boundary conditions. In the absence of external charges and currents, the elementary boundary conditions of macroscopic electrodynamics at an interface separating two media are $\hat{\mathbf{n}} \times (\mathbf{E}_2 - \mathbf{E}_1) = 0$ and $\hat{\mathbf{n}} \times (\mathbf{H}_2 - \mathbf{H}_1) = 0$, enforcing the continuity of the tangential component of the electric and magnetic fields, respectively. However, as we have seen in Sect. 8.1.1, the introduction of the d-parameters renders the above boundary conditions inapplicable. Hence, a revised set of boundary conditions which reflect the presence of the d-parameters is required. These can be encoded through (normally-oriented) surface polarization and (tangentially-oriented) surface current terms, herein denoted, respectively, by $\mathbf{P}(\mathbf{r}) = \boldsymbol{\pi}(\mathbf{r})\delta(\mathbf{r} - \mathbf{r}_{\partial\Omega})$ and $\mathbf{J}(\mathbf{r}) = \mathbf{K}(\mathbf{r})\delta(\mathbf{r} - \mathbf{r}_{\partial\Omega})$, which are nonzero only at the interface, i.e., for $\mathbf{r} = \mathbf{r}_{\partial\Omega}$. Specifically, the surface polarization and current densities can be defined in terms of the d-parameters as [17, 18, 21]

$$\boldsymbol{\pi} \equiv \epsilon_0 d_\perp \left[\hat{\mathbf{n}} \cdot (\mathbf{E}_2 - \mathbf{E}_1)\right] \hat{\mathbf{n}}\,, \tag{8.25a}$$

$$\mathbf{K} \equiv i\omega d_\parallel \left[\hat{\mathbf{n}} \times (\mathbf{D}_2 - \mathbf{D}_1) \times \hat{\mathbf{n}}\right]. \tag{8.25b}$$

After some manipulations, these surface polarizations and currents can be self-consistently absorbed into a revision of the conventional boundary conditions[5]

$$\begin{aligned} \hat{\mathbf{n}} \times (\mathbf{E}_2 - \mathbf{E}_1) &= \hat{\mathbf{n}} \times \left[-\epsilon_0^{-1} \nabla_s \left(\boldsymbol{\pi} \cdot \hat{\mathbf{n}}\right)\right] \\ &= -d_\perp \hat{\mathbf{n}} \times \left[\nabla \hat{\mathbf{n}} \cdot (\mathbf{E}_2 - \mathbf{E}_1)\right], \end{aligned} \tag{8.26a}$$

$$\hat{\mathbf{n}} \times (\mathbf{H}_2 - \mathbf{H}_1) = i\omega d_\parallel \left[\hat{\mathbf{n}} \times (\mathbf{D}_2 - \mathbf{D}_1) \times \hat{\mathbf{n}}\right] \tag{8.26b}$$

where ∇_s is the nabla operator that lives in the two-dimensional surface which defines the interface. It is more elucidating, nonetheless, to split the fields into parallel and perpendicular components (with respect to the interface), so that the previous equations become

[5] While the boundary conditions on the tangential parts of **E** and **H** is fully sufficient to uniquely couple solutions across the interface, the complementary set of boundary conditions—on the normal components of **B** and **D**—is occasionally more convenient. The boundary condition for the normal components of **B**, is unchanged from its classical counterpart, i.e. $\hat{\mathbf{n}} \cdot (\mathbf{B}_2 - \mathbf{B}_1) = 0$; for the normal components of **D**, it is $\hat{\mathbf{n}} \cdot (\mathbf{D}_2 - \mathbf{D}_1) = d_\parallel \nabla_\parallel \cdot \left[\hat{\mathbf{n}} \times (\mathbf{D}_2 - \mathbf{D}_1) \times \hat{\mathbf{n}}\right]$ [21].

$$\mathbf{E}_{2,\parallel} - \mathbf{E}_{1,\parallel} = -d_\perp \nabla_\parallel \left(E_{2,\perp} - E_{1,\perp}\right) , \tag{8.27a}$$
$$\mathbf{H}_{2,\parallel} - \mathbf{H}_{1,\parallel} = i\omega d_\parallel \left(\mathbf{D}_{2,\parallel} - \mathbf{D}_{1,\parallel}\right) \times \hat{\mathbf{n}} . \tag{8.27b}$$

Clearly, and as we have anticipated, the incorporation of the Feibelman d-parameters introduces discontinuities in the parallel components of the (classical, or asymptotic) electric and magnetic fields. The magnitude of such discontinuities is naturally proportional to the d-parameters.

Modified Fresnel coefficients with Feibelman d-parameters. In possession of the boundary conditions personified by Eqs. (8.27), we now have all that is required for deriving the nonclassical equivalents of Fresnel's reflection and transmission coefficients for a planar dielectric–metal interface. In what follows, we adopt the same coordinate system and conventions as in Sect. 8.1.1 (see also Fig. 8.3).

We seek transverse magnetic (TM) solutions, in the dielectric and metal half-spaces, of the form:

Dielectric half-space ($z > 0$):

$$\mathbf{H}_\mathrm{d} = \left(e^{-ik_{z,\mathrm{d}}z} + r_p e^{ik_{z,\mathrm{d}}z}\right) e^{i(qx-\omega t)}\, \hat{\mathbf{y}} , \tag{8.28a}$$
$$\mathbf{E}_\mathrm{d} = \left[E_{x,\mathrm{d}}(z)\, \hat{\mathbf{x}} + E_{z,\mathrm{d}}(z)\, \hat{\mathbf{z}}\right] e^{i(qx-\omega t)} , \tag{8.28b}$$

where $k_{z,\mathrm{d}} = \sqrt{\epsilon_\mathrm{d} k_0^2 - q^2}$, and

$$E_{x,\mathrm{d}}(z) = -\frac{k_{z,\mathrm{d}}}{\omega\epsilon_0\epsilon_\mathrm{d}} \left(e^{-ik_{z,\mathrm{d}}z} - r_p e^{ik_{z,\mathrm{d}}z}\right) , \tag{8.29a}$$
$$E_{z,\mathrm{d}}(z) = -\frac{q}{\omega\epsilon_0\epsilon_\mathrm{d}} \left(e^{-ik_{z,\mathrm{d}}z} + r_p e^{ik_{z,\mathrm{d}}z}\right) , \tag{8.29b}$$

as determined by Maxwell's equations.

Metal half-space ($z < 0$):

$$\mathbf{H}_\mathrm{m} = t_p e^{-ik_{z,\mathrm{m}}z} e^{i(qx-\omega t)}\, \hat{\mathbf{y}} , \tag{8.30a}$$
$$\mathbf{E}_\mathrm{m} = \left[E_{x,\mathrm{m}}(z)\, \hat{\mathbf{x}} + E_{z,\mathrm{m}}(z)\, \hat{\mathbf{z}}\right] e^{i(qx-\omega t)} , \tag{8.30b}$$

where $k_{z,\mathrm{m}} = \sqrt{\epsilon_\mathrm{m} k_0^2 - q^2}$, and

$$E_{x,\mathrm{m}}(z) = -\frac{k_{z,\mathrm{m}}}{\omega\epsilon_0\epsilon_\mathrm{m}} t_p e^{-ik_{z,\mathrm{m}}z} , \tag{8.31a}$$
$$E_{z,\mathrm{m}}(z) = -\frac{q}{\omega\epsilon_0\epsilon_\mathrm{m}} t_p e^{-ik_{z,\mathrm{m}}z} . \tag{8.31b}$$

Making use of the modified boundary conditions Eq. (8.27), which for a planar interface take the form

$$E_{\mathrm{x,d}}(0) - E_{\mathrm{x,m}}(0) = -iqd_{\perp}\left[E_{\mathrm{z,d}}(0) - E_{\mathrm{z,m}}(0)\right], \tag{8.32a}$$

$$H_{\mathrm{y,d}}(0) - H_{\mathrm{y,m}}(0) = -i\omega d_{\parallel}\left[D_{\mathrm{x,d}}(0) - D_{\mathrm{x,m}}(0)\right], \tag{8.32b}$$

one obtains the following linear system:

$$\frac{k_{\mathrm{z,d}}}{\epsilon_{\mathrm{d}}}\left(r_p - 1\right) + \frac{k_{\mathrm{z,m}}}{\epsilon_{\mathrm{m}}}t_p = iq^2 d_{\perp}\left[\frac{1}{\epsilon_{\mathrm{d}}}\left(1 + r_p\right) - \frac{1}{\epsilon_{\mathrm{m}}}t_p\right], \tag{8.33a}$$

$$1 + r_p - t_p = -id_{\parallel}\left[k_{\mathrm{z,d}}(r_p - 1) + k_{\mathrm{z,m}}t_p\right]. \tag{8.33b}$$

After some somewhat lengthy but straightforward algebraic manipulations, one finally arrives at the nonclassical reflection and transmission coefficients for p-polarization, reading [6, 7, 17, 20]

$$r_p = \frac{\epsilon_{\mathrm{m}}k_{\mathrm{z,d}} - \epsilon_{\mathrm{d}}k_{\mathrm{z,m}} + (\epsilon_{\mathrm{m}} - \epsilon_{\mathrm{d}})\left[iq^2 d_{\perp} - ik_{\mathrm{z,d}}k_{\mathrm{z,m}}d_{\parallel}\right]}{\epsilon_{\mathrm{m}}k_{\mathrm{z,d}} + \epsilon_{\mathrm{d}}k_{\mathrm{z,m}} - (\epsilon_{\mathrm{m}} - \epsilon_{\mathrm{d}})\left[iq^2 d_{\perp} + ik_{\mathrm{z,d}}k_{\mathrm{z,m}}d_{\parallel}\right]}, \tag{8.34a}$$

and

$$t_p = \frac{2\epsilon_{\mathrm{m}}k_{\mathrm{z,d}}}{\epsilon_{\mathrm{m}}k_{\mathrm{z,d}} + \epsilon_{\mathrm{d}}k_{\mathrm{z,m}} - (\epsilon_{\mathrm{m}} - \epsilon_{\mathrm{d}})\left[iq^2 d_{\perp} + ik_{\mathrm{z,d}}k_{\mathrm{z,m}}d_{\parallel}\right]}, \tag{8.34b}$$

where only terms up to linear-order in $qd_{\perp,\parallel}$ have been kept.

Consistently, in the electrostatic limit ($c \to \infty$) one has $k_{\mathrm{d,m}} \to iq$, thereby recovering the nonretarded reflection and transmission amplitudes for p-polarized waves given by Eqs. (8.19a) and (8.19b).

Proceeding in a similar fashion for s-polarized [or transverse electric (TE)] waves, the corresponding nonclassical reflection and transmission coefficients are given by [6, 7, 17]

$$r_s = \frac{k_{\mathrm{z,d}} - k_{\mathrm{z,m}} + (\epsilon_{\mathrm{m}} - \epsilon_{\mathrm{d}})ik_0^2 d_{\parallel}}{k_{\mathrm{z,d}} + k_{\mathrm{z,m}} - (\epsilon_{\mathrm{m}} - \epsilon_{\mathrm{d}})ik_0^2 d_{\parallel}}, \tag{8.35a}$$

$$t_s = \frac{2k_{\mathrm{z,d}}}{k_{\mathrm{z,d}} + k_{\mathrm{z,m}} - (\epsilon_{\mathrm{m}} - \epsilon_{\mathrm{d}})ik_0^2 d_{\parallel}}. \tag{8.35b}$$

Surface plasmon polariton dispersion with Feibelman d-parameters. The retarded dispersion relation of surface plasmon polaritons at a planar dielectric–metal interface can be fetched from the poles of the associated scattering coefficients—see

Eqs. (8.34). Thus, the nonclassical SPP spectrum within the framework of Feibelman's d-parameters stems from the implicit condition:

$$\epsilon_m k_{z,d} + \epsilon_d k_{z,m} - (\epsilon_m - \epsilon_d)\left[iq^2 d_\perp + i k_{z,d} k_{z,m} d_\parallel\right], \tag{8.36}$$

or, equivalently, while noting that $k_{z,(d,m)} = i\kappa_{d,m}$ with $\kappa_{d,m} = \sqrt{q^2 - \epsilon_{d,m} k_0^2}$,

$$\frac{\epsilon_d}{\kappa_d} + \frac{\epsilon_m}{\kappa_m} - (\epsilon_m - \epsilon_d)\left[\frac{q^2}{\kappa_d \kappa_m} d_\perp - d_\parallel\right] = 0\,. \tag{8.37}$$

Equation (8.37) determines the retarded dispersion relation of SPPs taking Feibelman's quantum surface corrections into account (to leading-order), namely nonlocal effects and damping of these collective excitations via their decay into electron-hole pairs [6, 7, 23]. Recall that in this formulation $\epsilon_m \equiv \epsilon_m(\omega)$ is still the local bulk dielectric function of the metal; therefore the above-mentioned nonclassical corrections enter via the d-parameters alone. Returning to Eq. (8.37), it is clear that the well-know classical result $\epsilon_d/\kappa_d + \epsilon_m/\kappa_m = 0$ [cf. Eq. (2.23)] is obtained by letting $d_{\perp,\parallel} \to 0$. Additionally, in the electrostatic limit one has $\kappa_{d,m} \to q$ and the nonretarded surface plasmon resonance, given by Eq. (8.21), is recovered.

Figure 8.5 shows the nonclassical spectral properties of plasmons in a planar vacuum–jellium interface, contrasting the retarded and nonretarded regimes, as well as the classical and nonclassical behaviors. Three (inverse) length scales characterize the plasmonic dispersion in the planar system: the free-space wavevector k_0, the plasmon wavevector q, and the inverse of the centroid of induced charge $[\mathrm{Re}\, d_\perp]^{-1}$.

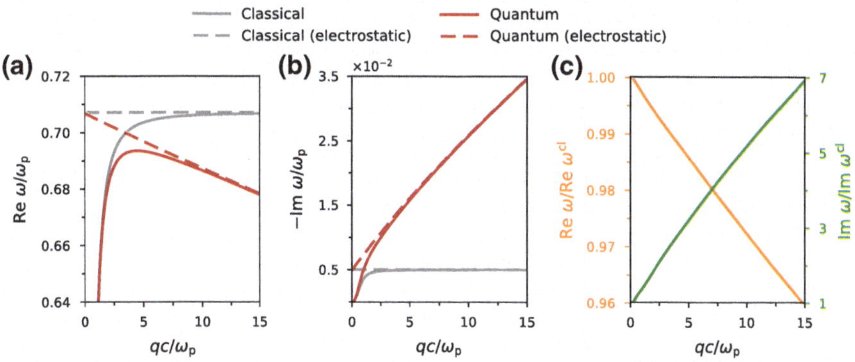

Fig. 8.5 Nonclassical spectral properties of SPPs in a planar vacuum–jellium interface computed via the d-parameters' formalism [corresponding to the roots of Eq. (8.37)]. The d-parameters' data is taken from TDDFT calculations performed by Christensen et al. [3, 28]. The material parameters are the same as in Fig. 8.4a. **a** Surface plasmon polariton dispersion relation and **b** Corresponding resonance widths. **c** Nonclassical spectral shifts and resonance broadenings (including retardation effects)

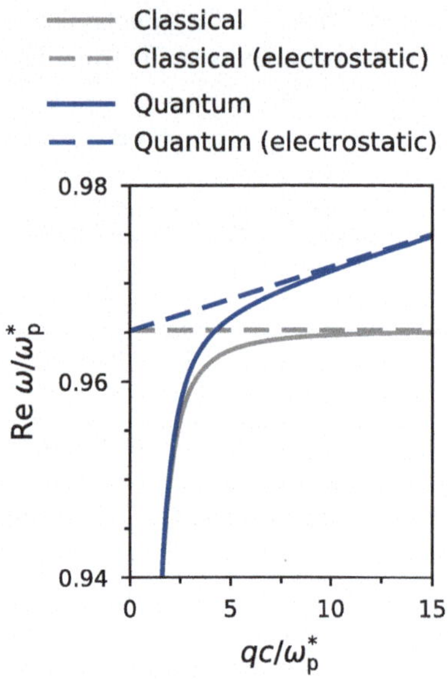

Fig. 8.6 Nonclassical dispersion relation of surface plasmon polaritons in a planar vacuum–silver interface. The *d*-parameters' data is taken from TDDFT calculations performed by Christensen et al. [3]. The material parameters are the same as in Fig. 8.4b

The plasmon dispersion, consequently, spans up to three distinct regimes, namely a classical, retarded regime $q \sim k_0 \ll |d_\perp|^{-1}$, a deeply nonclassical, quasi-static regime $q \sim |d_\perp|^{-1} \gg k_0$, and an intermediate regime. Figure 8.5a, b demonstrate that each of these regimes are well-realized in the planar $r_s = 3$ jellium: (i) at small wavevectors, nonclassical effects are negligible; (ii) at large wavevectors, they substantially redshift and broaden the plasmonic dispersion, manifesting the spill-out characteristic of simple metals (i.e., $\text{Re}\, d_\perp > 0$) and surface-enhanced Landau damping, respectively, consistent with earlier findings [7, 10, 11, 42, 43]; and (iii) at intermediate wavevectors, both retardation and nonclassical corrections are non-negligible, and therefore need to be taken into account simultaneously. Intriguingly, the existence of a well-defined intermediate regime demonstrates that the transition from classical to nonclassical response is intrinsically multiscale.

In the same spirit, we present in Fig. 8.6 the salient features of SPPs sustained at a vacuum–silver interface. In this case—and as we have seen when studying Fig. 8.4c—the incorporation of quantum surface corrections via the *d*-parameters' mesoscopic theory leads to a blueshift of the SPP frequency for moderate-to-larger wavevectors, courtesy of electronic spill-out (i.e., $\text{Re}\, d_{\text{eff}} < 0$). We further remark that in silver the dynamics of the (conducting) 5s electrons are strongly influenced by core electrons in the (filled) 4d band, which, semiclassically, amounts to a complex-valued, frequency-dependent background dielectric function $\epsilon_\infty \equiv \epsilon_\infty(\omega)$. Incidentally,[6] this affects

[6]On a more technical note, this also makes a TDDFT implementation strictly based on the jellium approximation inapplicable. This can be *partly* circumvented by accounting for *d*-band screening semiclassically—see, for instance, Refs. [3, 7]—, albeit it is reasonable to expect a loss in accuracy when comparing with the predictions for "true jellium metals". Hence, ideally, a TDDFT imple-

the magnitude of the nonclassical corrections to the SPPs spectral features depending simultaneously on the frequency-dependence of ϵ_∞ and d_{eff}.

8.2 Quantum Nonlocal Response of Plasmonic Spheres

8.2.1 Multipolar Polarizability of a Nanosphere Including Feibelman d-Parameters

Let us consider the response of a small (i.e., $R \ll \lambda$) spherical nanoparticle of radius R due to a quasi-static multipolar field; henceforth, all dynamic quantities are implicitly assumed to evolve as $e^{-i\omega t}$. The aim of the present section is to determine the multipolar polarizability of a metallic nanosphere, from which the system's optical response and associated localized surface plasmon resonances can be unambiguously calculated.

In the quasi-static (or nonretarded) limit, the scalar electrostatic potential is governed by Laplace's equation, $\nabla^2\Phi(\mathbf{r}) = 0$, and therefore admits an expansion in a series of multipoles [4], for instance

$$\Phi(r, \theta, \phi) = \sum_{lm} \varphi_l(r) Y_l^m(\theta, \phi), \tag{8.38}$$

where due to symmetry considerations one may to separate the electrostatic potential as a product of a radial part, $\varphi_l(r)$, and an angular part, $Y_l^m(\theta, \phi)$. The latter are the spherical harmonics while the former reads $\varphi_l(r) = A_l r^l + B_l r^{-l-1}$ [4]. Since the electric potential inside the nanosphere needs to remain finite, we have

$$\Phi_{\text{in}}(r, \theta, \phi) = \sum_{lm} c_l \left(\frac{r}{R}\right)^l Y_l^m(\theta, \phi), \tag{8.39}$$

whereas in the region outside the spherical particle we write

$$\Phi_{\text{out}}(r, \theta, \phi) = \sum_{lm} \left[\left(\frac{r}{R}\right)^l - a_l \left(\frac{R}{r}\right)^{l+1}\right] Y_l^m(\theta, \phi). \tag{8.40}$$

In the above expressions, the coefficients a_l and c_l play the role of "reflection" and "transmission" amplitudes for a given multipole of order l (they do not depend on m, however, due to azimuthal symmetry[7]). Once a_l and c_l are determined, the

---

mentation with a more sophisticated account of the atomic orbitals and atomic structure should be employed [8, 16].

[7]In fact, for this reason one could, alternatively, set $m = 0$ right from the start and describe the angular part of the potential using the Legendre polynomials alone [4].

electromagnetic response of the metal sphere—including its LSP resonances, optical cross-sections, LDOS, etc—can be readily computed.

In order to find the so far unknown a_l and c_l coefficients, one has to invoke the appropriate boundary conditions. Within the local response approximation, in the absence of external charges and currents, these are $\left(\mathbf{E}_{\parallel}^{\text{out}} - \mathbf{E}_{\parallel}^{\text{in}}\right)\big|_{\mathbf{r}\in\partial V} = 0$ and $\hat{\mathbf{n}} \cdot \left(\mathbf{D}_{\parallel}^{\text{out}} - \mathbf{D}_{\parallel}^{\text{in}}\right)\big|_{\mathbf{r}\in\partial V} = 0$, which simply state the continuity of the potential and the discontinuity of its derivative (weighted by the permittivities on each side) across an interface, ∂V. However—and as we have already seen in Sect. 8.1—within the mesoscopic formalism of the Feibelman d-parameters these are modified and read instead (recall the discussion at the start of Sect. 8.1.2):

$$\mathbf{E}_{\parallel}^{\text{out}} - \mathbf{E}_{\parallel}^{\text{in}} = -d_{\perp}\boldsymbol{\nabla}_{\parallel}\left(E_r^{\text{out}} - E_r^{\text{in}}\right), \tag{8.41a}$$

$$\hat{\mathbf{r}} \cdot \left(\mathbf{D}^{\text{out}} - \mathbf{D}^{\text{in}}\right) = d_{\parallel}\boldsymbol{\nabla}_{\parallel} \cdot \left[\hat{\mathbf{r}} \times \left(\mathbf{D}^{\text{out}} - \mathbf{D}^{\text{in}}\right) \times \hat{\mathbf{r}}\right]. \tag{8.41b}$$

where the subscript "$\parallel$" simply denotes the components that are parallel to the nanosphere's surface at a given point (and whose normal unit vector is radially-oriented). Thus, $\boldsymbol{\nabla} \equiv (\nabla_r, \boldsymbol{\nabla}_{\parallel})$. Now, making use of the connection between the electric field and the corresponding potential, i.e., $\mathbf{E} = -\boldsymbol{\nabla}\Phi$, Eqs. (8.41) yield[8]

$$1 - a_l - c_l + l\frac{d_{\perp}}{R}\left[1 + \frac{l+1}{l}a_l - c_l\right] = 0, \tag{8.42a}$$

$$\epsilon_{\text{m}}c_l - \epsilon_{\text{d}}\left[1 + \frac{l+1}{l}a_l\right] + (l+1)\frac{d_{\parallel}}{R}\left[\epsilon_{\text{m}}c_l - (1-a_l)\epsilon_{\text{d}}\right] = 0. \tag{8.42b}$$

Solving for a_l, one obtains (to linear-order in $R^{-1}d_{\perp,\parallel}$)

$$a_l = \frac{(\epsilon_{\text{m}} - \epsilon_{\text{d}})\left[1 + l d_{\perp}R^{-1} + (l+1)d_{\parallel}R^{-1}\right]}{\epsilon_{\text{m}} + (\epsilon_{\text{m}} - \epsilon_{\text{d}})(l+1)d_{\parallel}R^{-1} + (1 + 1/l)\left[\epsilon_{\text{d}}(1 + l d_{\perp}R^{-1}) - \epsilon_{\text{m}} l d_{\perp}R^{-1}\right]}. \tag{8.43}$$

Finally, the sough-after multipolar polarizability then reads ($\alpha_l = 4\pi R^{2l+1}a_l$)

$$\alpha_l = 4\pi R^{2l+1}\frac{(\epsilon_{\text{m}} - \epsilon_{\text{d}})\left[1 + \frac{l}{R}\left(d_{\perp} + \frac{l+1}{l}d_{\parallel}\right)\right]}{\epsilon_{\text{m}} + \frac{l+1}{l}\epsilon_{\text{d}} - (\epsilon_{\text{m}} - \epsilon_{\text{d}})\frac{l+1}{R}\left(d_{\perp} - d_{\parallel}\right)}. \tag{8.44}$$

---

[8] Here it is useful to recall the spherical harmonic differential equation [44]:

$$\frac{1}{\sin\theta}\frac{\partial}{\partial\theta}\left(\sin\theta\frac{\partial Y_l^m}{\partial\theta}\right) + \frac{1}{\sin^2\theta}\frac{\partial^2 Y_l^m}{\partial\phi^2} = -l(l+1)Y_l^m.$$

This expression therefore includes quantum surface corrections incorporated via Feibelman's mesoscopic theory. Equation (8.44) can then be straightforwardly employed to compute the corresponding nonclassical cross-sections, LDOS, and LSP dispersion, in the nonretarded regime. Furthermore, it is apparent from this equation that the classical multipolar polarizability [45, 46],

$$\alpha_l^{\mathrm{cl}} = 4\pi R^{2l+1} \frac{\epsilon_{\mathrm{m}} - \epsilon_{\mathrm{d}}}{\epsilon_{\mathrm{m}} + \dfrac{l+1}{l}\epsilon_{\mathrm{d}}}, \tag{8.45}$$

is reinstated [cf. Eq. (2.33)] in the limit of vanishing d-parameters. Lastly, the well-known Clausius–Mossotti dipolar polarizability [2, 47], that is, $\alpha_{\mathrm{CM}} = 4\pi R^3 \frac{\epsilon_{\mathrm{m}}-\epsilon_{\mathrm{d}}}{\epsilon_{\mathrm{m}}+2\epsilon_{\mathrm{d}}}$, immediately follows from the previous expression upon setting $l = 1$.

Nonretarded LSPs: quantum corrections. In the planar configuration, we have obtained the surface plasmon dispersion condition from the poles of the corresponding reflection coefficient for p-polarized waves. Likewise, for a metallic nanosphere, its localized surface plasmon resonances follow from the poles of the associated multipolar polarizability (8.44). Hence, the dispersion of LSP resonances—parameterized by the angular momentum l—stem from the implicit condition:

$$\epsilon_{\mathrm{m}} + \frac{l+1}{l}\epsilon_{\mathrm{d}} - (\epsilon_{\mathrm{m}} - \epsilon_{\mathrm{d}}) \frac{l+1}{R} \left(d_\perp - d_\parallel\right). \tag{8.46}$$

Moreover, if the metal's optical response in the asymptotic region is given by a Drude-like dielectric function of the form $\epsilon_{\mathrm{m}} = \epsilon_\infty - \omega_{\mathrm{p}}^2/(\omega^2 + i\omega\gamma)$, one finds

$$\omega(\omega + i\gamma) = \omega_{\mathrm{p}}^2 \frac{1 - (l+1)(d_\perp - d_\parallel)R^{-1}}{\epsilon_\infty + \epsilon_{\mathrm{d}}(l+1)/l - (\epsilon_\infty - \epsilon_{\mathrm{d}})(l+1)(d_\perp - d_\parallel)R^{-1}} \tag{8.47}$$

which, nonetheless, remains an *implicit* equation for the LSP dispersion because the d-parameters are frequency-dependent (and, potentially, ϵ_∞ and/or ϵ_{d} may be too).

In the same spirit of Eqs. (8.24), it is instructive to determine from Eq. (8.46) the approximate expressions for the real and imaginary parts of the first-order spectral correction, $\omega_l^{(1)} \equiv \omega_l^{(0)} + \delta\omega_l + \mathcal{O}([\omega_l - \omega_l^{(0)}]^2)$, which, for the particular case where $\epsilon_{\mathrm{d}} = \epsilon_\infty = 1$, read [3, 17]

$$\mathrm{Re}\ \omega_l^{(1)} \simeq \mathrm{Re}\ \omega_l^{(0)} \left[1 - \frac{l+1}{2R}\,\mathrm{Re}\left(d_\perp^{(0)} - d_\parallel^{(0)}\right)\right], \tag{8.48a}$$

$$-\mathrm{Im}\ \omega_l^{(1)} \simeq \frac{\gamma}{2} + \mathrm{Re}\ \omega_l^{(0)} \frac{l+1}{2R}\,\mathrm{Im}\left(d_\perp^{(0)} - d_\parallel^{(0)}\right), \tag{8.48b}$$

where $\mathrm{Re}\ \omega_l^{(0)} = \omega_{\mathrm{p}}/\sqrt{1 + (l+1)/l}$ is the lth-order classical nonretarded LSP frequency and $d_\alpha^{(0)} \equiv d_\alpha(\mathrm{Re}\ \omega_l^{(0)})$ for $\alpha = \{\perp, \parallel\}$. Equations (8.48a) and (8.48b) capture the leading-order nonclassical LSP frequency shift and broadening, respectively, for LSP frequencies around $\omega_l^{(0)}$. In addition, by comparing Eqs. (8.48) with their planar counterparts (8.24) it becomes apparent that $(l+1)/R$ plays the role of an effective wavevector; for large l, this effective wavevector tends to $\sim l/R$ as expected from simple standing-wave arguments. Furthermore, it is interesting to note that the incorporation of quantum mechanical effects breaks the scale-invariance that usually characterizes the classical nonretarded limit, wherein plasmon resonances $\omega^{\rm cl}$ are scale-independent (e.g., Re $\omega^{\rm cl} = \omega_{\rm p}/\sqrt{1+\epsilon_{\rm d}}$ and Re $\omega^{\rm cl} = \omega_{\rm p}/\sqrt{1+2\epsilon_{\rm d}}$ for the surface and dipole plasmon of a planar and spherical jellium interface, respectively). Here, the introduction of the electronic length scale(s) associated with $d_{\perp,\parallel}$ breaks this scale-invariance, producing finite-size corrections parameterized by either $qd_{\perp,\parallel}$ or $d_{\perp,\parallel}/R$,

In the quasi-static regime the l-dependent LSP resonances for deeply subwavelength metal nanospheres follow from the solutions of Eq. (8.46). Figure 8.7a shows the calculated LSP resonances—associated with the first five multipoles—of an $r_s = 4$ jellium sphere with diameter of $2R = 6$ nm and embedded in a homogeneous host medium with $\epsilon_{\rm d} = 1$. The difference between the classical and quantum predictions is striking in a number of ways. Firstly, the account of quantum mechanical spill-out leads to LSP resonances that are systematically redshifted with respect to the classical predictions. Secondly, the frequency of the classical LSP resonances increases monotonically with the eigenindex l—from the dipolar LSP at Re $\omega^{\rm cl}_{l=1} = \omega_{\rm p}/\sqrt{3}$ up to the asymptotic limit Re $\omega^{\rm cl}_{l\to\infty} = \omega_{\rm p}/\sqrt{2}$ where they "pile up"—, contrasting with the nonmonotonic behavior exhibited by the nonclassical LSPs. Lastly, while the classical framework heralds a fixed, l-independent broadening (and, perhaps more disturbing, also size-independent) amounting to – Im $\omega_l^{\rm cl} = \gamma/2$, the quantum mechanical theory predicts increased broadening for higher-order LSP multipoles, in agreement with experimental observations [39, 48].

Figure 8.7b, c portrays the size-dependent nonclassical shift and broadening of the dipolar ($l = 1$) LSP resonance for silver spheres of varying radii. The latter is consequence of the inward spill (Re $d_{\rm eff} < 0$) of the induced density whereas the latter reflects the presence of a new decay channel (namely, surface-assisted electron-hole pair creation).

8.2.2 Generalized Mie Theory with Feibelman d-Parameters

In the preceding section we have analyzed the quasi-static response of small (i.e., $R \ll \lambda_0$) spheres taking into account nonlocal and quantum surface corrections incorporated via the d-parameters' formalism. Here, we extend our previous treatment to spheres of arbitrary size by developing a generalized Mie theory that incorporates mesoscopic length-scales of the plasmon-supporting electron gas—that are omitted in a classical, LRA-description—through the Feibelman d-parameters.

While the mesoscopic reflection coefficients of the planar system have been determined by Feibelman [6], the corresponding scattering coefficients of the spherically

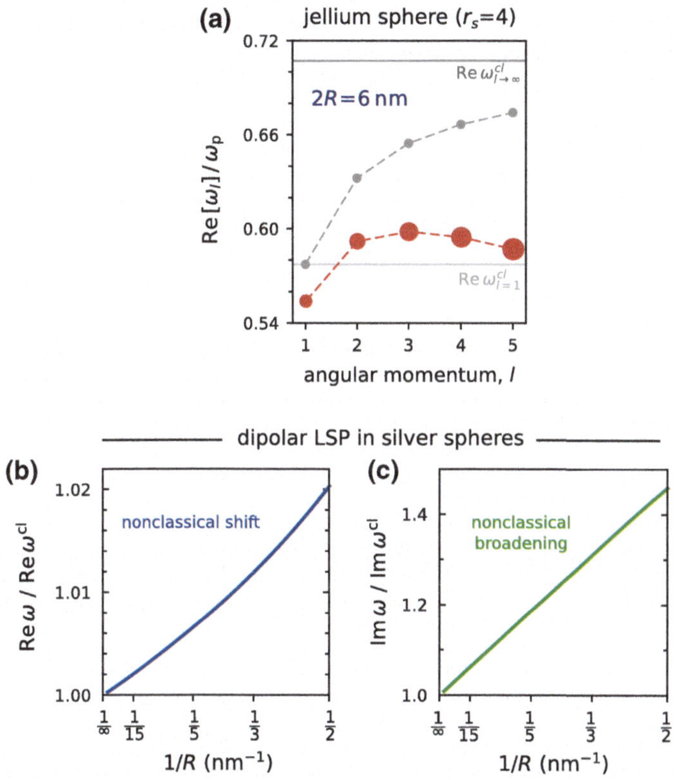

Fig. 8.7 Spectral properties of nonretarded LSPs in metal nanospheres calculated via Eq. (8.46). The d-parameters' data was obtained from TDDFT calculations performed by Christensen et al. [3]. The nanoparticles are assumed to be embedded in a host environment with dielectric constant $\epsilon_d = 1$. **a** Classical (gray circles) and quantum (red circles) multipolar LSP resonances of a $2R = 6$ nm metal nanosphere (specifically, an $r_s = 4$ jellium; $\hbar\omega = 5.89$ eV, $\epsilon_\infty = 1$, $\hbar\gamma = 0.1$ eV, and $d_\parallel = 0$). The circle's radius is proportional to the resonance's width. **b** [**c**] Nonclassical shift [broadening] of the dipolar LSP resonance for silver nanospheres of varying radii. We describe the bulk dielectric function of silver by a Drude-type expression of the form $\epsilon_m(\omega) = \epsilon_\infty(\omega) - \omega_p/(\omega^2 + i\omega\gamma)$ with $\hbar\omega_p = 9.02$ eV and $\hbar\gamma = 22$ meV and $\epsilon_\infty(\omega)$ is taken from Johnson and Christy's experimental data [29] (by subtracting the aforementioned Drude contribution)

symmetric system—the so-called Mie coefficients a_l^{TM} and b_l^{TE} [49]—have remained unknown until recently (see Publication A), despite their significant practical utility, e.g., for calculating the sphere's LSP resonances, optical cross-sections, Purcell enhancements, near-field interactions, etc.

Mesoscopic Mie coefficients. We have derived the mesoscopic generalization of Mie's theory by incorporating the Feibelman d-parameters through a generalization of the usual electromagnetic boundary conditions [17, 21] [cf. Eqs. (8.27)]. The mathematical details are somewhat lengthy and are thus provided in Appendix C (and also in the Supplementary Information of Publication A). As such, we shall not repeat them here and we will rather focus our discussion on the observables akin to the optical response of metal spheres.

Specifically, for a metallic sphere of radius R, the generalized, nonclassical TM and TE Mie coefficients are[9] [17]:

$$a_l^{\mathrm{TM}} = \frac{\epsilon_{\mathrm{m}} j_l(x_{\mathrm{m}})\Psi_l'(x_{\mathrm{d}}) - \epsilon_{\mathrm{d}} j_l(x_{\mathrm{d}})\Psi_l'(x_{\mathrm{m}}) + \Delta_{\perp,\parallel}^{\mathrm{num}}}{\epsilon_{\mathrm{m}} j_l(x_{\mathrm{m}})\xi_l'(x_{\mathrm{d}}) - \epsilon_{\mathrm{d}} h_l^{(1)}(x_{\mathrm{d}})\Psi_l'(x_{\mathrm{m}}) + \Delta_{\perp,\parallel}^{\mathrm{den}}}, \tag{8.49a}$$

where

$$\Delta_{\perp,\parallel}^{\mathrm{num}} = (\epsilon_{\mathrm{m}} - \epsilon_{\mathrm{d}}) \left\{ j_l(x_{\mathrm{d}}) j_l(x_{\mathrm{m}}) \left[l(l+1)\right] d_\perp + \Psi_l'(x_{\mathrm{d}})\Psi_l'(x_{\mathrm{m}})\, d_\parallel \right\} R^{-1}, \tag{8.49b}$$

$$\Delta_{\perp,\parallel}^{\mathrm{den}} = (\epsilon_{\mathrm{m}} - \epsilon_{\mathrm{d}}) \left\{ h_l^{(1)}(x_{\mathrm{d}}) j_l(x_{\mathrm{m}}) \left[l(l+1)\right] d_\perp + \xi_l'(x_{\mathrm{d}})\Psi_l'(x_{\mathrm{m}})\, d_\parallel \right\} R^{-1}, \tag{8.49c}$$

and

$$b_l^{\mathrm{TE}} = \frac{j_l(x_{\mathrm{m}})\Psi_l'(x_{\mathrm{d}}) - j_l(x_{\mathrm{d}})\Psi_l'(x_{\mathrm{m}}) + \left(x_{\mathrm{m}}^2 - x_{\mathrm{d}}^2\right) j_l(x_{\mathrm{d}}) j_l(x_{\mathrm{m}})\, d_\parallel R^{-1}}{j_l(x_{\mathrm{m}})\xi_l'(x_{\mathrm{d}}) - h_l^{(1)}(x_{\mathrm{d}})\Psi_l'(x_{\mathrm{m}}) + \left(x_{\mathrm{m}}^2 - x_{\mathrm{d}}^2\right) h_l^{(1)}(x_{\mathrm{d}}) j_l(x_{\mathrm{m}})\, d_\parallel R^{-1}}, \tag{8.49d}$$

with dimensionless wavevectors $x_j \equiv k_j R$, spherical Bessel and Hankel functions of the first kind $j_l(x)$ and $h_l^{(1)}(x)$, and the Riccati–Bessel functions $\Psi_l(x) \equiv x j_l(x)$ and $\xi_l(x) \equiv x h_l^{(1)}(x)$; primed functions denote their derivatives.

Notice that the quantum mechanical TE Mie coefficient b_l^{TE} is independent of the perpendicular Feibelman d-parameter, $d_\perp$ (a featured naturally shared with its planar kin, the TE reflection coefficient r_s). This should not really constitute a surprise, since for a TE wave the electric field is purely tangential to the particle's surface and therefore has no perpendicular component capable of inducing a displacement of the charge density along the normal to the sphere's surface. On the other hand, the TM Mie coefficient a_l^{TM} describes TM waves, whose electric field in general possesses both radial and tangential components, and, consequently, both $d_\perp$ and $d_\parallel$ are present in these coefficients.

Finally, we comment that in the small-radius limit, $x_j \ll 1$, a small-argument expansion of spherical Bessel and Hankel functions produces the nonretarded equivalent of the TM Mie coefficient a_l^{TM}, i.e., the mesoscopic multipolar polarizability [cf. Eq. (8.44)] previously derived in Sect. 8.2.1.

Quantum corrections to the optical response of plasmonic spheres. A particularly alluring feature of the theoretical formalism set forth here, is that the mathematical structure of the usual expressions for the cross-sections, LDOS, etc, is unchanged; that is, the nonclassical optical response is obtained using the very same textbook expressions, simply by replacing the classical Mie coefficients by the quan-

[9] The two remaining mesoscopic Mie coefficients c_l^{TE} and d_l^{TM}, associated with the internal (or transmitted) fields, can be found in Appendix C (and in the Supplementary Information of Publication A).

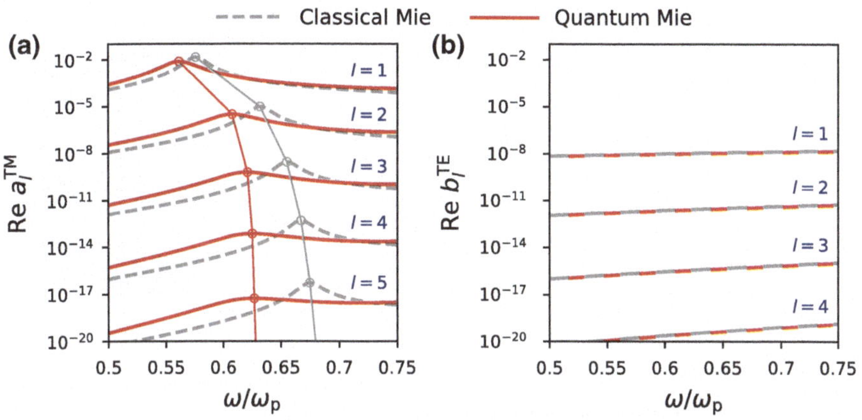

Fig. 8.8 Real parts of the Mie coefficients for a $R = 5$ nm jellium sphere with $r_s = 4$. The metal particle is assumed to be embedded in vacuum (ϵ_d=1). The d-parameters are obtained using the TDDFT data provided by Christensen et al. [3], and a Drude-type broadening of $\hbar\gamma = 0.1$ eV is assumed. **a** Real part of the Mie coefficient associated with TM polarization, a_l^{TM}, for different angular momenta, l (the line-connected circles indicate their respective maxima). **b** Real part of the Mie coefficient associated with TM polarization, b_l^{TE}, for different angular momenta, l

tum mechanical Mie coefficients presented in Eqs. (8.49). For instance, the extinction cross-section for a metal sphere illuminated by a monochromatic plane-wave is therefore given by

$$\sigma_{ext} = \frac{2\pi}{k_d^2} \sum_{l=1}^{\infty} (2l + 1) \operatorname{Re}\{a_l^{TM} + b_l^{TE}\}. \tag{8.50}$$

For relatively small metal spheres, the optical response is dominated by a series of peaks originating from the excitation of LSP resonances. The latter occur at frequencies for which a_l^{TM} has pole, as illustrated in Fig. 8.8. A LSP resonance is said to be of electric dipole character for $l = 1$, of electric quadrupole character for $l = 2$, and so on. It is also apparent from Fig. 8.8 that the dipole resonance ($l = 1$) contributes the most (at least for subwavelength spheres) for the cross-section, with the successively higher-order resonances becoming increasingly negligible (note the logarithmic scale). The TE Mie coefficients b_l^{TE} are essentially featureless and their contribution for the extinction cross-section is several orders of magnitude smaller than the one associated with TM Mie coefficients a_l^{TM} of the same order. The impact of quantum nonlocal effects, i.e., a redshift owing to spill-out ($\operatorname{Re} d_\perp > 0$) and broadening due to surface-enabled Landau damping, can be clearly observed in Fig. 8.8 as well. It is also interesting to note that the TE Mie coefficients are unchanged by such effects, which can be understood by the fact that here $d_\parallel = 0$ [check Eq. (8.49d)].

Spectral features of LSPs and quantum nonlocal effects. Figure 8.9 outlines the plasmonic features of metal spheres as a function of their radii. In most respects,

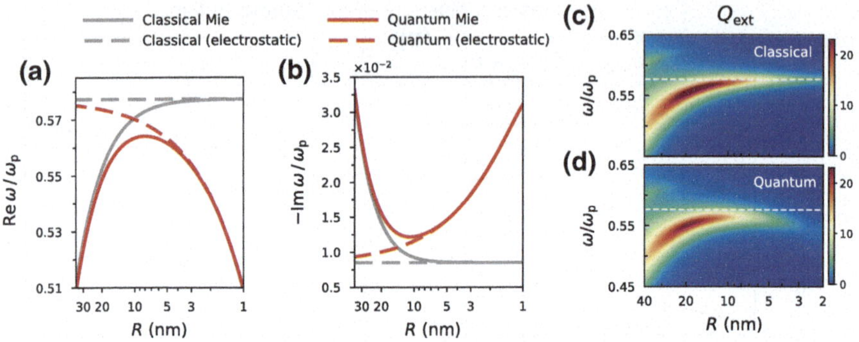

Fig. 8.9 Nonclassical spectral properties of LSPs calculated taking into account quantum nonlocal surface corrections. **a** Dipole LSP resonance frequency, and **b** Associated resonance widths, for metal spheres of varying radii (notice the inverted scale). **c** Classical and **d** Quantum extinction cross-sections for plasmonic spheres in air normalized to their geometrical cross-sections, $Q_{\text{ext}} \equiv \sigma_{\text{ext}}/(\pi R^2)$. The white dashed lines mark the classical nonretarded dipole resonance at $\omega_{\text{p}}/\sqrt{3}$. Material parameters: jellium metal ($r_s = 4$ and $\hbar\gamma = 0.1$ eV) and $\epsilon_{\text{d}} = 1$

they mirror the qualitative conclusions drawn for the planar case (see Fig. 8.5), but with the inverse radius R^{-1} playing the role of an effective wavevector.[10] Concretely, and focusing on the dipole LSP, Fig. 8.9a, b plainly show the shortcomings of the classical theory for jellium spheres with dimensions below $2R \sim 20$ nm. For extremely small spheres, the nonretarded limit (Sect. 8.2.1) reproduces the nonclassical redshift and broadening well. Again, we observe an intermediate region where both retardation and nonclassical effects are of comparable magnitude. Incidentally, this region corresponds to the regime probed by several experiments that investigated nonclassical plasmons [12, 13, 39, 50]. Our results therefore underscore the necessity of simultaneously accounting for retardation and quantum effects when interpreting experimental data. Finally, in Fig. 8.9c, d we present the normalized extinction cross-sections of jellium spheres under plane-wave illumination. Besides reproducing the main features already observed in Fig. 8.9a, b, they also exhibit extra resonances due to higher-order LSP modes (chiefly, the $l = 2$). The cross-section of these higher-order LSPs, however, fall off rapidly with decreasing radii owing to the realization of the dipole limit. In the nonclassical case this reduction is amplified further, as higher-order LSPs are progressively more impacted by surface-induced Landau damping [cf. Eq. (8.48b)]. These predictions are in good agreement with experimental data [7, 10, 11, 39, 43, 48].

The formalism and results presented in the preceding sections—i.e., Sects. 8.1 and 8.2—establish the foundations governing plasmon-enhanced nanophotonic phenomena in the mesoscopic regime. In the following, we exploit this understanding to assess plasmon–emitter interactions at the nanoscale.

[10] Increased losses at large radii are due to radiation damping.

8.3 Plasmon–Emitter Interactions at the Nanoscale

The exceedingly rapid growth of plasmonics has been largely motivated by the ability of surface plasmons to squeeze electromagnetic fields into deep subwavelength dimensions [51], thereby boosting the strength of light–matter interactions in nanoscale environments [52–54]. Plasmon-enabled phenomena include strong field enhancements [15, 55, 56], the modification of the emission/absorption of emitters via the Purcell effect [57–62], surface-enhanced Raman scattering [63, 64], ultrasensitive biochemical sensing [65–67], and plasmon-driven chemistry [68, 69], just to mention a few.

Indeed, the interaction between light and matter in free-space is an intrinsically weak process. Strikingly, this interaction strength can be enormously enhanced near material interfaces. This is especially true in plasmonics [51–54, 60, 67]: for an emitter separated from an interface by a subwavelength distance h, the decay rate is increased by a factor $\propto h^{-3}$ in the classical, macroscopic theory [recall Box 2.2]. However, as the separation—or the characteristic dimensions of the plasmonic system itself—is reduced to the nanoscale ($\lesssim 10 - 20\,\text{nm}$), the classical theory is rendered invalid due to its neglect of all intrinsic quantum mechanical lengths scales in the plasmonic material. Thus, to ascertain the ultimate limits of plasmon-mediated light–matter interactions, the classical theory must be augmented. Here, we apply the mesoscopic formalism introduced in Sects. 8.1 and 8.2 and investigate the impact of leading-order quantum mechanical corrections in a broad range of prominent plasmon-mediated light–matter interaction phenomena. First, we study two narrowband single-emitter phenomena: the Purcell enhancement of a dipole emitter [61, 62, 70, 71], associated with the enhancement of the local density of states (LDOS), and the enhancement of dipole-forbidden multipolar transitions [72–74]. Finally, and going beyond the single-emitter setting, we investigate plasmon-mediated energy transfer between two emitters [75–77], which can be either narrow- or broadband. Finally, we consider a particularly broadband effect: the enhancement of two-photon processes for an emitter near a metal surface [78–80].

8.3.1 *Nonclassical LDOS: Purcell Enhancement*

A hallmark of plasmonics is its ability to support extreme field enhancements and correspondingly large Purcell factors [61, 62, 67], enabling unprecedented control over the emission properties of emitters. At its core, this is a manifestation of the reshaping of the LDOS spectrum, which is enhanced near plasmon resonances [81–84]. Importantly, the Purcell enhancement is generally maximized at short emitter–surface separations, i.e., exactly where nonlocality and quantum effects become important. Thus, as we show in what follows, a rigorous description of the governing electrodynamics that incorporates nonclassical effects is not only necessary, but ultimately essential.

The LDOS, $\rho^{\mathrm{E}}_{\hat{\mathbf{n}}}$, experienced by an emitter with orientation $\hat{\mathbf{n}}$ (and incorporating both radiative and nonradiative contributions) is proportional to the imaginary part of the system's Green's dyadic, which in turn is expandable in the previously introduced scattering coefficients (recall Sect. 2.3) [47, 85, 86]. The expressions are succinctly presented in Box 8.2. We exploit this fact to directly incorporate nonclassical surface corrections into the LDOS, by simply adopting the mesoscopic scattering coefficients, Eqs. (8.34)–(8.35) or (8.49), instead of their classical equivalents.

Box 8.2—LDOS expressions in terms of the system's scattering coefficients

The LDOS experienced by a point-like dipole emitter embedded in a dielectric medium with dielectric constant ϵ_{d} and located at a distance h above a metal half-space is given by (derived in Eq. (2.53) and reproduced here again for the convenience of the reader) [47, 87]

$$\frac{\rho^{\mathrm{E}}_{\perp}}{\rho^{\mathrm{E}}_{0}} = 1 + \frac{3}{2}\,\mathrm{Re}\int_0^\infty \frac{u^3}{\sqrt{1-u^2}}\, r_p\, e^{2ik_{\mathrm{d}}h\sqrt{1-u^2}}\,\mathrm{d}u,$$

$$\frac{\rho^{\mathrm{E}}_{\parallel}}{\rho^{\mathrm{E}}_{0}} = 1 + \frac{3}{4}\,\mathrm{Re}\int_0^\infty \frac{u}{\sqrt{1-u^2}}\Big[r_s - (1-u^2)r_p\Big] e^{2ik_{\mathrm{d}}h\sqrt{1-u^2}}\,\mathrm{d}u,$$

for an emitter with its dipole moment oriented perpendicularly ($\perp$) or tangentially ($\parallel$), respectively, to the dielectric–metal interface (here assumed to be defined by the $z = 0$ plane). The perpendicularly oriented dipole only couples to TM modes, whereas the dipole in the parallel configuration couples to both TM and TE modes. At short emitter–metal separations, however, the TM contribution dominates, regardless of orientation. Moreover, since plasmonic excitations are TM polarized, the TM contribution is the main quantity of interest for plasmon-enhanced LDOS.

For an emitter at a distance h from the surface of a metallic sphere of radius R, the LDOS can be evaluated via [46, 88]

$$\frac{\rho^{\mathrm{E}}_{\perp}}{\rho^{\mathrm{E}}_{0}} = 1 + \frac{3}{2}\frac{1}{y^2}\sum_{l=1}^{\infty}(2l+1)l(l+1)\,\mathrm{Re}\Big\{-a_l^{\mathrm{TM}}\Big[h_l^{(1)}(y)\Big]^2\Big\},$$

$$\frac{\rho^{\mathrm{E}}_{\parallel}}{\rho^{\mathrm{E}}_{0}} = 1 + \frac{3}{4}\frac{1}{y^2}\sum_{l=1}^{\infty}(2l+1)\,\mathrm{Re}\Big\{-a_l^{\mathrm{TM}}\big[\xi_l'(y)\big]^2 - b_l^{\mathrm{TE}}\big[\xi_l(y)\big]^2\Big\},$$

for an emitter with its dipole oriented along the radial ($\perp$) or tangential ($\parallel$) directions, respectively. Additionally, we have introduced the dimensionless radial emitter position $y \equiv k_{\mathrm{d}}(R+h)$ for brevity of notation.

The above expressions also highlight a key feature of this formalism for mesoscale electrodynamics: conveniently, in order to calculate the quantum mechanically corrected LDOS within the d-parameters framework one only needs to replace the standard Mie coefficients by their generalized nonclassical counterparts, Eqs. (8.49). The same also holds for the familiar Fresnel reflection coefficients and their nonclassical counterparts, Eqs. (8.34)–(8.35).

In Fig. 8.10a, b we show the classical and quantum LDOS, normalized to the free-space LDOS, ρ_0^E, near a planar metal interface as a function of the emitter–metal separation h, for a normally-oriented emitter (see the Supplementary Information of Publication A for the parallel and orientation-averaged cases). The enhancement of the LDOS near the surface plasmon frequency is markedly sharper in the classical case and less pronounced in the nonclassical one at shorter separations. This observation is particularly evident in Fig. 8.10b, which shows the plasmon-enhanced LDOS for different emitter–metal separations. In the classical formulation, the peak in the LDOS remains relatively sharp, approaching the nonretarded plasmon frequency $\omega_p/\sqrt{2}$ at small separations. Contrasting this, in the nonclassical framework the LDOS peak redshifts (consistent with the spill-out characteristic of jellium metals) with decreasing h, and evolves into a broad spectral feature at very small emitter–metal distances. This behavior reflects the nonclassical corrections to the plasmonic spectrum outlined in Sect. 8.1. Evidently, the most significant impact of nonclassicality here is the substantial reduction (notice the logarithmic scale) of the maximum-attainable LDOS in the nonclassical case, particularly for $h \lesssim 10\,\text{nm}$. Lastly, it is interesting to observe the emergence of a broad spectral peak at frequencies above $\omega_p/\sqrt{2}$ that is absent in the classical setting. This feature is a manifestation of the so-called surface-multipole plasmon or Bennet mode [89] that originates due to the finite-size of the inhomogeneous surface region [7]; mathematically, it corresponds to a pole in $d_\perp(\omega)$; physically, it represents an out-of-plane oscillation confined to the surface region.

Figure 8.10c, d show the LDOS of a radially-oriented emitter placed at a distance h from the surface of an $R = 5\,\text{nm}$ metal sphere. The LDOS enhancement in the spherical geometry is richer in features, partly because the sphere, unlike the plane, has an intrinsic length scale (its radius R), and partly because it hosts a series of l-dependent multipolar LSPs. The LDOS enhancement is centered around these LSP frequencies. In the nonclassical case, we again observe redshifted and broadened spectral features relative to their classical counterparts. The impact of Landau damping is amplified by the order of the LSP mode, cf. Eq. (8.48); as a result, only the dipole and quadrupole modes are discernible in the nonclassical case (in the classical case, a faint $l = 3$ LSP remains identifiable). Next, in Fig. 8.10e we investigate the LDOS enhancement's dependence on the sphere's radius R for a fixed emitter–sphere separation of $h = 10\,\text{nm}$. In particular, the impact of nonclassical effects—particularly its reduction of the maximum LDOS—is more pronounced at smaller radii, in agree-

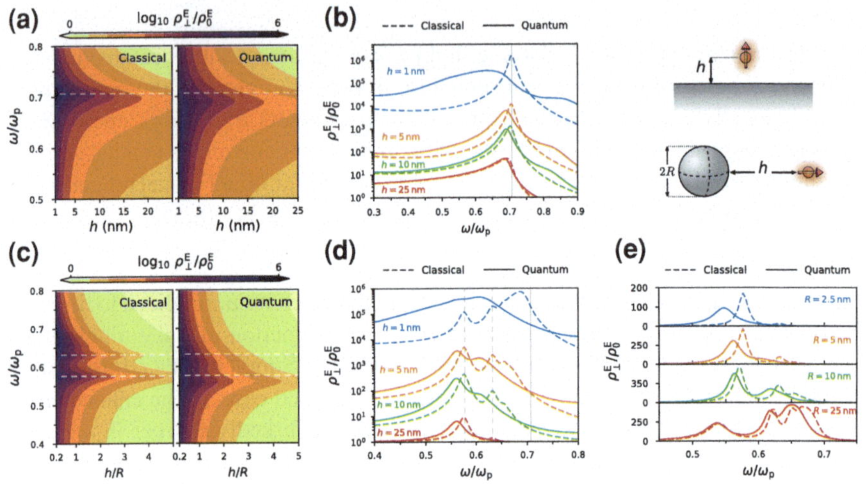

Fig. 8.10 Quantum corrections to the Purcell enhancement in nanoplasmonics. Normalized LDOS, $\rho^{\mathrm{E}}/\rho_0^{\mathrm{E}}$, experienced by a normally-oriented emitter near a planar metal surface (**a, b**), and a metal sphere (**c–e**). **a** Normalized LDOS map as a function of the emitter's frequency and separation from the planar jellium surface. The horizontal line marks the classical nonretarded surface plasmon frequency, $\omega_p/\sqrt{2}$. **b** Plasmon-mediated LDOS enhancement at different emitter–surface separations. The vertical line marks the position of $\omega_p/\sqrt{2}$. **c** Density plot of the normalized LDOS map near an $R = 5$ nm jellium sphere as a function of the emitter's frequency and h/R ratio. The horizontal lines marks the classical nonretarded dipole and quadrupole LSP frequencies, $\omega_p/\sqrt{1 + (l + 1)/l}$, for $l \in \{1, 2\}$. **d** Plasmon-mediated LDOS enhancement near an $R = 5$ nm nanosphere for distinct emitter–surface distances. Vertical lines mark the $l = 1$, $l = 2$, and $l = \infty$ classical nonretarded LSPs. **e** Normalized LDOS for plasmonic spheres with various radii at a fixed $h = 10$ nm emitter–surface separation. Material parameters as in Fig. 8.9

ment with the $\propto(l + 1)R^{-1}$ scaling previously derived in Eq. (8.48). In fact, for very small metal spheres, only the LDOS enhancement associated with the dipole plasmon remains identifiable in the nonclassical case, due to surface-enabled Landau damping. Crucially, although deviations from classicality are most pronounced for spheres with radii $\lesssim$10 nm, even relatively large spheres (that would otherwise be considered within the classical regime, e.g. $2R = 50$ nm) exhibit significant nonclassical corrections at small emitter–metal separations. Indeed, this constitutes an example of a multiscale regime where both retardation (a classical effect) and quantum effects must be addressed in concert.

8.3.2 *Dipole-Forbidden Electric Multipolar Transitions*

The set of optical transitions associated with the emission of radiation by atoms is in practice limited due to the mismatch between the atom's size and the wavelength of the radiation emitted by it. This fact leads to the selection rules for dipole-allowed transitions that originate from the so-called dipole approximation [90]. Such transitions, however, constitute only a fraction of a much richer spectrum. Nevertheless, transition rates other than the dipole-allowed are simply too slow (by several orders of magnitude) to be accessible in practice and are consequently termed "forbidden" [91, 92]. Recently, it has been shown that it is possible to increase the effective light–matter coupling strength for such transitions by exploiting the shrinkage of the wavelength of light brought about by surface plasmons [17, 74, 93]. Notwithstanding this, a satisfactory framework for describing the impact of nonclassical effects in the plasmonic enhancement of forbidden transitions remains somewhat elusive. Below, we pursue a remedy for this by extending our formalism of mesoscopic electrodynamics to the class of dipole-forbidden transitions of electric multipolar character [17], which can be exploited to probe even larger plasmon momenta. These are transitions in which the orbital angular momentum of the emitter changes by more than one; hereafter denoted En with $n = 2, 3, 4, \ldots$ (thus, E1 denotes a dipole transition, E2 a quadrupole transition, and so on). It should be emphasized that although in the following we consider hydrogenic systems—for the sake of definiteness—the theory presented here can be readily applied to any point-like emitter (e.g., atoms, quantum dots, nitrogen-vacancy centers, or dyes).

Let us consider an emitter located at a distance h from a planar metal surface (Fig. 8.11a), and treat the light–matter interaction in its vicinity using a formulation of macroscopic quantum electrodynamics (detailed in Appendix D) which enables a rigorous account of the quantum nature of the emitter and of the plasmon, as well as the inherent presence of loss [94, 95]. Within this framework, the multipolar decay rates, $\Gamma_{\mathrm{E}n}$, can be evaluated as [17, 74] (see Appendix D)

$$\Gamma_{\mathrm{E}n} = 2\alpha^3\omega_0 \left[\frac{(k_0 a_{\mathrm{B}})^{n-1}}{(n-1)!}\right]^2 \Xi \int_0^\infty u^{2n} e^{-2uk_0h}\, \mathrm{Im}\, r_p\, du\,, \tag{8.51}$$

where $u \equiv q/k_0$, a_{B} denotes the Bohr radius, α is the fine-structure constant, and the dimensionless quantity Ξ is related to the matrix element associated with the transition. In the previous expression, the quasi-static limit is assumed, valid for $k_0 h \ll 1$. Nonetheless, in our calculations we use the *retarded* reflection coefficient to accurately incorporate the plasmon pole's spectral position. Moreover, in this limit $\Gamma^{\mathrm{tot}}_{\mathrm{E}n} = \Gamma^0_{\mathrm{E}n} + \Gamma_{\mathrm{E}n} \simeq \Gamma_{\mathrm{E}n}$ since the free-space contribution $\Gamma^0_{\mathrm{E}n}$ is many orders of magnitude smaller [91, 92].

In Fig. 8.11b we plot the En decay rates associated with the 6{p, d, f, g, h} $\to$ 2s transition series in hydrogen (δ-transitions of the Balmer series). While at relatively large distances from the metal the spontaneous emission rates of higher-order multipolar transitions are several orders of magnitude slower than E1, this difference is dramatically reduced at smaller emitter–metal separations. Interestingly, at nanometric separations the higher-order multipolar rates can exceed the E1 free-space rate, signaling a breakdown of traditional dipole-allowed selection rules. In addition, the inclusion of nonclassical effects via d-parameters *increases* the multipolar decay rates relative to the classical predictions (Fig. 8.11b, inset) by roughly one order of magnitude at the smallest separations. To understand the physical mechanism for this additional enhancement, we show in Fig. 8.11c–e the integrand of Eq. (8.51) for the first three multipolar orders, each evaluated at three distinct atom–metal separations. Two main contributions can be readily identified: (i) a sharp, resonant contribution corresponding to the plasmon pole embodied in $\mathrm{Im}\, r_p$ at the transition frequency (i.e., at the intersection of the blue and red lines in Fig. 8.11a), associated with emission into plasmons; and (ii) a broad, nonresonant contribution associated with quenching by lossy channels in the metal, e.g. Landau damping, disorder, phonons, etc. The relative contribution of (i) and (ii) to the overall decay rate depends strongly on the emitter–metal separation (due to the $u^{2n} e^{-2uk_0 h}$ scaling of the integrand), with loss-related quenching dominating over plasmon emission at very small emitter–metal separations. This effect is more pronounced for higher-order multipolar transitions since the integrand of Eq. (8.51) initially grows with u^{2n}. With the above-mentioned considerations in mind, the additional nonclassical enhancement can now be understood: it is a direct result of an increased non-resonant, loss-related contribution due to surface-enabled Landau damping. Finally, the dotted lines in Fig. 8.11b, f indicate regions in which a significant fraction of $\Gamma_{\mathrm{E}n}$ is accumulated at very large wavevectors where the condition $q\, \mathrm{Re}\, d_\perp \ll 1$ is only approximately valid; evidently, at the smallest separations and at large transitions orders n, our mesoscopic framework is pushed beyond its range of validity.

Figure 8.11f considers a similar transition in a hydrogen-like atom, but now occurring at a higher frequency—i.e., closer to $\omega_{\mathrm{p}}/\sqrt{2}$—and therefore able to probe larger plasmon wavevectors. We assume, for simplicity, that the magnitude of the matrix elements in Eq. (8.51) still equal those in the 6{p, d, f, g, h} $\rightarrow$ 2s hydrogen series. The enhancement of the En rates is qualitatively similar to the previous case, albeit with some quantitative differences: for instance, as shown in Fig. 8.11g–i, the resonant plasmon contribution now peaks at larger u, a simple consequence of the increased plasmon momentum at this higher transition frequency. This is in principle beneficial because even a small increase in confinement can result in a huge increase of the decay rates due to the u^{2n} scaling of $d\Gamma_{\mathrm{E}n}/du$. However, plasmon losses tend to increase concomitantly with increasing confinement, resulting in broader plasmon peaks (cf. Fig. 8.11g–i). Lastly, we observe that the nonclassical multipolar decay rates no longer consistently exceed the classical predictions at this higher frequency, contrasting the findings shown in Fig. 8.11b. This difference reflects a more complicated and substantial nonclassical modification of the plasmonic response at such frequency, namely, that the overall impact on $\Gamma_{\mathrm{E}n}$ ultimately results from an nontrivial interplay between the modified scattering response $\mathrm{Im}\, r_p$ and the scaling $u^{2n} e^{-2uk_0 h}$.

In summary, these calculations demonstrate that quantum surface corrections substantially modify the multipolar decay rates from those predicted using classical electrodynamics. Radiation from such electric multipolar transitions can be delivered to the far-field by outcoupling the SPPs via gratings or antennas. Moreover, even in the regime dominated by nonresonant enhancement, the breakage of the conventional selection rules should still have clear experimental signatures [74], with potential implications for photovoltaic devices [68] or hot-electron physics [68, 69].

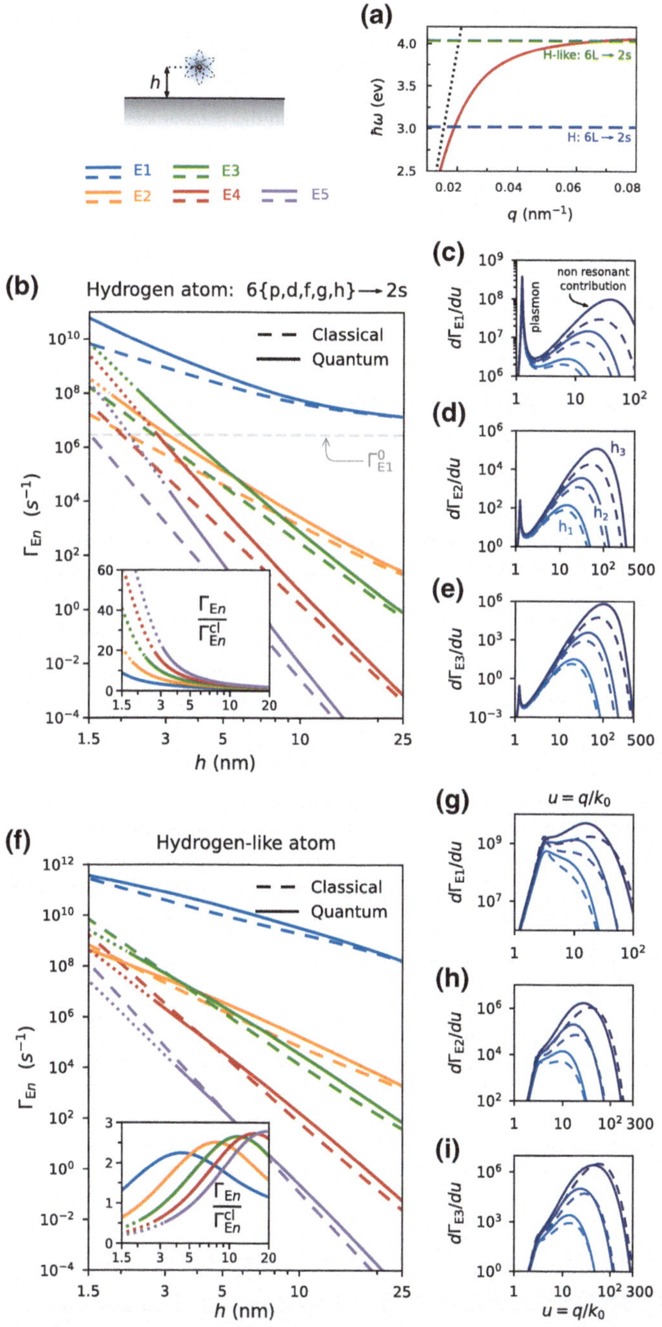

(a)
h
E1
E2
E3
E4
E5
ħω (ev)
H-like: 6L → 2s
H: 6L → 2s
q (nm⁻¹)
(b)
Hydrogen atom: 6{p,d,f,g,h} → 2s
Classical
Quantum
Γ_En (s⁻¹)
h (nm)
(c)
non resonant contribution
plasmon
dΓ_E1/du
(d)
dΓ_E2/du
h1
h2
h3
(e)
dΓ_E3/du
(f)
Hydrogen-like atom
Classical
Quantum
Γ_En (s⁻¹)
h (nm)
(g)
u = q/k0
dΓ_E1/du
(h)
dΓ_E2/du
(i)
dΓ_E3/du
u = q/k0

◂**Fig. 8.11** Enhancement of dipole-forbidden electric multipole transitions for an emitter near a metal surface. **a** Position of the transition frequencies corresponding to **b–e** (dashed blue) and to **f–i** (dashed green). **b** Rates for multipolar transitions, $\Gamma_{\mathrm{E}n}$, associated with the 6{p, d, f, g, h} → 2s series of transitions in hydrogen ($\hbar\omega_0 = 3.02\,\mathrm{eV}$). The dashed lines indicate the classical rates whereas the solid lines represent the predictions within the d-parameter formalism. The dashed gray horizontal line marks the E1 rate in free-space, for comparison. The inset shows the relative rates, $\Gamma_{\mathrm{E}n}/\Gamma^{\mathrm{cl}}_{\mathrm{E}n}$. **c–e** Differential rates $\mathrm{d}\Gamma_{\mathrm{E}n}/\mathrm{d}u$ [integrand of Eq. (8.51)] for various atom–surface separations: $h_1 = 10\,\mathrm{nm}$, $h_2 = 5\,\mathrm{nm}$, and $h_3 = 2.5\,\mathrm{nm}$. **f** and **g–i** are the respective equivalents of **b** and **c–e**, but now assuming that the transitions occur at $\omega_0 = 0.97\,\omega_\mathrm{p}/\sqrt{2}$. The dotted lines correspond to regions where our nonclassical formalism is extrapolated (we set the threshold for its validity when more than 10% of the contribution for the rates comes from wavevectors beyond $q_{\mathrm{th}}\,\mathrm{Re}\,d_\perp = 1/3$)

8.3.3 Energy Transfer Between Two Emitters Near a Metal Surface

The interaction between emitters in optical cavities or near plasmonic structures is instrumental to many scientific disciplines, ranging from quantum optics [96] to chemical physics and the life sciences [97, 98]. A prominent example is energy transfer (ET) between two fluorophores: the fundamental process by which an excited flourophore (the donor) lowers its energy by transferring it to another flourophore (the acceptor). The signature of this mechanism is the observation of fluorescence emitted by the acceptor. In free-space, the ET between the two emitters takes place primarily via dipole–dipoles interaction and is typically short-ranged; in this limit, it is commonly referred to as Förster resonant energy transfer (FRET). In this context, the integration of emitters with plasmonic nanostructures can enhance the emitter–emitter ET rate, herein denoted by Γ_{ET}, through the introduction of a new, plasmonic near-field channel between the donor (D) and the acceptor (A) [99–101].

The calculation of Γ_{ET} involves the system's Green's function [47, 75–77, 82], which in turn depends on the system's scattering coefficients. More concretely, the ET rate from a donor located at $\mathbf{r}_\mathrm{D}$ to an acceptor placed at $\mathbf{r}_\mathrm{A}$ can be determined via [47, 75–77, 82]

$$\Gamma_{\mathrm{ET}} = \int_0^\infty w_{\mathrm{ET}}(\mathbf{r}_\mathrm{D}, \mathbf{r}_\mathrm{A}, \omega)\, f_\mathrm{D}^{\mathrm{em}}(\omega)\, f_\mathrm{A}^{\mathrm{abs}}(\omega)\, \mathrm{d}\omega\,, \tag{8.52a}$$

where the kernel $w_{\mathrm{ET}}(\mathbf{r}_\mathrm{D}, \mathbf{r}_\mathrm{A}, \omega)$ is the energy transfer amplitude that governs the medium-assisted interaction, and it is given by

$$w_{\mathrm{ET}}(\mathbf{r}_\mathrm{D}, \mathbf{r}_\mathrm{A}, \omega) = \frac{2\pi}{\hbar^2}\left(\frac{\omega^2}{\epsilon_0 c^2}\right)^2 \left|\boldsymbol{\mu}_\mathrm{A}^* \cdot \overleftrightarrow{\mathbf{G}}(\mathbf{r}_\mathrm{D}, \mathbf{r}_\mathrm{A}; \omega) \cdot \boldsymbol{\mu}_\mathrm{D}\right|^2. \tag{8.52b}$$

Here, $f_\mathrm{D}^{\mathrm{em}}$ ($f_\mathrm{A}^{\mathrm{abs}}$) stands for the donor's emission (acceptor's absorption) spectrum, and $\boldsymbol{\mu}_\mathrm{D}$ ($\boldsymbol{\mu}_\mathrm{A}$) the corresponding dipole moment. Notice that all the information about the nanophotonic environment is embodied in the Green's dyadic. Therefore, if the

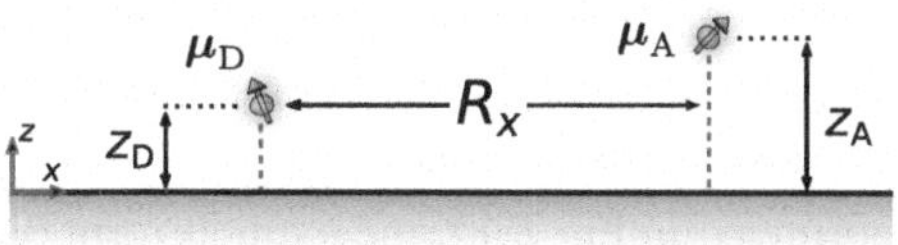

Fig. 8.12 Illustration of two dipoles in the vicinity of a planar metal surface. The dipole–surface separation is given by z_D (z_A) for the donor (acceptor), and its dipole moment is characterized by μ_D (μ_A). Without loss of generality, we have chosen our coordinate system so that $R_x = |x_D - x_A|$

Green's tensor is either known or can be calculated for the system under consideration, then the rate of energy transfer can be determined by evaluating Eq. (8.52). For the half-plane and sphere plasmonic structures, the nonclassical energy transfer rates are then straightforwardly obtained by substituting the classical scattering coefficients by their corresponding mesoscopic counterparts (namely, Eqs. (8.34) and (8.35) for the planar interface, and Eqs. (8.49) for the sphere).

At this point it is also instructive to note that, contrary to the computation of the LDOS where the (imaginary part of the) Green's dyadic is evaluated at the same point ($\mathbf{r} = \mathbf{r}' = \mathbf{r}_0$), the rate of energy transfer depends on the (absolute square of the) Green's function taken at the positions of the donor and of the acceptor. Additionally, the total (broadband) energy transfer rate depends on the overlap between the emission spectrum of the donor, $f_D^{em}(\omega)$, and the absorption spectrum of the acceptor, $f_A^{abs}(\omega)$.

Planar interface. We now consider two electric point-dipoles (e.g., each a generic two-level system) above a planar metal surface, as illustrated in Fig. 8.12. Therefore, according to Eqs. (8.52) the enhancement (with respect to the energy transfer rate in same medium but without the metal half-space, Γ_{ET}^0) of the energy transfer rate between the donor and the acceptor can be calculated as

$$\frac{\Gamma_{ET}}{\Gamma_{ET}^0} = \frac{\int_0^\infty d\omega\, \omega^4 \left|\hat{\mathbf{n}}_A \cdot \left[\overleftrightarrow{\mathbf{G}}_0(\mathbf{r}_D, \mathbf{r}_A; \omega) + \overleftrightarrow{\mathbf{G}}_{ref}(\mathbf{r}_D, \mathbf{r}_A; \omega)\right] \cdot \hat{\mathbf{n}}_D\right|^2 f_D^{em}(\omega) f_A^{abs}(\omega)}{\int_0^\infty d\omega\, \omega^4 \left|\hat{\mathbf{n}}_A \cdot \overleftrightarrow{\mathbf{G}}_0(\mathbf{r}_D, \mathbf{r}_A; \omega) \cdot \hat{\mathbf{n}}_D\right|^2 f_D^{em}(\omega) f_A^{abs}(\omega)}, \tag{8.53}$$

where $\overleftrightarrow{\mathbf{G}}_0(\mathbf{r}_D, \mathbf{r}_A; \omega)$ and $\overleftrightarrow{\mathbf{G}}_{ref}(\mathbf{r}_D, \mathbf{r}_A; \omega)$ are the Green's dyadics associated with the homogeneous medium and with the reflected part owing to the metal surface, correspondingly. The explicit form of the former has been written in Eq. (2.42) whereas the components of the latter—for the geometry considered here—are presented in Box 8.3. In Eq. (8.53), $\hat{\mathbf{n}}_A$ and $\hat{\mathbf{n}}_D$ are unit vectors denoting, respectively, the orientation of the acceptor and donor dipole moments.

Box 8.3—Components of the $\overleftrightarrow{\mathbf{G}}_{\text{ref}}(\mathbf{r}_\text{D},\mathbf{r}_\text{A};\omega)$ dyadic

For the geometry portrayed in Fig. 8.12, the explicit form of the elements of the Green's tensor $\overleftrightarrow{\mathbf{G}}_{\text{ref}}(\mathbf{r}_\text{D}, \mathbf{r}_\text{A}; \omega)$ is given by:

$$\left[\overleftrightarrow{\mathbf{G}}_{\text{ref}}(\mathbf{r}_\text{D}, \mathbf{r}_\text{A}; \omega)\right]_{zz} = \frac{i}{4\pi k_\text{d}^2} \int_0^\infty dq \, \frac{q^3}{k_z} r_p \, J_0(qR_x) \, e^{ik_z(z_D+z_A)} ,$$

$$\left[\overleftrightarrow{\mathbf{G}}_{\text{ref}}(\mathbf{r}_\text{D}, \mathbf{r}_\text{A}; \omega)\right]_{\substack{xx \\ yy}} = \frac{i}{8\pi} \int_0^\infty dq \, \frac{q}{k_z} r_s \, [J_0(qR_x) \pm J_2(qR_x)] \, e^{ik_z(z_D+z_A)}$$
$$- \frac{i}{8\pi k_\text{d}^2} \int_0^\infty dq \, q k_z \, r_p \, [J_0(qR_x) \mp J_2(qR_x)] \, e^{ik_z(z_D+z_A)} ,$$

$$\left[\overleftrightarrow{\mathbf{G}}_{\text{ref}}(\mathbf{r}_\text{D}, \mathbf{r}_\text{A}; \omega)\right]_{\substack{xz \\ zx}} = \pm \frac{1}{4\pi k_\text{d}^2} \int_0^\infty dq \, q^2 \, r_p \, J_1(qR_x) \, e^{ik_z(z_D+z_A)} ,$$

where $k_\text{d} = \sqrt{\epsilon_\text{d}}\omega/c$ and $k_z = \sqrt{k_\text{d}^2 - q^2}$, and J_n designate the nth order Bessel function of the first kind. All the remaining elements vanish for the configuration under consideration (Fig. 8.12).

Having established the theoretical foundations describing the energy transfer between two electric dipoles, we now investigate the impact of nonclassical corrections to plasmon-mediated ET between two emitters near a planar metal surface. In what follows, we restrict our analysis to the specific configuration depicted in Fig. 8.13a.

Figure 8.13b–e show the ET amplitude $w_\text{ET}(R, \omega)$ (evaluated at $z_\text{A} = z_\text{D} \equiv h$ with a donor–acceptor separation $R \equiv |\mathbf{r}_\text{A} - \mathbf{r}_\text{D}| = R_x$) normalized to its value in free-space $w_\text{ET}^0(R, \omega)$. The advantage of such procedure is that this ratio is emitter-independent, facilitating a discussion on the impact of the plasmonic response (also, for aligned narrowband emitters where $f_\text{D}^\text{em}(\omega) f_\text{A}^\text{abs}(\omega) \sim \delta(\omega - \omega_0)$, this simply amounts to the total ET rate enhancement $\Gamma_\text{ET}/\Gamma_\text{ET}^0$; we shall return to this point below). Our results demonstrate that the omission of quantum mechanical effects leads to a significant overestimation of the normalized ET amplitudes, across a broad parameter space. This discrepancy is particularly pronounced for emitter–metal separations of about $h \lesssim 10 - 15$ nm, and spans a wide range of donor–acceptor separations, R. The ET dependence on R is particularly interesting and spans several distinct regimes: (i) for large R relative to the SPP's propagation length, L_p, the metal's impact is negligible [the emitters are simply too far away for the ET to be mediated by SPPs (i.e., a SPP excited by the donor will be dissipated long before it reaches the acceptor)]; (ii) for $R \sim L_\text{p}$, the ET *enhancement* reaches its maxi-

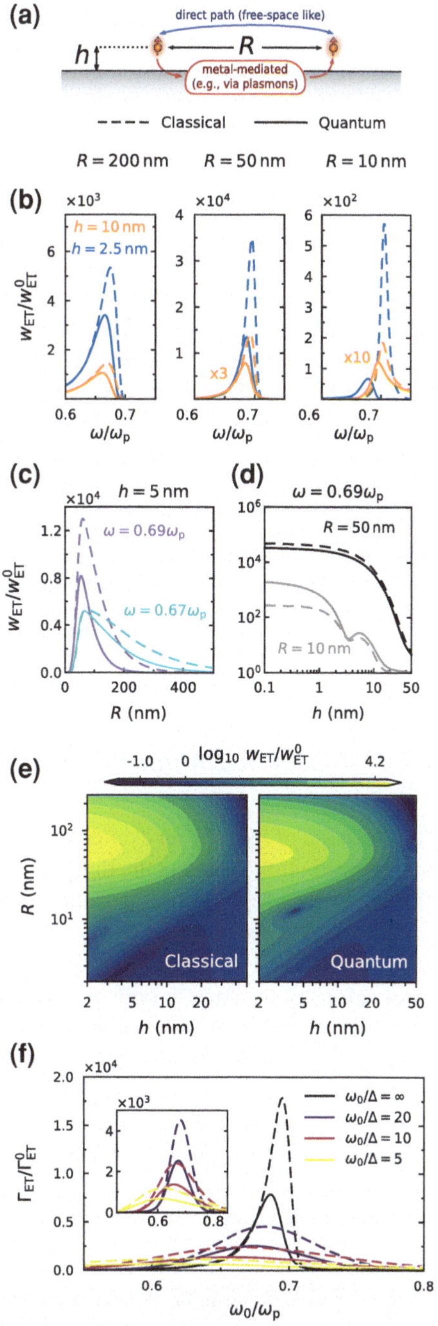

Fig. 8.13 Quantum corrections to plasmon-mediated energy transfer near a planar metal. **a** Two emitters transfer energy through free-space and metal-mediated channels (mutual separation $R \equiv |\mathbf{r}_A - \mathbf{r}_D|$, metal–interface offset $h = z_D = z_A$, and normally-oriented, i.e., $\boldsymbol{\mu}_A \parallel \boldsymbol{\mu}_D \parallel \hat{\mathbf{z}}$). **b** Spectral dependence of the (normalized) ET amplitude, w_{ET}/w^0_{ET}, for varying emitter–emitter distances. **c, d** Normalized transfer amplitude as a function of R (h), for fixed h (R) (see labels). **e** Contours of the classical and nonclassical energy transfer amplitudes at $\omega_0 = 0.97\,\omega_p/\sqrt{2}$. **f** Plasmonic enhancement of ET rates ($h = 5$ nm and $R = 50$ nm) for broadband emitters of varying $Q = \omega_0/\Delta$ as a function of (joint) emitter frequency ω_0 (inset: finite-Q emitters only)

mum, whose position and value are dictated by the spectral properties of the SPP, and therefore is affected both by the nonclassical spectral shift and broadening; and (iii) for $R \ll h$, the interaction is dominated by the free-space channel, rendering the metal's impact negligible again.

Furthermore, for emitters of sufficient spectral width, the ET can assume a broadband[11] aspect: we explore this in Fig. 8.13f by computing $\Gamma_{\mathrm{ET}}/\Gamma_{\mathrm{ET}}^{0}$ for a Gaussian donor–acceptor overlap $f_{\mathrm{D}}^{\mathrm{em}}(\omega) f_{\mathrm{A}}^{\mathrm{abs}}(\omega) = e^{-(\omega-\omega_0)^2/2\Delta^2}/\sqrt{2\pi}\,\Delta$, centered at ω_0 and with a (joint) width Δ and quality factor $Q \equiv \omega_0/\Delta$. Figure 8.13f shows the normalized classical and nonclassical broadband integrated ET rates for several Q as a function of the center frequency ω_0. Clearly, the maximum of $\Gamma_{\mathrm{ET}}/\Gamma_{\mathrm{ET}}^{0}$ decreases with Q, with a concomitant broadening and redshifting of the central peak. Interestingly, though the highest ET rate enhancements are obtained at large Q, and for ω_0 near the SPP's resonance, this shows that spectrally misaligned emitters can benefit from small Q factors, as this extends their spectral tails into the resonant region. More importantly, our results show that nonclassicality remains important even in the case of broadband emitters, and that nonclassical deviations persist (after being broadband integrated) even when the joint spectral width is larger that the nonclassical plasmon resonance shift.

Finally, Fig. 8.13 plainly demonstrates the importance of accounting for nonclassical effects when calculating energy transfer rates, as they impose limits to the maximum-attainable plasmon-enhanced energy transfer rate between emitters.

8.3.4 Plasmon-Enhanced Two-Photon Emission

The emission of light by an excited emitter is generally very well-described by first-order perturbation theory in the light–matter interaction of quantum electrodynamics (this includes every process considered so far). However, at higher order in the interaction, the possibility of two- and multi-photon spontaneous emission emerges. While two-photon spontaneous emission was predicted in the 1930s [103], it eluded observation for decades in both atomic and solid-state systems [78, 79], due to the exacerbated weakness of the interaction at second order. Despite this, two-photon emission is in principle an attractive process due to the correlated nature of the emitted photons (entangled, for instance, in energy and in angular momentum). In this context too, the extreme nanoscale confinement of plasmons in metals provides new opportunities to enhance two-photon emission dramatically [80] (in the form of two-plasmon emission), with recent work identifying potential opportunities to enhance two-photon emission to be as strong [74], or even stronger [104], than single-photon

[11] When $f_{\mathrm{D}}^{\mathrm{em}}(\omega) f_{\mathrm{A}}^{\mathrm{abs}}(\omega)$ in Eq. (8.53) extends above ω_{p}, it becomes necessary to evaluate $d_{\perp}(\omega)$ for $\omega > \omega_{\mathrm{p}}$ as well. Direct calculation via TDDFT is cumbersome above ω_{p}, since the induced potential extends into the bulk; instead, following Refs. [19, 102], we extrapolate $d_{\perp}(\omega)$ for $\omega > \omega_{\mathrm{p}}$ by enforcing exact sum-rules and asymptotic limits on a fit of $d_{\perp}(\omega)$ over frequencies below ω_{p} (see Supporting Information of Publication A for further details).

emission. Nevertheless, with these enticing possibilities being enabled essentially by extreme nanoscale confinement, it is legitimate to anticipate a sizable impact of nonclassical effects.

A minimal model of two-photon spontaneous emission is shown in Fig. 8.14a, where we illustrate an emitter at a distance h from a semi-infinite metallic interface. To isolate the parts of two-photon emission that depend on the metallic interface, as opposed to the detailed atomic level structure, we consider two-photon transitions between the s-states of a simple hydrogenic atom. This subgrouping includes the most prominent example of two-photon emission: the 2s $\rightarrow$ 1s transition in hydrogen, with level separation $\omega_0 = \omega_{2s} - \omega_{1s} \approx 10.2\,\text{eV}$. The level separation ω_0 restricts the frequencies of the two emitted photons to $\omega \in]0, \omega_0[$ and $\omega' \equiv \omega_0 - \omega$ (reflecting energy conservation) but otherwise leaves their difference unconstrained. The emission process is consequently inherently broadband, with the total rate Γ_{TPE} resulting from the sum over all energy-allowed (ω, ω')-partitions[12]: $\Gamma_{\text{TPE}} = \int_0^{\omega_0} (d\Gamma_{\text{TPE}}/d\omega)\, d\omega$, where $d\Gamma_{\text{TPE}}/d\omega$ is the differential decay rate for two-photon emission into frequencies ω and $\omega_0 - \omega$. As an example, for the 2s $\rightarrow$ 1s transition of hydrogen in free-space, $d\Gamma^0_{\text{TPE}}/d\omega$, exhibits a broad peak around the equal $\omega = \omega' = \omega_0/2$ splitting, as shown in Fig. 8.14b. Its integral, corresponding to the decay rate, is about $8.3\,\text{s}^{-1}$, nearly eight orders of magnitude slower than the 2p $\rightarrow$ 1s dipole-allowed single-photon transition ($\approx 6.3 \times 10^8\,\text{s}^{-1}$) [91].

In the presence of a metallic interface, the situation can change quite drastically due to a strongly enhanced LDOS. Importantly, two-photon emission benefits "twice" from an enhancement of the LDOS, as can be appreciated in the following nonretarded expression [104] describing the enhancement of the differential decay rate $d\Gamma_{\text{TPE}}/d\omega$ for an s $\rightarrow$ s transition in a hydrogenic atom, relative to its free-space value $d\Gamma^0_{\text{TPE}}/d\omega$ (see Supporting Information of Publication A):

$$\frac{d\Gamma_{\text{TPE}}/d\omega}{d\Gamma^0_{\text{TPE}}/d\omega} = \frac{1}{2}\left(\frac{\rho^{\text{E}}_{\perp}(\omega)}{\rho^{\text{E}}_0(\omega)}\right)\left(\frac{\rho^{\text{E}}_{\perp}(\omega_0 - \omega)}{\rho^{\text{E}}_0(\omega_0 - \omega)}\right). \tag{8.54}$$

Each fraction is a Purcell factor; thus, the order of magnitude two-photon differential enhancement is roughly the square of the one-photon enhancement (Fig. 8.10). To be more precise, the differential two-photon enhancement is directly and simply related to the one-photon enhancement: it is (half) the Purcell enhancement at ω multiplied by its reflection about $\omega_0/2$.

Figure 8.14c, d contrast the classical and nonclassical predictions of the differential two-photon emission enhancement near a metal surface for different values of the (hydrogen-like emitter's) transition frequency, its separation from the surface, and emission frequency. For separations $\gtrsim 10\,\text{nm}$ nonclassical effects modify the physics quantitatively (notice the logarithmic scale), but not qualitatively. Deviations from classicality substantially increase at the separation of 5 nm, with clear signatures of nonclassical broadening in particular. At a 1 nm separation, the classical and non-

[12]For $\omega_0 > \omega_p$, $\Gamma_{\text{TPE}} = \int_0^{\omega_0} (d\Gamma_{\text{TPE}}/d\omega)\, d\omega$ requires the evaluation of $d_\perp(\omega)$ for frequencies above ω_p, and thus here we also employ the strategy described in Footnote 11.

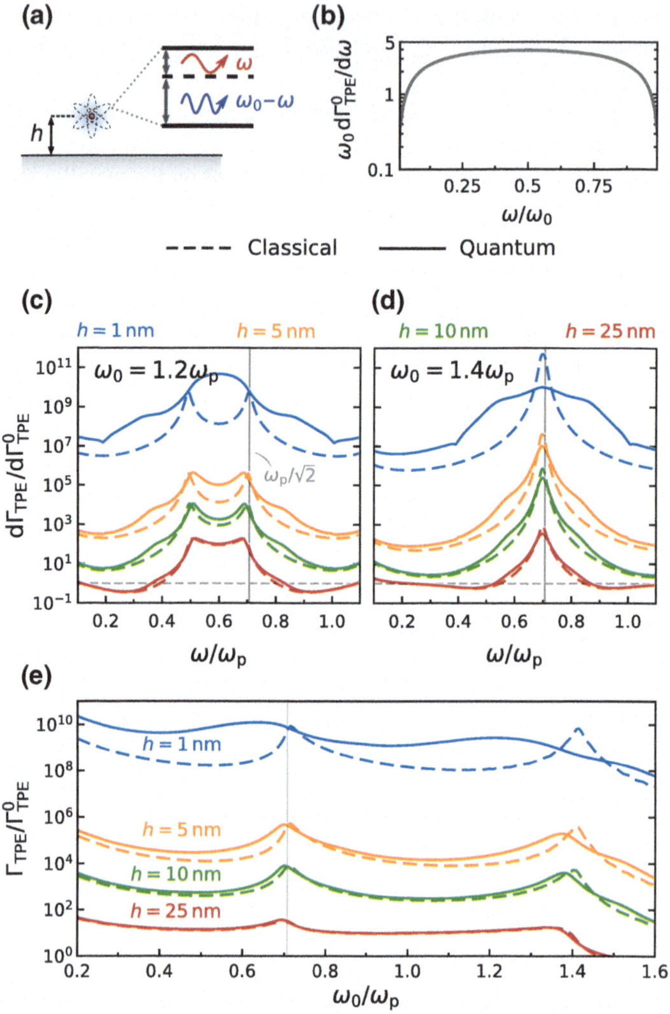

Fig. 8.14 Nonclassical corrections to two-photon emission enhancement. **a** Two-photon emission (s $\rightarrow$ s transition) of a hydrogenic atom above a planar air–metal interface. **b** Free-space differential decay rate for the 2s $\rightarrow$ 1s two-photon transition in hydrogen, as a function of ω/ω_0. **c** [**d**] Enhancement of the differential decay rate $\mathrm{d}\Gamma_{\mathrm{TPE}}/\mathrm{d}\omega$, Eq. (8.54), as a function of frequency for different emitter–interface separations at a transition frequency of $\omega_0 = 1.2\omega_\mathrm{p}$ [$1.4\omega_\mathrm{p}$]. Solid lines correspond to our quantum predictions; dashed lines to classical predictions. **e** Enhancement of the integrated two-photon decay rate Γ_{TPE}, as a function of transition frequency and emitter–surface separation

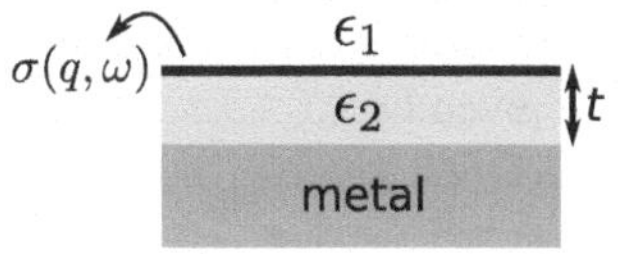

Fig. 8.15 Schematic representation of the planar dielectric–graphene–dielectric–metal structure considered in the text. The graphene–metal separation is controlled by the thickness of the dielectric in the inner region with thickness t

classical predictions differ qualitatively: at the transition frequency $\omega_0 = 1.2\omega_p$ (Fig. 8.14c) the peak-structure and position is mostly dissimilar (as can be understood and expected from Fig. 8.10b, where the LDOS peak is similarly displaced from the classical prediction); at $\omega_0 = 1.4\omega_p$ (Fig. 8.14d), classical and nonclassical peak positions still coincide but the nonclassical spectrum is far broader.

Finally, the impact of nonclassicality on the enhancement of the total (i.e., integrated) two-photon decay rate is shown in Fig. 8.14e. For small separations, the classical prediction can be quantitatively inaccurate by an order of magnitude. However, as also seen in the case of the LDOS, the classical prediction does not necessarily lead to an overestimation of the decay rates: for some transition frequencies, the nonclassical decay rate is higher, due to a redistribution of LDOS into regions in which the classical LDOS was low. Due to the quadratic dependence of two-photon emission enhancement on the LDOS, this process is much more sensitive to deviations from classicality. The considerations about two-photon emission provided in this section constitutes yet another example of the substantial impact of nonclassical effects in nanoscale plasmon–emitter interactions.

8.4 Graphene–Dielectric–Metal Structures: Application of the d-Parameters' Formalism

Before concluding the present chapter, we return to the dielectric–graphene–dielectric–metal (DGDM) heterostructure studied in Chap. 7—and, more concretely, in Sect. 7.3—and apply the d-parameter formalism developed above to account for the quantum nonlocal response of the metal substrate. This will allow us to extend our previous hydrodynamic-theory considerations to a significantly broader scope, wherein quantum mechanical surface corrections can be readily incorporated through the Feibelman d-parameters' formalism. For instance—and as we have already seen in the preceding sections—the latter includes surface-assisted Landau damping, which is overlooked in the HDM, and electronic spill-out, which is inherently forbidden in the standard HDM treatment (due to the its "hard-wall" character).

The spectrum of the plasmonic excitations supported by the DGDM heterostructure (Fig. 8.15) is obtainable by analyzing the poles of the system's scattering coefficients for p-polarized waves. For the three-layered (two interfaces) geometry considered here, these are given by

$$1 - r_p^{2|G|1}\, r_p^{2M}\, e^{i2k_{z,2}t} = 0\,, \tag{8.55}$$

where $r_p^{2|G|1}$ and r_p^{2M} are, respectively, the p-polarization reflection coefficients associated with the reflection from the graphene (for waves incoming from the dielectric medium with ϵ_2) and with the reflection from the metal. The complex exponential naturally takes into account multiple reflections within the slab of finite thickness t. Substituting the aforementioned reflection amplitudes one arrives at the implicit condition:

$$\begin{aligned}&\left[\frac{\epsilon_1}{\kappa_1}+\frac{\epsilon_2}{\kappa_2}+\frac{i\sigma}{\omega\epsilon_0}\right]\left[\epsilon_m\kappa_2+\epsilon_2\kappa_m-\left(\epsilon_m-\epsilon_2\right)\left(q^2d_\perp-\kappa_2\kappa_m d_\parallel\right)\right]=\\&\left[\frac{\epsilon_1}{\kappa_1}-\frac{\epsilon_2}{\kappa_2}+\frac{i\sigma}{\omega\epsilon_0}\right]\left[\epsilon_m\kappa_2-\epsilon_2\kappa_m+\left(\epsilon_m-\epsilon_2\right)\left(q^2d_\perp+\kappa_2\kappa_m d_\parallel\right)\right]e^{-2\kappa_2 t}\,,\end{aligned} \tag{8.56}$$

where, as before, $\kappa_j = \sqrt{q^2 - \epsilon_j k_0^2}$ for $j \in \{1, 2, \mathrm{m}\}$.

For jellium metals, where $d_\parallel = 0$ due to charge-neutrality [7], and for the particular case where $\epsilon_d \equiv \epsilon_1 = \epsilon_2$, the previous expression reduces to

$$\begin{aligned}\left[1+\frac{2\epsilon_d}{\kappa_d}\frac{\omega\epsilon_0}{i\sigma}\right]\left[\epsilon_m\kappa_d+\epsilon_d\kappa_m-\left(\epsilon_m-\epsilon_d\right)q^2d_\perp\right]&=\\\left[\epsilon_m\kappa_d-\epsilon_d\kappa_m+\left(\epsilon_m-\epsilon_d\right)q^2d_\perp\right]&e^{-2\kappa_d t}\,.\end{aligned} \tag{8.57}$$

The dispersion relation associated with the system's plasmonic excitations stems from the solutions of Eq. (8.56) [or its simpler sibling Eq. (8.57)]. As we have seen in Chap. 7, the screening exerted by the metal leads to graphene plasmons that are acoustic-like and that emerge at lower frequencies (for the same wavevector) than the ones associated with conventional graphene plasmons. Importantly, we note that the frequency regime that is relevant for graphene plasmonics (cf. Chap. 4) typically spans the THz and the mid-IR spectral window; this corresponds to frequencies well below the plasma frequency of 3D metals.

Conveniently, in the $\omega \ll \omega_p$ regime the $d_\perp$-parameter can be well approximated by an expression of the form

$$\boxed{d_\perp(\omega) \simeq \zeta + i\frac{\omega}{\omega_p}\xi \qquad (\text{for } \omega \ll \omega_p)}\,, \tag{8.58}$$

as shown by Persson et al. [19, 102] by exploiting Krammers–Kronig relations and the response's asymptotic limits. In the jellium approximation, the parameters $\zeta \equiv \zeta(r_s)$

and $\xi \equiv \xi(r_s)$ are constants that depend uniquely on the density parameter r_s. Hence, Eq. (8.58) states that, for $\omega \ll \omega_p$, Re $d_\perp$ is constant and that Im $d_\perp$ grows linearly with frequency. Indeed, this behavior can be seen empirically upon inspecting the TDDFT data for the (jellium) $d_\perp$-parameters shown in Fig. 8.2. In the following, we pursue this understanding to calculate the dispersion of acoustic-like graphene plasmons (AGPs), which can then be used to ascertain the impact of quantum mechanical effects affecting the metal's electromagnetic response.

Figure 8.16 exhibits the spectrum of AGPs calculated using three distinct models (contrasting in levels of complexity) for the system's response. In brief, the figure is built in the same vein as Fig. 7.6, but now with the metal's response described within the Feibelman d-parameter formalism instead of the HDM. In particular, Fig. 8.16a shows that—for a fixed wavevector—the AGP's resonance redshifts when describing the nonlocal response of the metal using the mesoscopic framework of the d-parameters, which is in accordance with fact that Re $d_\perp > 0$ (electronic spill-out) for jellium metals. This is behavior is the opposite of the one predicted by the HDM where the result is always a blue-shift due to the neglect of spill-out effects. On the other hand, Fig. 8.16b depicts the imaginary part of the AGP's wavevector (that characterizes the mode's propagation length); it shows that the incorporation of Feibelman d-parameters consistently leads to larger values for the imaginary part, presumably due to the inclusion of Landau damping in the metal's response. However, the modification of Im q is not independent of the corresponding Re q and thus an isolated analysis of the former alone can be misleading (this is the case, for instance, when comparing the "fully LRA" curve with the graphene described at the RPA level). Hence, a perhaps more fair comparison of the effect of the overall losses is to do so by defining a quality factor $Q \equiv \frac{\text{Re } q}{\text{Im } q}$. We pursue this approach in Fig. 8.16c where the "fully classical" case unsurprisingly delivers the largest quality factors because it only accommodates Drude-type damping. The incorporation of quantum mechanical effects, first in the graphene alone, and then both on the graphene and metal, consecutively reduces the quality factors exhibited by the AGPs. Nonetheless, the impact of metal-related losses is barely significant; indeed, this is highlighted by the fact that the black dashed and red solid curves are almost indistinguishable in Fig. 8.16c. The overall behavior of the AGP's dispersion upon including quantum surface corrections via Feibelman d-parameters can be understood by noting that (i) for $\omega \ll \omega_p$, i.e., the regime considered here, Re $d_\perp$ tends to a finite, non-negligible value and thus leads to a redshift (blueshift) if Re $d_\perp > 0$ (Re $d_\perp < 0$) of the AGP's dispersion; and (ii) in the $\omega \ll \omega_p$ limit, Im $d_\perp$ is very small (even going to zero in the static limit) and therefore only incurs in a negligible increase to Im q, which remains dominated by losses intrinsic to the graphene sheet alone. Quite remarkably, this line of reasoning is able to explain the experimental observations reported by Iranzo et al. [56] where AGPs were excited using a metallic grating, and the graphene–metal separation was only about $t = 0.7$ nm (approximate thickness of an one-atom-thick hBN layer).

The investigations set forth in the present section entail a particularly interesting feature which is ought to be experimentally-relevant: assuming that all other experimental parameters are known, by measuring the AGP's dispersion one could

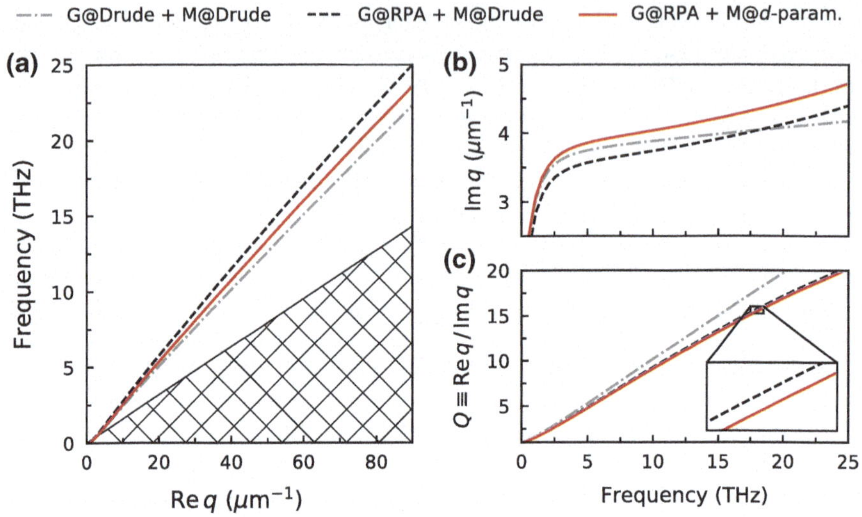

Fig. 8.16 Influence of quantum surface corrections, arising from the metal's response, on the dispersion relation of acoustic-like graphene plasmons. The latter is calculated considering three different levels of sophistication: (i) both materials are assumed to exhibit a response of the Drude kind [gray dot-dashed line]; (ii) the graphene's response is modeled at the nonlocal RPA level, while the metal is still considered to exhibit a Drude-type response [black dashed line]; and (iii) graphene remains treated at the level of the nonlocal RPA whereas the metal's response is modeled within the d-parameters' formalism described in the text [red solid line]. The values for the parameters ζ and ξ figuring in Eq. (8.58) are taken from Ref. [19]; explicitly, $\zeta = 1.41$ Å and $\xi = 0.33$ Å, corresponding to a density parameter of $r_s \approx 3.97$ ($\hbar\omega \simeq 5.95$ eV). **a** Dispersion diagram [$\omega/(2\pi)$ versus Re q] of AGPs computed within three distinct approximations. **b** Imaginary part of the AGP's wavevector corresponding to the dispersion plotted in **a**. **c** AGPs' quality factor, defined as $Q \equiv \text{Re}\ q/\,\text{Im}\ q$. The inset shows a $\times 10$ zoom of the region spanned by the small rectangle indicated in the figure. Remaining setup parameters: $E_F = 0.2$ eV and $\hbar\gamma = 8$ meV for graphene, and $\hbar\gamma_{\text{m}} = 0.1$ eV for the metal; the graphene–metal separation is taken as $t = 1$ nm, and $\epsilon_{\text{d}} = 1$ [the latter is assumed here in order to be consistent with the d-parameter data (that has been calculated assuming vacuum next to the metal surface [19])]

indirectly assess the low-frequency behavior of the metal's nonclassical response, and, chiefly, reveal whether it leads to an outward or to an inward spill of the induced electron density about the metal's surface region. The concept we propose is illustrated in Fig. 8.17; it plainly shows the impact of $d_\perp(0) \equiv \text{Re}\ d_\perp(\omega \ll \omega_{\text{p}})$, that is, the position of the centroid of the metal's induced charge with respect to the classically-sharp metal boundary, on the dispersion relation of AGPs. Specifically, the magnitude of the shift imparted to the AGP's dispersion is proportional to the absolute value of $d_\perp(0)$, whereas the direction of the shift depends on its sign (towards the red if positive and towards the blue if negative). Similarly, the impact of Im $d_\perp$ could in principle be inferred by measuring Im q at different frequencies to determine the slope associated with the imaginary part of Eq. (8.58).

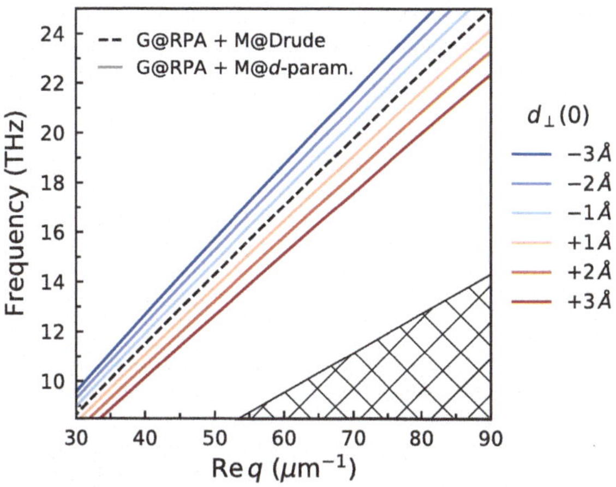

Fig. 8.17 Impact of $d_\perp(0) \equiv \mathrm{Re}\, d_\perp(\omega \ll \omega_p)$ on the dispersion exhibited by acoustic-like graphene plasmons. The parameters [with the exception of $d_\perp(0)$] are the same as in Fig. 8.16

References

1. Gonçalves PAD, Peres NMR (2016) An introduction to graphene plasmonics, 1st edn. World Scientific, Singapore. https://doi.org/10.1142/9948
2. Maradudin AA, Barnes WL, Sambles JR (eds) (2014) Modern plasmonics, 1st edn. Elsevier, Amsterdam
3. Christensen T, Yan W, Jauho AP, Soljačić M, Mortensen NA (2017) Phys Rev Lett 118:157402. https://doi.org/10.1103/PhysRevLett.118.157402
4. Jackson JD (1998) Classical electrodynamics, 3rd edn. Wiley, New York
5. Duffy DG (2015) Green's functions with applications, 2nd edn. CRC Press, Boca Raton
6. Feibelman PJ (1982) Prog Surf Sci 12(4):287. https://doi.org/10.1016/0079-6816(82)90001-6
7. Liebsch A (1997) Electronic excitations at metal surfaces. Springer, New York
8. Varas A, García-González P, Feist J, García-Vidal FJ, Rubio A (2016) Nanophotonics 5(3):409. https://doi.org/10.1515/nanoph-2015-0141
9. Zhu W, Esteban R, Borisov AG, Baumberg JJ, Nordlander P, Lezec HJ, Aizpurua J, Crozier KB (2016) Nat Commun 7:11495. https://doi.org/10.1038/ncomms11495
10. Reiners T, Ellert C, Schmidt M, Haberland H (1995) Phys Rev Lett 74:1558. https://doi.org/10.1103/PhysRevLett.74.1558
11. Mandal S, Wang J, Winans RE, Jensen L, Sen A (2013) J Phys Chem C 117(13):6741. https://doi.org/10.1021/jp310514z
12. Scholl JA, Koh AL, Dionne JA (2012) Nature 483(7390):421. https://doi.org/10.1038/nature10904
13. Campos A, Troc N, Cottancin E, Pellarin M, Weissker HC, Lermé J, Kociakand M, Hillenkamp M (2018) Nat Phys https://doi.org/10.1038/s41567-018-0345-z
14. Dias EJC, Iranzo DA, Gonçalves PAD, Hajati Y, Bludov YV, Jauho AP, Mortensen NA, Koppens FHL, Peres NMR (2018) Phys Rev B 97:245405. https://doi.org/10.1103/PhysRevB.97.245405
15. Bozhevolnyi SI, Mortensen NA (2017) Nanophotonics 6:1185. https://doi.org/10.1515/nanoph-2016-0179
16. Marques MA, Ullrich CA, Nogueira F, Rubio A, Burke K, Gross EKU (2006) Time-dependent density functional theory. Lecture notes in physics. Springer, New York
17. Gonçalves PAD, Christensen T, Rivera N, Jauho AP, Mortensen NA, Soljačić M (2020). Plasmon-emitter interactions at the nanoscale. Nat Commun 11:366. https://doi.org/10.1038/s41467-019-13820-z

18. Yan W, Wubs M, Asger Mortensen N (2015) Phys Rev Lett 115:137403. https://doi.org/10.1103/PhysRevLett.115.137403
19. Persson BNJ, Apell P (1983) Phys Rev B 27:6058. https://doi.org/10.1103/PhysRevB.27.6058
20. Jin D, Hu Q, Neuhauser D, von Cube F, Yang Y, Sachan R, Luk TS, Bell DC, Fang NX (2015) Phys Rev Lett 115:193901. https://doi.org/10.1103/PhysRevLett.115.193901
21. Yang Y, Di Z, Yan W, Agarwal A, Zheng M, Joannopoulos JD, Lalanne P, Christensen T, Berggren KK, Soljačić M (2019) A general theoretical and experimental framework for nanoscale electromagnetism. Nature 576(7786):248–252. https://doi.org/10.1038/s41586-019-1803-1
22. Gradshteyn IS, Ryzhik IM (2007) Table of integrals, series, and products, 7th edn. Academic, London
23. Apell P (1981) Phys Scr 24(4):795. https://doi.org/10.1088/0031-8949/24/4/019
24. Ford GW, Weber WH (1984) Phys Rep 113(4):195. https://doi.org/10.1016/0370-1573(84)90098-X
25. Liebsch A (1987) Phys Scr 35(3):354. https://doi.org/10.1088/0031-8949/35/3/022
26. Mukhopadhyay G (1987) Phys Scr 36(4):676. https://doi.org/10.1088/0031-8949/36/4/011
27. Apell P, Monreal R, Flores F (1984) Solid State Commun 52(12):971. https://doi.org/10.1016/0038-1098(84)90490-3
28. Christensen T, Yan W (2018) Private communication
29. Johnson PB, Christy RW (1972) Phys Rev B 6:4370. https://doi.org/10.1103/PhysRevB.6.4370
30. Tsuei KD, Plummer EW, Feibelman PJ (1989) Phys Rev Lett 63:2256. https://doi.org/10.1103/PhysRevLett.63.2256
31. Tsuei KD, Plummer EW, Liebsch A, Kempa K, Bakshi P (1990) Phys Rev Lett 64:44. https://doi.org/10.1103/PhysRevLett.64.44
32. Tsuei KD, Plummer EW, Liebsch A, Pehlke E, Kempa K, Bakshi P (1991) Surf Sci 247(2):302. https://doi.org/10.1016/0039-6028(91)90142-F
33. Sprunger PT, Watson GM, Plummer EW (1992) Surf Sci 269–270:551. https://doi.org/10.1016/0039-6028(92)91307-W
34. Bréchignac C, Cahuzac P, Kebaïli N, Leygnier J, Sarfati A (1992) Phys Rev Lett 68:3916. https://doi.org/10.1103/PhysRevLett.68.3916
35. Bréchignac C, Cahuzac P, Leygnier J, Sarfati A (1993) Phys Rev Lett 70:2036. https://doi.org/10.1103/PhysRevLett.70.2036
36. Kim BO, Lee G, Plummer EW, Dowben PA, Liebsch A (1995) Phys Rev B 52:6057. https://doi.org/10.1103/PhysRevB.52.6057
37. Moresco F, Rocca M, Hildebrandt T, Zielasek V, Henzler M (1998) EPL 43(4):433. https://doi.org/10.1209/epl/i1998-00377-0
38. Rocca M, Valbusa U (1993) Surf Sci 287–288:770. https://doi.org/10.1016/0039-6028(93)91069-2. Proceedings of the 8th international conference on solid surfaces
39. Raza S, Kadkhodazadeh S, Christensen T, Di Vece M, Wubs M, Mortensen NA, Stenger N (2015) Nat Commun 6:8788. https://doi.org/10.1038/ncomms9788
40. Lipparini E, Pederiva F (1993) Z Phys D: At Mol Clust 27(3):281. https://doi.org/10.1007/BF01436545
41. Feibelman PJ (1993) Surf Sci 282(1):129. https://doi.org/10.1016/0039-6028(93)90616-R
42. Teperik TV, Nordlander P, Aizpurua J, Borisov AG (2013) Phys Rev Lett 110:263901. https://doi.org/10.1103/PhysRevLett.110.263901
43. Liebsch A (1993) Phys Rev B 48:11317. https://doi.org/10.1103/PhysRevB.48.11317
44. Zwillinger D (1997) Handbook of differential equations, 3rd edn. Academic, London
45. Pelton M, Bryant GW (2013) Introduction to metal-nanoparticle plasmonics. Wiley, New York
46. Christensen T, Yan W, Raza S, Jauho AP, Mortensen NA, Wubs M (2014) ACS Nano 8(2):1745. https://doi.org/10.1021/nn406153k

47. Novotny L, Hecht B (2012) Principles of nano-optics, 2nd edn. Cambridge University Press, Cambridge
48. Kreibig U, Vollmer M (1995) Optical properties of metal clusters. Springer, Berlin
49. Bohren CF, Huffman DR (1983) Absorption and scattering of light by small particles. Wiley, New York
50. Genzel L, Martin TP, Kreibig U (1975) Z Phys B Condens Matter 21(4):339. https://doi.org/10.1007/BF01325393
51. Gramotnev DK, Bozhevolnyi SI (2010) Nat Photonics 4(2):83. https://doi.org/10.1038/nphoton.2010.282, https://www.nature.com/articles/nphoton.2009.282
52. Fernández-Domínguez AI, García-Vidal FJ, Martín-Moreno L (2017) Nat Photonics 11(1):8. https://www.nature.com/articles/nphoton.2016.258
53. Flick J, Rivera N, Narang P (2018) Nanophotonics 7(9):1479. https://doi.org/10.1515/nanoph-2018-0067
54. Fernández-Domínguez AI, Bozhevolnyi SI, Mortensen NA (2018) ACS Photonics 5(9):3447. https://doi.org/10.1021/acsphotonics.8b00852
55. Ciracì C, Hill RT, Mock JJ, Urzhumov Y, Fernández-Domínguez AI, Maier SA, Pendry JB, Chilkoti A, Smith DR (2012) Science 337(6098):1072. https://doi.org/10.1126/science.1224823
56. Alcaraz Iranzo D, Nanot S, Dias EJC, Epstein I, Peng C, Efetov DK, Lundeberg MB, Parret R, Osmond J, Hong JY, Kong J, Englund DR, Peres NMR, Koppens FHL (2018) Science 360(6386):291. https://doi.org/10.1126/science.aar8438
57. Purcell EM (1946) Phys Rev 69:681
58. Anger P, Bharadwaj P, Novotny L (2006) Phys Rev Lett 96:113002. https://doi.org/10.1103/PhysRevLett.96.113002
59. Kinkhabwala A, Yu Z, Fan S, Avlasevich Y, Müllen K, Moerner W (2009) Nat Photonics 3(11):654. https://doi.org/10.1038/nphoton.2009.187
60. Vasa P, Wang W, Pomraenke R, Lammers M, Maiuri M, Manzoni C, Cerullo G, Lienau C (2013) Nat Photonics 7(2):128. https://doi.org/10.1038/nphoton.2012.340
61. Akselrod GM, Argyropoulos C, Hoang TB, Ciracì C, Fang C, Huang J, Smith DR, Mikkelsen MH (2014) Nat Photonics 8:835. https://doi.org/10.1038/nphoton.2014.228
62. Pelton M (2015) Nat Photonics 9:427. https://doi.org/10.1038/nphoton.2015.103
63. Kneipp K, Wang Y, Kneipp H, Perelman LT, Itzkan I, Dasari RR, Feld MS (1997) Phys Rev Lett 78:1667. https://doi.org/10.1103/PhysRevLett.78.1667
64. Haes AJ, Haynes CL, McFarland AD, Schatz GC, Van Duyne RP, Zou S (2005) MRS Bull 30(5):368–375. https://doi.org/10.1557/mrs2005.100
65. Mayer KM, Hafner JH (2011) Chem Rev 111(6):3828. https://doi.org/10.1021/cr100313v
66. Brolo AG (2012) Nat Photonics 6:709. https://doi.org/10.1038/nphoton.2012.266
67. Chikkaraddy R, de Nijs B, Benz F, Barrow SJ, Scherman OA, Rosta E, Demetriadou A, Fox P, Hess O, Baumberg JJ (2016) Nature 535(7610):127. https://doi.org/10.1038/nature17974
68. Brongersma ML, Halas NJ, Nordlander P (2015) Nat Nanotechnol 10(1):25. https://doi.org/10.1038/nnano.2014.311
69. Zhou L, Swearer DF, Zhang C, Robatjazi H, Zhao H, Henderson L, Dong L, Christopher P, Carter EA, Nordlander P et al (2018) Science 362(6410):69. https://doi.org/10.1126/science.aat696
70. Russell KJ, Liu TL, Cui S, Hu EL (2012) Nat Photonics 6:459. https://doi.org/10.1038/nphoton.2012.112
71. Koenderink AF (2010) Opt Lett 35(24):4208. https://doi.org/10.1364/OL.35.004208
72. Kern AM, Martin OJF (2012) Phys Rev A 85:022501. https://doi.org/10.1103/PhysRevA.85.022501
73. Filter R, Mühlig S, Eichelkraut T, Rockstuhl C, Lederer F (2012) Phys Rev B 86:035404. https://doi.org/10.1103/PhysRevB.86.035404
74. Rivera N, Kaminer I, Zhen B, Joannopoulos JD, Soljačić M (2016) Science 353(6296):263. https://doi.org/10.1126/science.aaf6308

75. Dung HT, Knöll L, Welsch DG (2002) Phys Rev A 65:043813. https://doi.org/10.1103/PhysRevA.65.043813
76. Wubs M, Vos WL (2016) New J Phys 18(5):053037. https://doi.org/10.1088/1367-2630/18/5/053037
77. Ren J, Wu T, Yang B, Zhang X (2016) Phys Rev B 94:125416. https://doi.org/10.1103/PhysRevB.94.125416
78. Cesar CL, Fried DG, Killian TC, Polcyn AD, Sandberg JC, Yu IA, Greytak TJ, Kleppner D, Doyle JM (1996) Phys Rev Lett 77:255. https://doi.org/10.1103/PhysRevLett.77.255
79. Hayat A, Ginzburg P, Orenstein M (2008) Nat Photonics 2(4):238. https://doi.org/10.1038/nphoton.2008.28
80. Nevet A, Berkovitch N, Hayat A, Ginzburg P, Ginzach S, Sorias O, Orenstein M (2010) Nano Lett 10(5):1848. https://doi.org/10.1021/nl1005806
81. Barnett SM, Loudon R (1996) Phys Rev Lett 77:2444. https://doi.org/10.1103/PhysRevLett.77.2444
82. Blum C, Zijlstra N, Lagendijk A, Wubs M, Mosk AP, Subramaniam V, Vos WL (2012) Phys Rev Lett 109:203601. https://doi.org/10.1103/PhysRevLett.109.203601
83. Sanders S, Manjavacas A (2018) ACS Photonics 5:2437. https://doi.org/10.1021/acsphotonics.8b00225
84. Shim H, Fan L, Johnson SG, Miller OD (2019) Phys Rev X 9:011043. https://doi.org/10.1103/PhysRevX.9.011043
85. Chew WC (1995) Waves and fields in inhomogenous media. Wiley-IEEE Press
86. Prigogine I, Rice SA (eds) (1978) Advances in chemical physics, vol XXXVII. Wiley, New York
87. Carminati R, Cazé A, Cao D, Peragut F, Krachmalnicoff V, Pierrat R, Wilde YD (2015) Surf Sci Rep 70(1):1. https://doi.org/10.1016/j.surfrep.2014.11.001
88. Chew H (1987) J Chem Phys 87(2):1355. https://doi.org/10.1063/1.453317
89. Bennett AJ (1970) Phys Rev B 1:203. https://doi.org/10.1103/PhysRevB.1.203
90. Cohen-Tannoudji C, Diu B, Laloe F (1978) Quantum mechanics, vol 2. Wiley, New York
91. Wiese WL, Smith MW, Glennon BM (1966) Atomic transition probabilities, vol 1. NIST. https://www.govinfo.gov/app/details/GOVPUB-C13-d53059923643c9d4d7265eb9e770b013
92. Wiese WL, Fuhr JR (2009) J Phys Chem Ref Data 38(3):565. https://doi.org/10.1063/1.3077727
93. Andersen ML, Stobbe S, Sørensen AS, Lodahl P (2011) Nat Phys 7(3):215. https://doi.org/10.1038/nphys1870
94. Peřina J (ed) (2001) Coherence and statistics of photons and atoms. Wiley, New York
95. Scheel S, Buhmann S (2008) Acta Phys Slovaca 58(5):675
96. Brandes T (2005) Phys Rep 408(5):315. https://doi.org/10.1016/j.physrep.2004.12.002
97. Selvin PR (2000) Nat Struct Biol 7:730–734. https://doi.org/10.1038/78948
98. Lakowicz JR (ed) (2000) Principles of fluorescence spectroscopy, 3rd edn. Springer, New York
99. Andrew P, Barnes WL (2004) Science 306(5698):1002. https://doi.org/10.1126/science.1102992
100. Lunz M, Gerard VA, Gun'ko YK, Lesnyak V, Gaponik N, Susha AS, Rogach AL, Bradley AL (2011) Nano Lett 11(8):3341. https://doi.org/10.1021/nl201714y
101. Govorov AO, Lee J, Kotov NA (2007) Phys Rev B 76:125308. https://doi.org/10.1103/PhysRevB.76.125308
102. Persson BNJ, Zaremba E (1984) Phys Rev B 40:5669. https://doi.org/10.1103/PhysRevB.30.5669
103. Göppert-Mayer M (1931) Ann Phys 401(3):273. https://doi.org/10.1002/andp.19314010303
104. Rivera N, Rosolen G, Joannopoulos JD, Kaminer I, Soljačić M (2017) Proc Natl Acad Sci USA 114(52):13607. https://doi.org/10.1073/pnas.1713538114

Chapter 9
Conclusions and Outlook

Throughout this thesis we have investigated and discussed the salient features of plasmonics and light–matter interactions in various contexts and material platforms. Our theoretical studies have been concerned with three- and two-dimensional plasmonic materials and their nanostructures, and our considerations have comprised a diverse toolkit of classical, semiclassical, and quantum mechanical treatments. Special emphasis has been given to the incorporation of quantum surface corrections for rigorously modeling the emergence of nonclassical behavior in the nanoscopic limit.

In closing, we shall summarize the content of each chapter, followed by a brief discussion about new and potentially interesting opportunities arising from this work.

Chapter 2 | **Classical Electrodynamics of Solids**

The central purpose of this chapter was to provide a general and self-contained (as far as possible) introduction to the core elements that form the theoretical foundations that will be either exploited or further elaborated upon in the subsequent chapters.

Chapter 3 | **Electronic and Optical Properties of Graphene**

Like Chapter 2, this chapter served an essentially introductory purpose wherein the fundamentals of graphene physics—focusing on its electronic and optical properties—have been detailed in a cohesive fashion. In particular, and from a purely practical standpoint, the different models describing the conductivity of graphene constitute key quantities for unraveling the plasmonic response of graphene-based systems.

Chapter 4 | **Fundamentals of Graphene Plasmonics**

Here we have focused our efforts on the description of plasmons in graphene, starting from elementary considerations and then evolving towards more sophisticated descriptions. We have seen that plasmons in two-dimensional (2D) graphene differ from plasmons in conventional three-dimensional (3D) conductors in a number of ways; typically, such differences can be ascribed either to graphene's reduced dimensionality or to its curious conical bandstructure (exhibiting a linear energy-

P. A. D. Gonçalves, *Plasmonics and Light–Matter Interactions in Two-Dimensional Materials and in Metal Nanostructures*, Springer Theses,
https://doi.org/10.1007/978-3-030-38291-9_9

momentum dispersion in the relevant energy range for graphene plasmonics). After analyzing the plasmonic excitations in continuous, unpatterned graphene we have explored the effect of nanostructuring the graphene on the same. To that end, we have developed a nonretarded semi-analytical framework for calculating plasmon resonances and their spectral features in arbitrarily-shaped 2D nanostructures (made of graphene or otherwise); we then illustrated this by explicitly considering two archetypal graphene nanostructures (i.e., the ribbon and disk geometries). We have found that the nanostructure's shape can be used as an extra tuning knob—on top of the one provided by the material's electrical tunability—to tailor 2D plasmon resonances even further. Moreover, using the approach we have just described, we have theoretically studied plasmon coupling and ensuing hybridization in 2D nanoslits (i.e., the inverse structure of a nanoribbon). We have shown that this system supports two plasmonic modes of opposite parity which result from symmetric and antisymmetric hybridizations between the two edge plasmons of the constituent half-sheets. Finally, we have abandoned a graphene-centered view and considered the case of 2D nanoslits based on anisotropic atomically-thin crystals (latter specifying on monolayer black phosphorus for definiteness).

We believe our work on hybridized plasmons in 2D nanoslits could catalyze future studies based on this system. One possibility is to exploit the 2D nanoslit's plasmon resonances for subwavelength waveguiding, biochemical sensing [1], optical manipulation [2] and trapping of nanosized particles [3], or even for quantum technologies based on quantum optics [4] and quantum information [5]. Several interesting open questions arise from this work; here we enumerate two: the first is pertained with the role of edge effects associated with the edge terminations of the slit [6–8]; the second is concerned with the breakdown of Babinet's principle for complementary structures and whether or not a sum rule relating the resonant frequencies of complementary 2D structures is derivable (in 3D, it has been shown that such a sum rule does exist [9]). Lastly, we note that many extensions can readily follow from our theoretical approach, such as the exploration of complex coupled resonances stemming from the pairing of 2D slits with other 2D nanostructures (e.g., ribbons, other slits, etc) or the study of magnetoplasmons in 2D nanoslits (see, for instance, Publication C for a study of magnetoplasmons in a half-sheet with anisotropic conductivity).

Chapter 5 | Two-Dimensional Channel Plasmons in Nonplanar Geometries

In this chapter we have proposed the exploitation of the folding of an otherwise planar graphene sheet—or any other plasmon-supporting 2D material—into an (out-of-plane) triangular-like shape as a mean to achieve deep subwavelength waveguiding and light localization using the plasmon modes guided along the apex of the structure. We have developed two distinct theoretical techniques for calculating the spectral properties of the system's 2D channel plasmons: one is a rigorous method based on a modification of the semi-analytical framework presented in Sect. 4.2.1, whereas the other relies on a perturbative approach known as effective-index method (EIM). Strikingly, we have found that the dispersion relation of 2D channel plasmons propagating along a Λ- (V-)shape wedge (groove) follows the same functional dependence as their planar-graphene plasmon counterparts, but now scaled by a (purely) geometric factor in which all the information about the system's geometry is contained. Moreover, when computing the dispersion the associated field distributions are concurrently obtained. Both methods are extremely computationally efficient, in stark contrast with commercially available full-wave numerical codes which tend to be computationally demanding and time-consuming.

In this context, the *semi-analytical* investigation of 2D-material-based wedges with other types of cross-sections could also be potentially interesting and it remains largely unexplored. This could in principle be achieved by adapting our formalism with minor extensions [10, 11] or, alternatively, by employing techniques from transformation optics [12–14].

Chapter 6 | Electrodynamics of Metals Beyond the Local-Response Approximation: Nonlocal Effects

We describe the electrodynamic response of 3D metals and metal-based nanostructures beyond the conventional local-response approximation (LRA). We start by introducing one of the most prominent frameworks for incorporation nonlocal effects in the metal's optical response, that is, the hydrodynamic model (HDM) of the free-electron gas. We analyze its implications for the nonlocal plasmonic response of a planar dielectric–metal interface and of metallic spheres. In both platforms, the HDM predicts a blueshift of the plasmon resonances that is parameterized by the hydrodynamic velocity parameter β and it is proportional to the surface plasmon's wavevector (inverse radius) in the planar (spherical) configuration. Recognizing the limitations of the HDM, e.g., the omission of Landau damping, we move towards an account of the optical response of the metal based on the full, nonlocal Lindhard dielectric function of the uniform electron gas). We have seen that the latter includes the full q-dependence of the dielectric function, and, importantly, the presence of Landau damping.

Based on the theoretical foundations set forth in this chapter, it should be relatively straightforward to extend or approach to other closely-related nonlocal models, e.g., the quantum infinite and finite barrier models [15] or the image potential model proposed by Chulkov and co-workers [16–18]. We further note that although many of this pioneering work has been known since the 1980s and/or 1990s, only very recently those theoretical frameworks have been applied to describe nonequilibrium emitter–surface interactions in the presence of spatial dispersion [19–21] and to heterostructures based on 2D materials and 3D metals [22].

Chapter 7 | Quantum Nonlocal Effects Probed by Ultraconfined Graphene Plasmons

Here we have theoretically investigated the intricate electromagnetic interaction of plasmons supported by a graphene sheet placed in the close vicinity of a metal substrate. We have found that the strong screening exhibited by metals in the terahertz (THz) regime leads to the emergence of acoustic-like graphene plasmons (AGPs) resembling the acoustic plasmon mode in double-layer graphene. We have seen that AGPs are able to render ultradeep subwavelength field confinement, reflecting the large plasmon wavevectors that characterize them. For this reason, they constitute formidable candidates to investigate simultaneously the frequency- and wavevector-dependence of the system's response (or, at the distance of a Fourier transform, the full spatiotemporal response). In agreement with a previous report [23], our results—based on the theory introduced in Ref. [23]—have shown that the AGP's dispersion can only be well described if nonlocal effects in the graphene are included, and that, at low carrier densities, many-body effects must be accounted for on top of those. Going beyond previous work, we have refined prior theoretical descriptions by considered nonlocality not only in graphene but also including it in the metal's response. Our approach was based on the HDM model and it revealed that AGPs can indeed be used to probe the nonlocal response of the underlying metal substrate when the graphene–metal separation is small (i.e., $\sim$ few-nm).

Further research can spur from the work detailed here. One example is to consider the same heterostructure but where the metal has been replaced by a superconductor, and investigate if and how can we use the extraordinary properties of AGPs to assess indirectly the electrodynamic response of the superconductor. Other extensions that have been recently proposed include the exploitation of other 2D conductors [24] and the exploration of AGPs for ultrasensitive sensing capabilities [25, 26].

Chapter 8 | Quantum Corrections in Plasmonics and Plasmon–Emitter Interactions

In this chapter we have considered the impact of nonclassical corrections in nanoplasmonics using a mesoscopic scattering framework that incorporates quantum nonlocal effects via quantum-informed surface-response functions known as Feibelman d-parameters. Specifically, we developed and applied our theoretical formalism for calculating leading-order-accurate quantum mechanical corrections in the spectral properties of surface plasmons in metal nanostructures and in a varied range of plasmon-enhanced light–matter interaction processes. The former include the prototypical planar and spherical systems (but can be generalized to arbitrary geometries via the d-parameter-modified mesoscopic boundary conditions presented here) whereas the latter include spontaneous dipole and multipole emission, energy transfer between emitters, and spontaneous two-photon emission. Our findings elucidate and contextualize the main physical mechanisms responsible for deviations from the classical response in light–matter interactions at the nanoscale: spectral shifting and surface-enabled Landau damping, manifesting the joint impact of spill-out and nonlocality. For deeply nanoscale emitter–surface separations, the deviations can be order-of-magnitude, thereby completely invalidating any quantitative prediction of the classical approach. Crucially, our formalism can be reliably employed from the classical regime (wavelength scales, local response) all the way down to the quantum mechanical domain (electronic length scales), and provides an unifying link across the blurred boundary between classical and quantum realms. In particular, we derived *analytical* expressions for the nonclassical scattering coefficients through which quantum mechanical effects—namely, nonlocality, electronic spill-out/spill-in, and surface-assisted Landau damping—are conveniently incorporated. These coefficients constitute the building blocks for determining the dynamics governing a plethora of nanophotonic phenomena, making the theory set forth here particularly well-suited for rigorously taking into account nonclassical effects in nanoplasmonics.

We believe that there are several enticing opportunities and open questions arising from this work: Firstly, while here we have employed d-parameters calculated from jellium TDDFT, other models—simpler or more sophisticated—can be readily applying using the very same mesoscopic scattering coefficients presented here. An example of an approach beyond the jellium approximation would be the of use atomistic TDDFT for investigating the role of atomic structure and valence-bound screening (which should be particularly relevant for noble metals like silver or gold). In this vein, it would be extremely valuable to compile databases of d-parameters at various dielectric–metal interfaces, analogous to the existing databases of local bulk permittivities [27] or of 2D materials [28, 29]. Note that our equations are general, and thus remain valid irrespective of the method employed in the computation (or of the experimentally-measured value [30]) of the d-parameters. Secondly, our approach can be readily extended to other prominent light–matter interaction processes, such as near-field radiative heat transfer [31], electron energy-loss spectroscopy [32, 33], or van der Waals [34, 35] and Casimir–Polder interactions [36]. Thirdly, we anticipate the extremely interesting—from both a fundamental and an experimental viewpoint—application of our mesoscopic d-parameter formalism for light–matter interactions beyond the point-dipole approximation of the emitter; this should not pose significant extra difficulties, based on what is known in the context of quantum dots integrated in nanophotonic systems [37–43]. Lastly, we speculate on a possible extension of the d-parameter formalism to two-dimensional systems (e.g., for describing edge plasmons and for a facilitated treatment of edge states).

Realizing the promise of plasmon-enhanced light–matter interactions inevitably involves multiscale plasmonic architectures, combining both macroscopic and nanoscale features in synergy. The development of the next generation of nanoscale optical devices consequently requires theoretical tools that incorporate the salient features of both the classical and quantum domains in a tractable manner. It is the author's hope that the theoretical frameworks presented in this thesis may constitute such tools, and, potentially, be valuable in the pursuit of the aforementioned developments.

References

1. Rodrigo D, Limaj O, Janner D, Etezadi D, García deAbajo FJ, Pruneri V, Altug H (2015) Science 349(6244):165. https://science.sciencemag.org/content/349/6244/165, https://doi.org/10.1126/science.aab2051
2. Kawata S, Tani T (1996) Opt Lett 21(21):1768 (1996). http://ol.osa.org/abstract.cfm?URI=ol-21-21-1768, https://doi.org/10.1364/OL.21.001768
3. Zhang J, Liu W, Zhu Z, Yuan X, Qin S (2016) Sci Rep 6:38086. https://doi.org/10.1038/srep38086

4. Chang DE, Sørensen AS, Hemmer PR, Lukin MD (2006) Phys Rev Lett 97:053002. https://doi.org/10.1103/PhysRevLett.97.053002
5. Calafell IA, Cox JD, Radonjić M, Saavedra JRM, García de Abajo FJ, Rozema LA, Walther P (2019) npj Quantum Inf. 5:37. https://doi.org/10.1038/s41534-019-0150-2
6. Thongrattanasiri S, Manjavacas A, García de Abajo FJ (2012) ACS Nano 6(2): 1766. https://doi.org/10.1021/nn204780e
7. Christensen T, Wang W, Jauho AP, Wubs M, Mortensen NA (2014) Phys Rev B 90:241414. https://doi.org/10.1103/PhysRevB.90.241414
8. Wedel KO, Mortensen NA, Thygesen KS, Wubs M (2018) Phys Rev B 98:155412. https://doi.org/10.1103/PhysRevB.98.155412
9. Apell SP, Echenique PM, Ritchie RH Ultramicroscopy 65(1):53. http://www.sciencedirect.com/science/article/pii/S0304399196000551, https://doi.org/10.1016/S0304-3991(96)00055-1
10. Garcia-Molina R, Gras-Marti A, Ritchie RH (1985) Phys Rev B 31:121. https://doi.org/10.1103/PhysRevB.31.121
11. Boardman AD, Garcia-Molina R, Gras-Marti A, Louis E (1985) Phys Rev B 32:6045.https://doi.org/10.1103/PhysRevB.32.6045
12. Chen H (2009) J Opt A Pure Appl Opt 11(7):075102. https://doi.org/10.1088/1464-4258/11/7/075102
13. Luo Y, Pendry JB, Aubry A (2010) Nano Lett 10(10):4186. https://doi.org/10.1021/nl102498s
14. Kadic M, Guenneau S, Enoch S, Huidobro PA, Martín-Moreno L, García-Vidal FJ, Renger J, Quidant R (2012) Nanophotonics 1:51. https://doi.org/10.1515/nanoph-2012-0011
15. Ford GW, Weber WH (1984) Phys Rep 113(4):195. http://www.sciencedirect.com/science/article/pii/037015738490098X, https://doi.org/10.1016/0370-1573(84)90098-X
16. Chulkov EV, Silkin VM, Echenique PM (1997) Surf Sci 391(1):L1217. http://www.sciencedirect.com/science/article/pii/S0039602897006535, https://doi.org/10.1016/S0039-6028(97)00653-5
17. Chulkov EV, Silkin VM, Echenique PM (1999) Surf Sci 437(3):330. http://www.sciencedirect.com/science/article/pii/S0039602899006688, https://doi.org/10.1016/S0039-6028(99)00668-8
18. Chulkov EV, Osma J, Sarrıa I, Silkin VM, Pitarke JM (1999) Surf Sci 433–435:882. http://www.sciencedirect.com/science/article/pii/S003960289900148X, https://doi.org/10.1016/S0039-6028(99)00148-X
19. Reiche D, Dalvit DAR, Busch K, Intravaia F (2017) Phys Rev B 95:155448. https://doi.org/10.1103/PhysRevB.95.155448
20. Oelschläger M, Busch K, Intravaia F (2018) Phys Rev A 97:062507. https://doi.org/10.1103/PhysRevA.97.062507
21. Reiche D, Oelschläger M, Busch K, Intravaia F (2019) J Opt Soc Am B 36(4):C52. http://josab.osa.org/abstract.cfm?URI=josab-36-4-C52, https://doi.org/10.1364/JOSAB.36.000C52
22. Echarri AR, Cox JD, García de Abajo FJ (2019) Optica 6(5):630. http://www.osapublishing.org/optica/abstract.cfm?URI=optica-6-5-630, https://doi.org/10.1364/OPTICA.6.000630
23. Lundeberg MB, Gao Y, Asgari R, Tan C, Van Duppen B, Autore M, Alonso-González P, Woessner A, Watanabe K, Taniguchi T, Hillenbrand R, Hone J, Polini M, Koppens FHL (2017) Science 357(6347):187. http://science.sciencemag.org/content/357/6347/187, https://doi.org/10.1126/science.aan2735
24. Lee IH, Martin-Moreno L, Mohr DA, Khaliji K, Low T, Oh SH (2018) ACS Photonics 5(6):2208. https://doi.org/10.1021/acsphotonics.8b00062
25. Chen S, Autore M, Li J, Li P, Alonso-González P, Yang Z, Martin-Moreno L, Hillenbrand R, Nikitin AY (2017) ACS Photonics 4(12):3089. https://doi.org/10.1021/acsphotonics.7b00654
26. Lee IH, Yoo D, Avouris P, Low T, Oh SH (2019) Nat Nanotechnol 14(4):313. https://doi.org/10.1038/s41565-019-0363-8
27. Palik ED (1997) Handbook of optical constants of solids, vol. 1–5. Academic Press

28. Haastrup S, Strange M, Pandey M, Deilmann T, Schmidt PS, Hinsche NF, Gjerding MN, Torelli D, Larsen PM, Riis-Jensen AC, Gath J, Jacobsen KW, Mortensen JJ, Olsen T, Thygesen KS (2018) 2D Mater. 5(4):042002. http://c2db.fysik.dtu.dk, https://doi.org/10.1088/2053-1583/aacfc1
29. Mounet N, Gibertini M, Schwaller P, Campi D, Merkys A, Marrazzo A, Sohier T, Castelli IE, Cepellotti A, Pizzi G, Marzari N (2018) Nat Nanotechnol 13:246.https://doi.org/10.1038/s41565-017-0035-5, https://doi.org/10.24435/materialscloud:2017.0008/v3
30. Yang Y, Di Z, Yan W, Agarwal A, Zheng M, Joannopoulos JD, Lalanne P, Christensen T, Berggren KK, Soljačić M (2019) A general theoretical and experimental framework for nanoscale electromagnetism. Nature 576(7786):248–252. https://doi.org/10.1038/s41586-019-1803-1
31. Cuevas JC, García-Vidal FJ (2018) ACS Photonics 5(10):3896. https://doi.org/10.1021/acsphotonics.8b01031
32. Raza S, Kadkhodazadeh S, Christensen T, Di Vece M, Wubs M, Mortensen NA, Stenger N (2015) Nat Commun 6:8788. https://www.nature.com/articles/ncomms9788, https://doi.org/10.1038/ncomms9788
33. Campos A, Troc N, Cottancin E, Pellarin M, Weissker HC, Lermé J, Kociakand M, Hillenkamp M (2018) Nat Phys. https://doi.org/10.1038/s41567-018-0345-z
34. Apell P (1981) Phys Scr 24(4):795. https://doi.org/10.1088/0031-8949/24/4/019
35. Luo Y, Zhao R, Pendry JB (2014) Proc Natl Acad Sci USA 111(52):18422. https://www.pnas.org/content/111/52/18422, https://doi.org/10.1073/pnas.1420551111
36. Bordag M, Klimchitskaya GL, Mohideen U, Mostepanenko VM (2009) Advances in the casimir effect. Oxford University Press, New York
37. Tighineanu P, Sørensen AS, Stobbe S, Lodahl P (2017) Quantum dots for quantum information technologies. Springer, Berlin, pp 165–198. https://doi.org/10.1007/978-3-319-56378-7_5
38. Andersen ML, Stobbe S, Sørensen AS, Lodahl P (2011) Nat Phys 7(3):215. https://www.nature.com/articles/nphys1870, https://doi.org/10.1038/nphys1870
39. Jun Ahn K, Knorr A (2003) Phys Rev B 68:161307. https://doi.org/10.1103/PhysRevB.68.161307
40. Rukhlenko ID, Handapangoda D, Premaratne M, Fedorov AV, Baranov AV, Jagadish C (2009) Opt Express 17(20):17570. http://www.opticsexpress.org/abstract.cfm?URI=oe-17-20-17570, https://doi.org/10.1364/OE.17.017570
41. Stobbe S, Kristensen PT, Mortensen JE, Hvam JM, Mørk J, Lodahl P (2012) Phys Rev B 86:085304. https://doi.org/10.1103/PhysRevB.86.085304
42. Zheng K, Žídek K, Abdellah M, Zhu N, Chábera P, Lenngren N, Chi Q, Pullerits T (2014) J Am Chem Soc 136(17):6259. https://doi.org/10.1021/ja411127w
43. Neuman T, Esteban R, Casanova D, García-Vidal FJ, Aizpurua J (2018) Nano Lett 18(4):2358. https://doi.org/10.1021/acs.nanolett.7b05297

Appendix A
Lindhard Dielectric Function of a Homogeneous 3DEG

The longitudinal dielectric function of the non-interacting free-electron gas in three-dimensions, within the random-phase approximation (RPA), is given by [1, 2]

$$\epsilon^{\mathrm{RPA}}(\mathbf{q},\omega) = 1 - \frac{e^2}{\epsilon_0 q^2}\chi_0(\mathbf{q},\omega)\,, \tag{A.1}$$

where $\chi_0(\mathbf{q},\omega)$ is the non-interacting polarizability (also known as density-density correlation function or the bare pair-bubble diagram) and follows from [1–3]

$$\chi_0(\mathbf{q},\omega) = \frac{2}{V}\sum_{\mathbf{k}} \frac{f_0(E_{\mathbf{k}}) - f_0(E_{\mathbf{k}+\mathbf{q}})}{\hbar\omega - E_{\mathbf{k}+\mathbf{q}} + E_{\mathbf{k}} + i\eta}, \tag{A.2}$$

where $E_{\mathbf{k}} = \frac{\hbar^2 k^2}{2m_e}$ are the single-particle energies, $f_0(E)$ the Fermi-Dirac distribution function, and η is an infinitesimally small factor to ensure the analytical properties of $\chi_0(\mathbf{q},\omega)$ in the complex ω plane (i.e., here $\chi_0(\mathbf{q},\omega)$ is a retarded function [2]).

The RPA dielectric function of a non-interacting 3DEG is also commonly referred to as the Lindhard dielectric function, owing to its derivation by Lindhard in 1954 [4].

Explicit form of the RPA dielectric function for a 3DEG

The explicit calculation of Eqs. (A.1) and (A.2)—which can be found in several many-body condensed matter physics textbooks; see, for instance Refs. [1–3]—gives

$$\begin{aligned}\mathrm{Re}\,\epsilon^{\mathrm{RPA}}(\mathbf{q},\omega) = 1 + \frac{q_{\mathrm{TF}}^2}{2q^2}\Bigg\{1 + \frac{k_F}{2q}\Bigg\{&\left[1 - \left(\frac{\hbar\omega + E_q}{\hbar v_F q}\right)^2\right]\ln\left|\frac{\hbar\omega + \hbar v_F q + E_q}{\hbar\omega - \hbar v_F q + E_q}\right| \\ &+ \left[1 - \left(\frac{\hbar\omega - E_q}{\hbar v_F q}\right)^2\right]\ln\left|\frac{\hbar\omega - \hbar v_F q - E_q}{\hbar\omega + \hbar v_F q - E_q}\right|\Bigg\}\Bigg\},\end{aligned} \tag{A.3a}$$

P. A. D. Gonçalves, *Plasmonics and Light–Matter Interactions in Two-Dimensional Materials and in Metal Nanostructures*, Springer Theses,
https://doi.org/10.1007/978-3-030-38291-9

for the real part, and

$$\mathrm{Im}\,\epsilon^{\mathrm{RPA}}(\mathbf{q},\omega)=\begin{cases}\frac{\pi}{2}\frac{\hbar\omega}{\hbar v_F q}\frac{q_{\mathrm{TF}}^2}{2q^2} & ,\ 0\le\hbar\omega<\hbar v_F q-E_q\ \wedge\ q\le 2k_F\\ \frac{\pi}{4}\frac{k_F}{q}\frac{q_{\mathrm{TF}}^2}{2q^2}\left[1-\left(\frac{\hbar\omega-E_q}{\hbar v_F q}\right)^2\right] & ,\ \hbar v_F q-E_q\le\hbar\omega\le\hbar v_F q+E_q\ \wedge\ q\le 2k_F\\ 0 & ,\ \hbar v_F q+E_q<\hbar\omega\ \wedge\ q\le 2k_F\\ 0 & ,\ 0\le\hbar\omega<E_q-\hbar v_F q\ \wedge\ q>2k_F\\ \frac{\pi}{4}\frac{k_F}{q}\frac{q_{\mathrm{TF}}^2}{2q^2}\left[1-\left(\frac{\hbar\omega-E_q}{\hbar v_F q}\right)^2\right] & ,\ E_q-\hbar v_F q\le\hbar\omega\le E_q+\hbar v_F q\ \wedge\ q>2k_F\\ 0 & ,\ E_q+\hbar v_F q<\hbar\omega\ \wedge\ q>2k_F\end{cases}, \tag{A.3b}$$

for the imaginary part. In the above, $q\equiv|\mathbf{q}|$, $E_q=\hbar^2q^2/(2m_e)$, and q_{TF} stands for the Thomas-Fermi wavevector, $q_{\mathrm{TF}}=\sqrt{\frac{e^2}{\epsilon_0}\frac{m_e k_F}{\pi^2\hbar^2}}$ [1].

Introducing the dimensionless variables $x=q/k_F$ and $y=\hbar\omega/E_F$, the previous expressions can be written as

$$\mathrm{Re}\,\epsilon^{\mathrm{RPA}}(x,y)=1+\mathcal{C}\frac{2}{\pi}\frac{1}{x^2}\left\{1+\frac{1}{2x}\left\{\left[1-\left(\frac{y+x^2}{2x}\right)^2\right]\ln\left|\frac{y+2x+x^2}{y-2x+x^2}\right|+\left[1-\left(\frac{y-x^2}{2x}\right)^2\right]\ln\left|\frac{y-2x-x^2}{y+2x-x^2}\right|\right\}\right\}, \tag{A.4a}$$

and

$$\mathrm{Im}\,\epsilon^{\mathrm{RPA}}(x,y)=\begin{cases}\mathcal{C}\frac{y}{x^3} & ,\ 0\le y<2x-x^2\ \wedge\ x\le 2\\ \mathcal{C}\frac{1}{x^3}\left[1-\left(\frac{y-x^2}{2x}\right)^2\right] & ,\ 2x-x^2\le y\le 2x+x^2\ \wedge\ x\le 2\\ 0 & ,\ 2x+x^2<y\ \wedge\ x\le 2\\ 0 & ,\ 0\le y<x^2-2x\ \wedge\ x>2\\ \mathcal{C}\frac{1}{x^3}\left[1-\left(\frac{y-x^2}{2x}\right)^2\right] & ,\ x^2-2x\le y\le x^2+2x\ \wedge\ x>2\\ 0 & ,\ x^2+2x<y\ \wedge\ x>2\end{cases}, \tag{A.4b}$$

where we have defined the dimensionless variable $\mathcal{C}=\dfrac{\alpha\, m_e c^2}{\hbar c\, k_F}$ for short-hand notation, with $\alpha\simeq 1/137.036$ denoting the fine-structure constant. Figure A.1 shows the

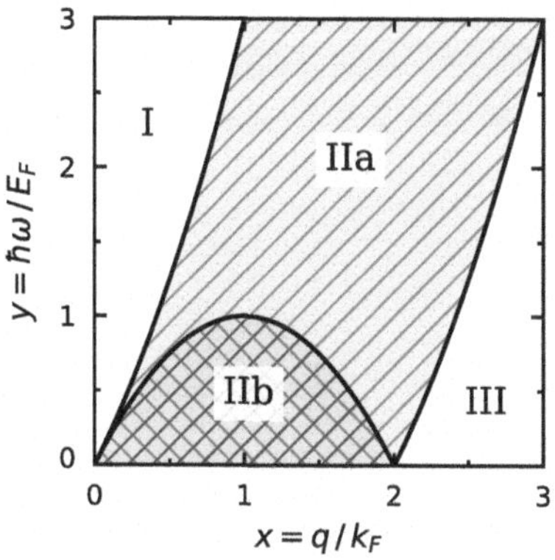

Fig. A.1 Electron-hole continuum of a non-interacting 3DEG. The hatched regions show the areas of the phase-space where Im $\epsilon^{\mathrm{RPA}}(x, y)$ is finite due to Landau damping

regions where Landau damping is present owing to single-particle particle excitations (i.e., regions where Im $\epsilon^{\mathrm{RPA}} \neq 0$).

The relaxation-time approximation in the RPA

In order to include electronic scattering within the relaxation-time approximation (RTA) in the RPA formalism, we employ Mermin's prescription (in order to maintain *local* particle number conservation), that is [3, 5]

$$\epsilon_{\gamma}^{\mathrm{RPA}}(q, \omega) = 1 + \frac{(\omega + i\gamma)\left[\epsilon^{\mathrm{RPA}}(q, \omega + i\gamma) - 1\right]}{\omega + i\gamma \dfrac{\epsilon^{\mathrm{RPA}}(q, \omega + i\gamma) - 1}{\epsilon^{\mathrm{RPA}}(q, 0) - 1}}, \tag{A.5}$$

where $\gamma = \tau^{-1}$ characterizes the scattering rate experienced by the charge carriers (and where τ is the corresponding relaxation-time).

References

1. Sólyom J (2010) Fundamentals of the physics of solids. Springer, Berlin
2. Bruus H, Flensberg K (2004) Many-body quantum theory in condensed matter physics: an introduction, Oxford Graduate Texts. Oxford University Press, Oxford
3. Giuliani G, Vignale G (2005) Quantum theory of the electron liquid. Cambridge University Press, Cambridge. https://doi.org/10.1017/CBO9780511619915
4. Lindhard J, Danske Videnskab Kgl (1954) Selskab Mat-Fys Medd 28
5. Mermin ND (1970) Phys Rev B 1:2362. https://doi.org/10.1017/CBO9780511619915; https://doi.org/10.1017/CBO9780511619915.2362

Appendix B
Derivation of the Specular Reflection Model for Nonlocal Plasmonics

Surface plasmon polaritons (SPPs) are surface electromagnetic waves resulting from collective excitations of the free-electron gas in conductors [1, 2]. The simplest system capable of sustaining SPPs is a planar dielectric–metal interface, with the dielectric and the metal filling each half-space, as depicted in Fig. B.1.

Here, we provide a detailed derivation of the the specular reflection model (SRM) for nonlocal plasmonics. In contrast to the simpler hydrodynamic model, here we employ the SRM together with Lindhard's dielectric function [3–5] in order to (i) incorporate the full wavevector dependence of the nonlocal dielectric function, and (ii) to account for Landau damping.

B.1 Reflection Coefficient for TM Waves in Terms of Surface Impedance

In the case of p-polarization—or transverse magnetic (TM) solutions—the electromagnetic field in the semi-infinite dielectric medium (filling the $z > 0$ half-space) can be written as

$$\begin{aligned} \mathbf{H}_{\mathrm{d}}(\mathbf{r}, t) &= H_{y,\mathrm{d}}(z)\, e^{i(qx-\omega t)}\, \hat{\mathbf{y}} \\ &= \left(e^{-ik_{z,\mathrm{d}}z} + r_p\, e^{ik_{z,\mathrm{d}}z}\right) e^{i(qx-\omega t)}\, \hat{\mathbf{y}}\,, \end{aligned} \tag{B.1a}$$

$$\begin{aligned} \mathbf{E}_{\mathrm{d}}(\mathbf{r}, t) &= \left[E_{x,\mathrm{d}}(z)\,\hat{\mathbf{x}} + E_{z,\mathrm{d}}(z)\,\hat{\mathbf{z}}\right] e^{i(qx-\omega t)} \\ &= -\frac{1}{\omega\epsilon_0\epsilon_{\mathrm{d}}} \left\{k_{z,\mathrm{d}}\left(e^{-ik_{z,\mathrm{d}}z} - r_p\, e^{ik_{z,\mathrm{d}}z}\right)\hat{\mathbf{x}} \right. \\ &\qquad \left. + q\left(e^{-ik_{z,\mathrm{d}}z} + r_p\, e^{ik_{z,\mathrm{d}}z}\right)\hat{\mathbf{z}}\right\} e^{i(qx-\omega t)}\,, \end{aligned} \tag{B.1b}$$

P. A. D. Gonçalves, *Plasmonics and Light–Matter Interactions in Two-Dimensional Materials and in Metal Nanostructures*, Springer Theses,
https://doi.org/10.1007/978-3-030-38291-9

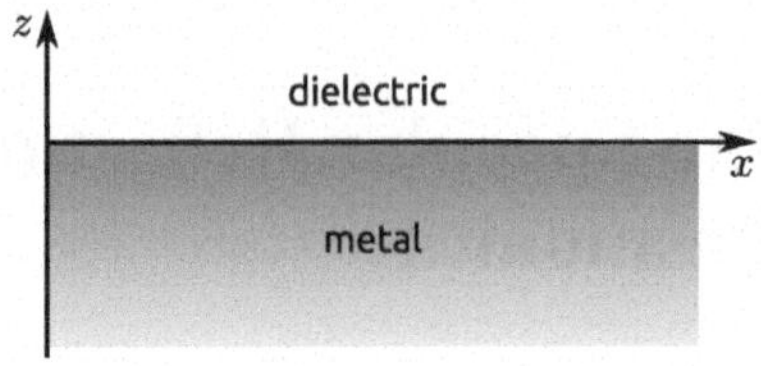

Fig. B.1 Illustration of a planar dielectric–metal interface, defined by $z = 0$ plane. The dielectric half-space is characterized by a relative permittivity, ϵ_{d}, whereas the electromagnetic properties of the metal are accounted for by a dielectric function of the form, $\epsilon_{\alpha\beta}(\mathbf{k}, \omega)$

where $k_{z,\mathrm{d}} = \sqrt{\epsilon_{\mathrm{d}} k_0^2 - q^2}$, and $k_0 = \omega/c$ denotes the wavevector of light in free-space.

Since we are interested in the case in which the metal is nonlocal, it is convenient to introduce the concept of surface impedance, Z_p, defined via [6–8]

$$Z_p = \frac{\mathbf{E}_{\mathrm{t}}}{\hat{\mathbf{n}}_{\mathrm{out}} \times \mathbf{H}_{\mathrm{t}}}, \tag{B.2}$$

where the subscript stands for the fields' tangential components (with respect to the interface), and $\hat{\mathbf{n}}_{\mathrm{out}}$ is a unit vector normal to the interface and pointing outwards (i.e., away from the metal). The surface impedance can also be regarded as a boundary condition, so that the electromagnetic field outside the metal can be determined without considering the electromagnetic field inside the metal [6, 8].

In particular, the surface impedance associated with the electromagnetic field (B.1), is given by

$$Z_p = -\left.\frac{E_x}{H_y}\right|_{z=0^-} = -\left.\frac{E_x}{H_y}\right|_{z=0^+}, \tag{B.3}$$

where in the last step we made explicit use of the fact that $\mathbf{E}_{\mathrm{t}}$ and $\mathbf{H}_{\mathrm{t}}$ are continuous across the interface.[1] Then, using Eqs. (B.1), we obtain an expression for the reflection coefficient for p-polarized waves in terms of the corresponding surface impedance, namely

$$r_p = \frac{\dfrac{k_{z,\mathrm{d}}}{\omega\epsilon_0\epsilon_{\mathrm{d}}} - Z_p}{\dfrac{k_{z,\mathrm{d}}}{\omega\epsilon_0\epsilon_{\mathrm{d}}} + Z_p}. \tag{B.4}$$

[1]That is, the tangential fields evaluated *just* inside the metal must be match the fields evaluated *just* outside the metal surface (see Refs. [6–8] for a more in-depth discussion).

Alternatively, the previous equation can also be cast as

$$r_p = \frac{\dfrac{k_{z,\mathrm{d}}}{\epsilon_\mathrm{d} k_0} - Z_p/Z_0}{\dfrac{k_{z,\mathrm{d}}}{\epsilon_\mathrm{d} k_0} + Z_p/Z_0}\,, \tag{B.5}$$

where the quantity $Z_0 = \sqrt{\mu_0/\epsilon_0}$ is the impedance of free-space.

B.2 Surface Impedance in a Nonlocal Medium: Specular Reflection Model

Let us consider a homogeneous nonlocal medium. We seek TM solutions, that can be written as

$$\mathbf{H}(\mathbf{r}) = H_y(z)\, e^{ik_x x}\,, \tag{B.6a}$$

$$\mathbf{E}(\mathbf{r}) = \left[E_x(z)\,\hat{\mathbf{x}} + E_z(z)\,\hat{\mathbf{z}}\right] e^{ik_x x}\,, \tag{B.6b}$$

where a harmonic time dependence of the form $e^{-i\omega t}$ is implicitly assumed herein. Using these together with the macroscopic Maxwell curl equations—i.e., Faraday's and Ampère's law—one finds

$$\frac{\partial}{\partial z} E_x(z) - ik_x E_z(z) = i\omega\mu_0 H_y(z)\,, \tag{B.7a}$$

$$\frac{\partial}{\partial z} H_y(z) = i\omega D_x(z)\,, \tag{B.7b}$$

$$k_x H_y(z) = -\omega D_z(z)\,. \tag{B.7c}$$

Now, combining Eqs. (B.7a) and (B.7c), yields

$$-k_x^2 E_z(z) - ik_x \frac{\partial}{\partial z} E_x(z) + \omega^2 \mu_0 D_z(z) = 0\,, \tag{B.8}$$

and differentiating Eq. (B.7a) with respect to the z-coordinate gives

$$\frac{\partial^2}{\partial z^2} E_x(z) - ik_x \frac{\partial}{\partial z} E_z(z) + \omega^2 \mu_0 D_x(z) = 0\,, \tag{B.9}$$

where we have made use of Eq. (B.7b).

We now introduce the central assumption behind the specular reflection model (SRM). As the name suggests, within this framework it is assumed that the electrons are specularly reflected at $z = 0$, implying that [9]

$$E_x(z) = E_x(-z) \quad \text{and} \quad E_z(z) = -E_z(-z)\,. \tag{B.10}$$

Hence, the fields in a metal half-space are identical to the ones in an infinite metal with a source current sheet at the surface of the former.[2] As it will become apparent later, since we are dealing with a nonlocal metal here, it is useful to work in the Fourier space (i.e., momentum space). To that end, we Fourier transform Eqs. (B.8) and (B.9); doing so,[3] we obtain (using integration by parts)

$$-k_x^2 E_z(k_z) + k_x k_z E_x(k_z) + \omega^2 \mu_0 D_z(k_z) = 0\,, \tag{B.12}$$

and

$$\begin{aligned} -k_z^2 E_x(k_z) + k_x k_z E_z(k_z) + \omega^2 \mu_0 D_x(k_z) &= -2 \left[\frac{\partial}{\partial z} E_x(z) - i k_x E_z(z) \right] \Bigg|_{z=0^-} \\ &= -2 i \omega \mu_0 H_y(0^-)\,, \end{aligned} \tag{B.13}$$

respectively. In the previous expression, the right-hand side arises from the discontinuities of $E_z(z)$ and $\partial E_x(z)/\partial z$ at $z = 0$ [cf. Eq. (B.10)], and in the last step we have used Eq. (B.7a).

The advantage of working in **k**-space now becomes apparent, since the constitutive relation (real-space representation) $\mathbf{D}(\mathbf{r}) = \epsilon_0 \int d\mathbf{r}' \overleftrightarrow{\epsilon}(\mathbf{r}, \mathbf{r}') \cdot \mathbf{E}(\mathbf{r}')$ reduces to a simple multiplication in momentum-space, specifically, $\mathbf{D}(\mathbf{k}) = \epsilon_0 \overleftrightarrow{\epsilon}(\mathbf{k}) \cdot \mathbf{E}(\mathbf{k})$. Therefore, for a p-polarized wave like the one represented by Eq. (B.6), we may write

$$D_x(k_z) = \epsilon_0 \epsilon_{xx} E_x(k_z) + \epsilon_0 \epsilon_{xz} E_z(k_z)\,, \tag{B.14a}$$
$$D_z(k_z) = \epsilon_0 \epsilon_{zx} E_x(k_z) + \epsilon_0 \epsilon_{zz} E_z(k_z)\,, \tag{B.14b}$$

where it is implicit that here $\epsilon_{\alpha\beta} \equiv \epsilon_{\alpha\beta}(\mathbf{k})$. Upon substituting these in Eqs. (B.13) and (B.12), one can solve the ensuing equations for $E_x(k_z)/H_y(0^+)$, which produces

$$\frac{E_x(k_z)}{H_y(0^-)} = -2 i \omega \mu_0 \frac{k_0^2 \epsilon_{zz} - k_x^2}{\left(k_0^2 \epsilon_{xx} - k_z^2\right)\left(k_0^2 \epsilon_{zz} - k_x^2\right) - \left(k_0^2 \epsilon_{xz} - k_x k_z\right)\left(k_0^2 \epsilon_{zx} - k_x k_z\right)}\,. \tag{B.15}$$

Instead of writing this relation in terms of the Cartesian components of the dielectric function, it is more useful to represent the latter in terms of longitudinal and transverse

---

[2]For a technical discussion on the SRM model, the reader is advised to consult Refs. [8, 9].

[3]We define the Fourier transform and its inverse, for an arbitrary function $F(z)$, as

$$F(k_z) = \int_{-\infty}^{\infty} dz F(z)\, e^{-i k_z z} \quad \text{and} \quad F(z) = \int_{-\infty}^{\infty} \frac{dk_z}{2\pi} F(k_z)\, e^{i k_z z}\,. \tag{B.11}$$

components. Noting that $\epsilon_{\alpha\beta}(\mathbf{k},\omega) = k_\alpha k_\beta/k^2 \epsilon_{\rm L}(\mathbf{k},\omega) + \left(\delta_{\alpha\beta} - k_\alpha k_\beta/k^2\right)\epsilon_{\rm T}(\mathbf{k},\omega)$ [1, 10], we have

$$\epsilon_{xx} = \frac{k_x^2}{k^2}\epsilon_{\rm L} + \frac{k_z^2}{k^2}\epsilon_{\rm T}\,, \tag{B.16a}$$

$$\epsilon_{zz} = \frac{k_z^2}{k^2}\epsilon_{\rm L} + \frac{k_x^2}{k^2}\epsilon_{\rm T}\,, \tag{B.16b}$$

$$\epsilon_{xz} = \epsilon_{zx} = \frac{k_x k_z}{k^2}\left(\epsilon_{\rm L} - \epsilon_{\rm T}\right)\,, \tag{B.16c}$$

where $k^2 = k_x^2 + k_z^2$. In possession of these identities, Eq. (B.15) becomes

$$\frac{E_x(k_z)}{H_y(0^-)} = -\frac{2i\omega\mu_0}{k^2 k_0^2}\left[\frac{k_x^2}{\epsilon_{\rm L}} + \frac{k_z^2}{\epsilon_{\rm T} - k^2/k_0^2}\right]\,, \tag{B.17}$$

which is almost the surface impedance (B.3). Indeed, all that it is left in order to determine Z_p is to Fourier transform Eq. (B.17) back to real-space and take the appropriate $z \to 0$ limit. Such a procedure then leads to following result for the surface impedance for p-polarization:

$$Z_p = -\frac{E_x(0^-)}{H_y(0^-)} = \frac{2i}{\pi}\frac{Z_0}{k_0}\int_0^\infty \frac{dk_z}{k^2}\left[\frac{k_x^2}{\epsilon_{\rm L}(\mathbf{k},\omega)} + \frac{k_z^2}{\epsilon_{\rm T}(\mathbf{k},\omega) - k^2/k_0^2}\right]\,, \tag{B.18}$$

where we have made explicit use of the fact that the integrand is an even function of k_z to write the integral only over the positive part.

We have therefore derived a closed-form (albeit non-trivial) expression for the surface impedance within the framework of the specular reflection model, describing a general nonlocal medium characterized by a general dielectric function of the form, $\epsilon_{\alpha\beta}(\mathbf{k},\omega) = k_\alpha k_\beta/k^2 \epsilon_{\rm L}(\mathbf{k},\omega) + \left(\delta_{\alpha,\beta} - k_\alpha k_\beta/k^2\right)\epsilon_{\rm T}(\mathbf{k},\omega)$. The explicit form for the longitudinal and transverse components of the dielectric function of the homogenous 3DEG can be found, for instance, in Ref. [4].

B.3 Reflection Coefficient and Nonlocal Plasmons Within the Specular Reflection Model

In possession of Eq. (B.18), the reflection coefficient for a nonlocal metal is then given by [see Eq. (B.5)]

$$r_p = \frac{k_{z,\rm d} - i\dfrac{2\epsilon_{\rm d}}{\pi}\displaystyle\int_0^\infty \frac{dk_z}{k^2}\left[\frac{k_x^2}{\epsilon_{\rm L}(\mathbf{k},\omega)} + \frac{k_z^2}{\epsilon_{\rm T}(\mathbf{k},\omega) - k^2/k_0^2}\right]}{k_{z,\rm d} + i\dfrac{2\epsilon_{\rm d}}{\pi}\displaystyle\int_0^\infty \frac{dk_z}{k^2}\left[\frac{k_x^2}{\epsilon_{\rm L}(\mathbf{k},\omega)} + \frac{k_z^2}{\epsilon_{\rm T}(\mathbf{k},\omega) - k^2/k_0^2}\right]}\,, \tag{B.19}$$

where we recall here that k has an explicit dependence on k_z, i.e., $k^2 = |\mathbf{k}|^2 = k_x^2 + k_z^2$. Also, in the language of Sect. B.1, we note that $k_x \equiv q$, whereas $k_{z,\mathrm{d}} = \sqrt{\epsilon_\mathrm{d} k_0^2 - q^2}$ (subjected to Im $k_{z,\mathrm{d}} \geq 0$).

The dispersion relation of plasmons supported by the dielectric–metal interface follows from the poles of r_p. Hence, the nonlocal plasmonic spectrum can be constructed out of the nonlocal plasmon condition:

$$1 + i\frac{2\epsilon_\mathrm{d}}{\pi} \frac{1}{\sqrt{\epsilon_\mathrm{d} k_0^2 - q^2}} \int_0^\infty \frac{dk_z}{k^2} \left[\frac{q^2}{\epsilon_\mathrm{L}(\mathbf{k}, \omega)} + \frac{k_z^2}{\epsilon_\mathrm{T}(\mathbf{k}, \omega) - k^2/k_0^2} \right] = 0 . \quad \text{(B.20)}$$

In the electrostatic limit, the above expression simplifies to

$$1 + q\,\frac{2\epsilon_\mathrm{d}}{\pi} \int_0^\infty \frac{dk_z}{q^2 + k_z^2} \frac{1}{\epsilon_\mathrm{L}\left(\sqrt{q^2 + k_z^2}, \omega\right)} = 0 . \quad \text{(B.21)}$$

References

1. Gonçalves PAD, Peres NMR (2016) An introduction to graphene plasmonics, 1st edn. World Scientific, Singapore. https://doi.org/10.1017/CBO9780511619915; https://doi.org/10.1017/CBO9780511619915ks/10.1142/9948
2. Maradudin AA, Barnes WL, Sambles JR (eds) (2014) Modern plasmonics, 1st edn. Elsevier
3. Lindhard J, Danske Videnskab Kgl (1954) Selskab Mat-Fys Medd 28
4. Sólyom J (2010) Fundamentals of the physics of solids. Springer, Berlin
5. Giuliani G, Vignale G (2005) Quantum theory of the electron liquid. Cambridge University Press, Cambridge. https://doi.org/10.1017/CBO9780511619915
6. Landau LD, Lifshitz EM (1960) Electrodynamics of continuous media. Pergamon Press, New York
7. Yuferev SV, Ida N (2009) Surface impedance boundary conditions: a comprehensive approach. CRC press
8. FordGW,WeberWH(1984)PhysRep113(4):195.http://www.sciencedirect.com/science/article/pii/037015738490098X, https://doi.org/10.1017/CBO9780511619915-1573(84)90098-X
9. Kliewer KL, Fuchs R (1968) Phys Rev 172:607. https://doi.org/10.1017/CBO9780511619915; https://doi.org/10.1017/CBO978051161991572.607
10. Wooten F (1972) Optical properties of solids. Academic, New York

Appendix C
Mie Theory with d-Parameters

In the present appendix, we provide a detailed account of the generalized Mie theory introduced in Sect. 8.2.2, and, in particular, the derivation of the mesoscopic Mie coefficients incorporating Feibelman d-parameters. These constitute the fundamental building-blocks for calculating the response of the scattering object—here a metallic sphere—to *any* external perturbation (in a basis of spherical vector waves).

Theoretical Framework

Wave equation in spherically symmetric systems. In a linear, isotropic, and homogeneous medium the electric and magnetic fields must satisfy their respective vector wave equations,

$$\nabla^2\mathbf{E}(\mathbf{r},\omega) + k_j^2\mathbf{E}(\mathbf{r},\omega) = 0 \quad \text{and} \quad \nabla^2\mathbf{H}(\mathbf{r},\omega) + k_j^2\mathbf{H}(\mathbf{r},\omega) = 0\,, \tag{C.1}$$

where $k_j \equiv \sqrt{\epsilon_j}k_0$. The electric and magnetic fields in spherical coordinates can be constructed in terms of vector harmonics, $\mathbf{M}_\nu(\mathbf{r})$ and $\mathbf{N}_\nu(\mathbf{r})$, which in turn are defined in terms of a pilot wave $\mathbf{c}$ and a generating scalar function $\psi_\nu(\mathbf{r})$ that obeys the (scalar) Helmholtz equation $\nabla^2\psi_\nu(\mathbf{r}) + k_j^2\psi_\nu(\mathbf{r}) = 0$ [1]. Strictly speaking, the pilot vector $\mathbf{c}$ should be a constant vector. However, and rather fortunately, in spherical coordinates and in this case alone, it turns out that we can take $\mathbf{c} = r\hat{\mathbf{r}}$; this enables us to construct a tangential solution $\mathbf{M}_\nu(\mathbf{r})$ (though $\mathbf{N}_\nu(\mathbf{r})$ is not purely normal) [2]. This is desirable because it allows us to associate each of the vector harmonics $\mathbf{M}_\nu(\mathbf{r})$ and $\mathbf{N}_\nu(\mathbf{r})$ with TE and TM waves, respectively.

Therefore, the vector harmonics $\mathbf{M}_\nu(\mathbf{r})$ and $\mathbf{N}_\nu(\mathbf{r})$—which are two independent solutions of the vector wave equation in spherical coordinates—may be written as [1, 2]

$$\mathbf{M}_{lm}^{e/o}(\mathbf{r}) \equiv \nabla \times \hat{\mathbf{r}}\, r\, \psi_{lm}^{e/o}(\mathbf{r})\,, \tag{C.2a}$$

$$\mathbf{N}_{lm}^{e/o}(\mathbf{r}) \equiv k_j^{-1}\nabla \times \nabla \times \hat{\mathbf{r}}\, r\, \psi_{lm}^{e/o}(\mathbf{r})\,, \tag{C.2b}$$

where the generation function is

P. A. D. Gonçalves, *Plasmonics and Light–Matter Interactions in Two-Dimensional Materials and in Metal Nanostructures*, Springer Theses,
https://doi.org/10.1007/978-3-030-38291-9

$$\psi_{lm}^{e/o}(r,\theta,\phi) \equiv z_l(k_j r) P_l^m(\cos\theta) \begin{Bmatrix} \cos(m\phi) \\ \sin(m\phi) \end{Bmatrix}, \tag{C.2c}$$

and where the radial part $z_l(k_j r)$ is either a spherical Bessel or Hankel function of the first kind, $j_l(k_j r)$ and $h_l^{(1)}(k_j r)$, respectively representing incoming and outgoing waves. Furthermore, the "quantum numbers" l and m are integers with values in the range $l \in [1, \infty[$ and $m \in [0, l]$, and P_l^m denote the associated Legendre polynomials [3]. The explicit form of the solenoidal (i.e., divergence-free) vector waves $\mathbf{M}_{lm}^{e/o}(\mathbf{r})$ and $\mathbf{N}_{lm}^{e/o}(\mathbf{r})$, obtained via Eqs. (C.2), reads[4]

$$\mathbf{M}_{lm}^{e/o}(\mathbf{r}) = \begin{bmatrix} 0 \\ m z_l(\rho_j) \dfrac{P_l^m(\mu)}{\sin\theta} \begin{Bmatrix} -\sin(m\phi) \\ \cos(m\phi) \end{Bmatrix} \\ -z_l(\rho) \dfrac{\partial P_l^m(\mu)}{\partial\theta} \begin{Bmatrix} \cos(m\phi) \\ \sin(m\phi) \end{Bmatrix} \end{bmatrix}, \tag{C.3a}$$

and

$$\mathbf{N}_{lm}^{e/o}(\mathbf{r}) = \begin{bmatrix} \dfrac{l(l+1)}{\rho_j} z_l(\rho_j) P_l^m(\mu) \begin{Bmatrix} \cos(m\phi) \\ \sin(m\phi) \end{Bmatrix} \\ \dfrac{1}{\rho_j} \left[\rho_j z_l(\rho_j)\right]' \dfrac{\partial P_l^m(\mu)}{\partial\theta} \begin{Bmatrix} \cos(m\phi) \\ \sin(m\phi) \end{Bmatrix} \\ \dfrac{m}{\rho_j} \left[\rho_j z_l(\rho_j)\right]' \dfrac{P_l^m(\mu)}{\sin\theta} \begin{Bmatrix} -\sin(m\phi) \\ \cos(m\phi) \end{Bmatrix} \end{bmatrix}, \tag{C.3b}$$

where $\rho_j \equiv k_j r$ and $\mu \equiv \cos\theta$, and the prime denotes differentiation with respect to ρ_j.

The spherical vector harmonics $\mathbf{M}_{lm}^{e/o}(\mathbf{r})$ and $\mathbf{N}_{lm}^{e/o}(\mathbf{r})$ may be regarded as fundamental solutions of the vector wave equation in spherical coordinates. Hence, the electric and magnetic fields can be constructed from an expansion in terms of such vector waves.

Incident plane-wave in terms of vector spherical harmonics. Let us assume an incident plane-wave traveling along the positive z-direction and polarized along $\hat{\mathbf{x}}$, namely[5]

$$\mathbf{E}^{\text{inc}} = E_0 e^{ik_d z} \hat{\mathbf{x}} = E_0 e^{ik_d r\cos\theta} \begin{bmatrix} \sin\theta\cos\phi \\ \cos\theta\cos\phi \\ -\sin\theta \end{bmatrix}. \tag{C.4}$$

We now want to cast the incident field, not as in Eq. (C.4), but rather in terms of the vector spherical waves introduced earlier, that is

[4]For the radial component of $\mathbf{N}_{lm}^{e/o}(\mathbf{r})$ we have made use of the Legendre differential equation [4] whose solutions are the associated Legendre polynomials.

[5]Note that $\hat{\mathbf{x}} = \sin\theta\cos\phi\,\hat{\mathbf{r}} + \cos\theta\cos\phi\,\hat{\boldsymbol{\theta}} - \sin\theta\,\hat{\boldsymbol{\phi}}$.

$$\mathbf{E}^{\text{inc}} = \sum_{l=1}^{\infty} \sum_{m=0}^{l} \sum_{\sigma=\{\text{e,o}\}} \left(A_{lm}^{\text{TM},\sigma} \mathbf{N}_{lm}^{\sigma} + A_{lm}^{\text{TE},\sigma} \mathbf{M}_{lm}^{\sigma} \right) , \tag{C.5}$$

where the expansion coefficients $A_{lm}^{\text{TM},\sigma}$ and $A_{lm}^{\text{TE},\sigma}$ follow from [1]

$$A_{lm}^{\text{TM},\sigma} = \frac{\int_0^{2\pi} \mathrm{d}\phi \int_0^{\pi} \mathrm{d}\theta \sin\theta \, \mathbf{E}^{\text{inc}} \cdot \mathbf{N}_{lm}^{\sigma}}{\int_0^{2\pi} \mathrm{d}\phi \int_0^{\pi} \mathrm{d}\theta \sin\theta \left|\mathbf{N}_{lm}^{\sigma}\right|^2} , \tag{C.6a}$$

$$A_{lm}^{\text{TE},\sigma} = \frac{\int_0^{2\pi} \mathrm{d}\phi \int_0^{\pi} \mathrm{d}\theta \sin\theta \, \mathbf{E}^{\text{inc}} \cdot \mathbf{M}_{lm}^{\sigma}}{\int_0^{2\pi} \mathrm{d}\phi \int_0^{\pi} \mathrm{d}\theta \sin\theta \left|\mathbf{M}_{lm}^{\sigma}\right|^2} . \tag{C.6b}$$

Carrying out the explicit calculations and exploiting the orthogonality relations of the sine and cosines functions, one finds that $A_{lm}^{\text{TM},o} = 0$ and $A_{lm}^{\text{TM},e} \propto \delta_{m1}$, and, similarly, that $A_{lm}^{\text{TE},e} = 0$ and $A_{lm}^{\text{TE},o} \propto \delta_{m1}$. The calculation of these coefficients then leads to

$$\mathbf{E}^{\text{inc}} = \sum_{l=1}^{\infty} E_0^l \left(-i\mathbf{N}_{l1}^{e\,[\text{d}]} + \mathbf{M}_{l1}^{o\,[\text{d}]} \right) , \tag{C.7}$$

where the extra superscript "[d]" indicates that the radial function is $z_l(\rho_{\text{d}}) \equiv j_l(\rho_{\text{d}})$ because the impinging field has to remain finite at the origin, and we have also defined $E_0^l = i^l \frac{2l+1}{l(l+1)} E_0$. From here, the corresponding incident magnetic field follows from Faraday's law, yielding

$$\mathbf{H}^{\text{inc}} = -\frac{k_{\text{d}}}{\omega\mu_0} \sum_{l=1}^{\infty} E_0^l \left(i\mathbf{N}_{l1}^{o\,[\text{d}]} + \mathbf{M}_{l1}^{e\,[\text{d}]} \right) , \tag{C.8}$$

where the properties of the vector spherical harmonics have been used [1, 2].

Equations (C.7) and (C.8) thereby reflect the expansion of the incident plane-wave in terms of vector spherical waves.

Internal and scattered fields. In possession of the incident electromagnetic field expansion, Eqs. (C.7) and (C.8), we now express in a similar fashion the internal and scattered fields using vector spherical waves. In this spirit, we write the fields inside the sphere as

$$\mathbf{E}^{\text{m}} = \sum_{l=1}^{\infty} E_0^l \left(-id_l^{\text{TM}} \mathbf{N}_{l1}^{e\,[\text{m}]} + c_l^{\text{TE}} \mathbf{M}_{l1}^{o\,[\text{m}]} \right) , \tag{C.9a}$$

$$\mathbf{H}^{\text{m}} = -\frac{k_{\text{m}}}{\omega\mu_0} \sum_{l=1}^{\infty} E_0^l \left(ic_l^{\text{TE}} \mathbf{N}_{l1}^{o\,[\text{m}]} + d_l^{\text{TM}} \mathbf{M}_{l1}^{e\,[\text{m}]} \right) . \tag{C.9b}$$

where now the superscript "[m]" highlights that the radial function is $z_l(\rho_\mathrm{m}) \equiv j_l(\rho_\mathrm{m})$ because it must to remain finite at the origin.

Outside the sphere, both spherical Bessel functions j_l and y_l are permitted and therefore we shall use the spherical Hankel function of the first kind for the radial part of the generating function. Specifically, we choose $z_l(\rho_\mathrm{d}) \equiv h_l^{(1)}(\rho_\mathrm{d})$ owing to its appropriate asymptotic behavior at large ρ_d [1], i.e., corresponding to an outgoing spherical wave (which is what we expect for the asymptotic scattered field). Within this reasoning, we therefore express the scattered fields as

$$\mathbf{E}^\mathrm{scat} = \sum_{l=1}^{\infty} E_0^l \left(i a_l^\mathrm{TM} \mathbf{N}_{l1}^{e\,[\mathrm{s}]} - b_l^\mathrm{TE} \mathbf{M}_{l1}^{o\,[\mathrm{s}]} \right), \tag{C.10a}$$

$$\mathbf{H}^\mathrm{scat} = \frac{k_\mathrm{d}}{\omega\mu_0} \sum_{l=1}^{\infty} E_0^l \left(i b_l^\mathrm{TE} \mathbf{N}_{l1}^{o\,[\mathrm{s}]} + a_l^\mathrm{TM} \mathbf{M}_{l1}^{e\,[\mathrm{s}]} \right), \tag{C.10b}$$

where this time the superscript "[s]" indicates that the radial function is $h_l^{(1)}(\rho_\mathrm{d})$ as we have just mentioned.

The amplitudes a_l^TM, b_l^TE, c_l^TE, and d_l^TM are the so-called Mie coefficients of the scattered and internal (i.e., transmitted) kind, in TM and TE flavors, respectively. These can be determined once the adequate boundary conditions are imposed (at the sphere's surface).

Nonclassical Mie coefficients

The Mie coefficients essentially entail all the relevant physics describing the electromagnetic response of a sphere of arbitrary size and material constitution. In the following, we derive closed-form expressions for the mesoscopic (i.e., nonclassical) Mie coefficients within the formalism of Feibelman d-parameters. In order to proceed, we need to invoke the appropriate boundary conditions, reading [cf. Eqs. (8.27b)], in spherical coordinates,

$$\left(\mathbf{E}_\Omega^\mathrm{inc} + \mathbf{E}_\Omega^\mathrm{scat} - \mathbf{E}_\Omega^\mathrm{m}\right)\Big|_{r=R} = -d_\perp \nabla_\Omega \left(E_r^\mathrm{inc} + E_r^\mathrm{scat} - E_r^\mathrm{m}\right)\Big|_{r=R}, \tag{C.11a}$$

$$\left(\mathbf{H}_\Omega^\mathrm{inc} + \mathbf{H}_\Omega^\mathrm{scat} - \mathbf{H}_\Omega^\mathrm{m}\right)\Big|_{r=R} = i\omega d_\parallel \left(\mathbf{D}_\Omega^\mathrm{inc} + \mathbf{D}_\Omega^\mathrm{scat} - \mathbf{D}_\Omega^\mathrm{m}\right) \times \hat{\mathbf{r}}\Big|_{r=R}, \tag{C.11b}$$

with $\Omega = \{\theta, \phi\}$ denoting the angular components (i.e., the components that are parallel to the sphere's surface). Equations (C.11), after some algebra, produce the following set of equations:

$$h_l^{(1)}(x_\mathrm{d}) b_l^\mathrm{TE} + j_l(x_\mathrm{m}) c_l^\mathrm{TE} = j_l(x_\mathrm{d}), \tag{C.12a}$$

$$\left[\xi_l'(x_\mathrm{d}) + \overline{d}_\perp h_l^{(1)}(x_\mathrm{d})\right] a_l^\mathrm{TM} + \left[\Psi_l'(x_\mathrm{m}) + \overline{d}_\perp j_l(x_\mathrm{m})\right] \frac{k_\mathrm{d}}{k_\mathrm{m}} d_l^\mathrm{TM} = \Psi_l'(x_\mathrm{d}) + \overline{d}_\perp j_l(x_\mathrm{d}), \tag{C.12b}$$

$$\left[h_l^{(1)}(x_\mathrm{d}) + \overline{d}_\parallel \xi_l'(x_\mathrm{d})\right] a_l^{\mathrm{TM}} + \left[j_l(x_\mathrm{m}) + \overline{d}_\parallel \Psi_l'(x_\mathrm{m})\right] \frac{k_\mathrm{m}}{k_\mathrm{d}} d_l^{\mathrm{TM}} = j_l(x_\mathrm{d}) + \overline{d}_\parallel \Psi_l'(x_\mathrm{d}), \tag{C.12c}$$

$$\left[\xi_l'(x_\mathrm{d}) - x_\mathrm{d}^2 \overline{d}_\parallel h_l^{(1)}(x_\mathrm{d})\right] b_l^{\mathrm{TE}} + \left[\Psi_l'(x_\mathrm{m}) - x_\mathrm{m}^2 \overline{d}_\parallel j_l(x_\mathrm{m})\right] c_l^{\mathrm{TE}} = \Psi_l'(x_\mathrm{d}) - x_\mathrm{d}^2 \overline{d}_\parallel j_l(x_\mathrm{d}), \tag{C.12d}$$

with $x_j = k_j R$. Additionally, the Riccati-Bessel functions $\Psi_l(x) = x j_l(x)$ and $\xi_l(x) = x h_l^{(1)}(x)$ have been employed, and the prime denotes the functions' derivatives with respect to their argument. Lastly, for shorthand notation, we have also introduced the dimensionless quantities

$$\overline{d}_\perp \equiv l(l+1)\frac{d_\perp}{R} \quad \text{and} \quad \overline{d}_\parallel \equiv \frac{d_\parallel}{R}. \tag{C.13a}$$

Solving the system of equations posed by Eqs. (C.12), we obtain the mesoscopic generalization of the Mie coefficients incorporating the Feibelman d-parameters:

$$a_l^{\mathrm{TM}} = \frac{\epsilon_\mathrm{m} j_l(x_\mathrm{m})\Psi_l'(x_\mathrm{d}) - \epsilon_\mathrm{d} j_l(x_\mathrm{d})\Psi_l'(x_\mathrm{m}) + (\epsilon_\mathrm{m} - \epsilon_\mathrm{d})\left[j_l(x_\mathrm{d}) j_l(x_\mathrm{m})\overline{d}_\perp + \Psi_l'(x_\mathrm{d})\Psi_l'(x_\mathrm{m})\overline{d}_\parallel\right]}{\epsilon_\mathrm{m} j_l(x_\mathrm{m})\xi_l'(x_\mathrm{d}) - \epsilon_\mathrm{d} h_l^{(1)}(x_\mathrm{d})\Psi_l'(x_\mathrm{m}) + (\epsilon_\mathrm{m} - \epsilon_\mathrm{d})\left[h_l^{(1)}(x_\mathrm{d}) j_l(x_\mathrm{m})\overline{d}_\perp + \xi_l'(x_\mathrm{d})\Psi_l'(x_\mathrm{m})\overline{d}_\parallel\right]}, \tag{C.14a}$$

$$b_l^{\mathrm{TE}} = \frac{j_l(x_\mathrm{m})\Psi_l'(x_\mathrm{d}) - j_l(x_\mathrm{d})\Psi_l'(x_\mathrm{m}) + (x_\mathrm{m}^2 - x_\mathrm{d}^2) j_l(x_\mathrm{d}) j_l(x_\mathrm{m})\overline{d}_\parallel}{j_l(x_\mathrm{m})\xi_l'(x_\mathrm{d}) - h_l^{(1)}(x_\mathrm{d})\Psi_l'(x_\mathrm{m}) + (x_\mathrm{m}^2 - x_\mathrm{d}^2) h_l^{(1)}(x_\mathrm{d}) j_l(x_\mathrm{m})\overline{d}_\parallel}, \tag{C.14b}$$

for the coefficients associated with the scattered outgoing wave, and

$$c_l^{\mathrm{TE}} = \frac{j_l(x_\mathrm{d})\xi_l'(x_\mathrm{d}) - h_l^{(1)}(x_\mathrm{d})\Psi_l'(x_\mathrm{d})}{j_l(x_\mathrm{m})\xi_l'(x_\mathrm{d}) - h_l^{(1)}(x_\mathrm{d})\Psi_l'(x_\mathrm{m}) + (x_\mathrm{m}^2 - x_\mathrm{d}^2) h_l^{(1)}(x_\mathrm{d}) j_l(x_\mathrm{m})\overline{d}_\parallel}, \tag{C.15a}$$

$$d_l^{\mathrm{TM}} = \frac{\sqrt{\epsilon_\mathrm{m}\epsilon_\mathrm{d}}\left[j_l(x_\mathrm{d})\xi_l'(x_\mathrm{d}) - h_l^{(1)}(x_\mathrm{d})\Psi_l'(x_\mathrm{d})\right]}{\epsilon_\mathrm{m} j_l(x_\mathrm{m})\xi_l'(x_\mathrm{d}) - \epsilon_\mathrm{d} h_l^{(1)}(x_\mathrm{d})\Psi_l'(x_\mathrm{m}) + (\epsilon_\mathrm{m} - \epsilon_\mathrm{d})\left[h_l^{(1)}(x_\mathrm{d}) j_l(x_\mathrm{m})\overline{d}_\perp + \xi_l'(x_\mathrm{d})\Psi_l'(x_\mathrm{m})\overline{d}_\parallel\right]}, \tag{C.15b}$$

for the coefficients associated with the internal fields. As in the planar case, we have only kept terms up to linear order in the d-parameters. We have also added TM and TE as superscripts in the Mie coefficients to highlight the character of the associated waves.

To the best of our knowledge, such quantum mechanical Mie coefficients with the incorporation of nonlocal and quantum surface corrections via Feibelman d-parameters, given above by Eqs. (C.14) and (C.15), have not been derived before. Thus, these results could open new perspectives for the investigation of nonlocal and quantum effects in the context of light–matter interactions in nanophotonics with metal spheres. With this in mind, they constitute a springboard to study phenomena beyond the classical regime, namely the influence of quantum nonlocal effects on the cross-sections, localized surface plasmon (LSP) resonances, electromagnetic LDOS, etc, for spherical metal particles of arbitrary size.

Quasi-static multipolar polarizability and nonretarded LSPs. In the electrostatic limit, the Mie coefficients akin to scattered outgoing waves reduce to a single quantity: the quasi-static multipolar polarizability, α_l. Since in the nonretarded regime we have $\omega R/c \to 0$, then one may perform small-argument expansions of $j_l(z)$, $h_l^{(1)}(z)$, and of the derivatives of the Riccati-Bessel functions, thereby yielding the following relation between the Mie coefficient a_l^{TM} and the multipolar polarizability:

$$\lim_{\omega R/c \to 0} a_l^{\text{TM}} = -i x_{\text{d}}^{2l+1} \frac{(l+1)(2l+1)}{l\,[(2l+1)!!]^2} \frac{\alpha_l}{4\pi R^{2l+1}} \,. \tag{C.16}$$

Taking the $\omega R/c \to 0$ limit, one finds

$$\alpha_l = 4\pi R^{2l+1} \frac{(\epsilon_{\text{m}} - \epsilon_{\text{d}}) \left[1 + \frac{l}{R}\left(d_\perp + \frac{l+1}{l} d_\parallel\right)\right]}{\epsilon_{\text{m}} + \frac{l+1}{l}\epsilon_{\text{d}} - (\epsilon_{\text{m}} - \epsilon_{\text{d}}) \frac{l+1}{R}\left(d_\perp - d_\parallel\right)} \,, \tag{C.17}$$

which agrees with the expression (8.44) obtained through a strictly nonretarded calculation (Sect. 8.2.1).

References

1. Bohren CF, Huffman DR (1983) Absorption and scattering of light by small particles. Wiley, New York
2. Stratton JA (1941) Electromagnetic theory. McGraw-Hill, New York
3. Abramowitz M, Stegun IA (1972) Handbook of mathematical functions: with formulas, graphs, and mathematical tables. Dover, New York. http://people.math.sfu.ca/~cbm/aands/
4. Zwillinger D (1997) Handbook of differential equations, 3rd edn. Academic, New York

Appendix D
Macroscopic Quantum Electrodynamics and Multipolar Decay Rates

Atom-field Hamiltonian. In the context of quantum electrodynamics (QED) we consider atom-field interactions whose dynamics are governed by the following non-relativistic Hamiltonian:

$$H = H_{\text{a}} + H_{\text{EM}} + H_{\text{int}} \,, \tag{D.1a}$$

$$H_{\text{a}} = \sum_n \left(\frac{p_n^2}{2m} - \frac{e^2}{4\pi \epsilon_0} \frac{1}{r_n} \right) , \tag{D.1b}$$

$$H_{\text{EM}} = \sum_{j=\{x,y,z\}} \int \mathrm{d}\mathbf{r} \int \mathrm{d}\omega \, \hbar\omega \left[f_j^\dagger(\mathbf{r}, \omega) f_j(\mathbf{r}, \omega) + \frac{1}{2} \right] , \tag{D.1c}$$

$$H_{\text{int}} = \sum_n \left[\frac{e}{2m} \left(\mathbf{p}_n \cdot \mathbf{A}(\mathbf{r}_n) + \mathbf{A}(\mathbf{r}_n) \cdot \mathbf{p}_n \right) + \frac{e}{2m} \mathbf{A}^2(\mathbf{r}_n) \right] , \tag{D.1d}$$

where H_{a} is the atom Hamiltonian (or, more generically, of an emitter, e.g., a quantum dot, etc), H_{EM} is the Hamiltonian of the electromagnetic (EM) field, and H_{int} stands for the Hamiltonian governing the atom-field interaction. As written above, the matter Hamiltonian H_{a} is treated at the Hartree-Fock level (consisting of a kinetic part together with the Coulomb potential), that is, electron-electron interactions, spin-orbit coupling, etc, will be neglected in what follows. Naturally, in any case these could in principle be incorporated here. Moreover, the Hamiltonian akin to the EM field is just like the one of a typical quantum harmonic oscillator, where the operator $f_j^\dagger(\mathbf{r}, \omega)$ ($f_j(\mathbf{r}, \omega)$) creates (annihilates) a field excitation at position $\mathbf{r}$ with frequency ω and oriented along the direction j. For the part of the Hamiltonian describing the interaction between the atom and the field, H_{int}, we take the

P. A. D. Gonçalves, *Plasmonics and Light–Matter Interactions in Two-Dimensional Materials and in Metal Nanostructures*, Springer Theses,
https://doi.org/10.1007/978-3-030-38291-9

so-called minimal-coupling interaction Hamiltonian.[6] Finally, $\mathbf{A}(\mathbf{r})$ is the vector potential associated with the EM field, from which both the electric and magnetic fields stem, in particular, $\mathbf{E}(\mathbf{r}) = -\frac{\partial}{\partial t}\mathbf{A}(\mathbf{r})$ and $\mathbf{H}(\mathbf{r}) = \mu_0^{-1}\nabla \times \mathbf{A}(\mathbf{r})$ [1, 2]. Notice that we have implicitly assumed that the scalar potential is identically zero (in our gauge). It should be noted that such vector and scalar potentials are simply useful mathematical entities/abstractions and, unlike $\mathbf{E}(\mathbf{r})$ and $\mathbf{H}(\mathbf{r})$, do not represent *per se* real physical quantities.

Below, we will be dealing with (linear) dispersive and dissipative media and therefore the common canonical quantization based on modal expansion is unsuitable. Ergo, we shall employ instead the formalism of macroscopic quantum electrodynamics [3, 4].

Dyadic Green's functions and macroscopic QED. From Maxwell's equations, the following wave equation is immediately apparent (assuming linear, isotropic, and nonmagnetic media) [1]:

$$\nabla \times \nabla \times \mathbf{E}(\mathbf{r}, \omega) - \frac{\omega^2}{c^2}\epsilon(\mathbf{r}, \omega)\mathbf{E}(\mathbf{r}, \omega) = i\omega\mu_0\mathbf{j}(\mathbf{r}, \omega)\,. \tag{D.2}$$

The solution of Eq. (D.2) can then be written in terms of the associated dyadic Green's function; in particular

$$\mathbf{E}(\mathbf{r}, \omega) = i\omega\mu_0\overleftrightarrow{\mathbf{G}}(\mathbf{r}, \mathbf{r}'; \omega) \cdot \mathbf{j}(\mathbf{r}, \omega)\,, \tag{D.3}$$

where the Green's dyadic satisfies

$$\nabla \times \nabla \times \overleftrightarrow{\mathbf{G}}(\mathbf{r}, \mathbf{r}'; \omega) - \frac{\omega^2}{c^2}\epsilon(\mathbf{r}, \omega)\overleftrightarrow{\mathbf{G}}(\mathbf{r}, \mathbf{r}'; \omega) = \overleftrightarrow{\mathbf{I}}\delta(\mathbf{r} - \mathbf{r}')\,. \tag{D.4}$$

Furthermore, we note the following useful properties of the Green's tensor [3]:

$$\overleftrightarrow{\mathbf{G}}^*(\mathbf{r}, \mathbf{r}'; \omega) = \overleftrightarrow{\mathbf{G}}(\mathbf{r}, \mathbf{r}'; -\omega^*)\,, \tag{D.5a}$$

$$\overleftrightarrow{\mathbf{G}}(\mathbf{r}', \mathbf{r}; \omega) = \overleftrightarrow{\mathbf{G}}^T(\mathbf{r}, \mathbf{r}'; \omega)\,, \tag{D.5b}$$

$$\frac{\omega^2}{c^2}\int \mathrm{d}\mathbf{x}\, \mathrm{Im}\{\epsilon(\mathbf{x}, \omega)\}\overleftrightarrow{\mathbf{G}}(\mathbf{r}, \mathbf{x}; \omega) \cdot \overleftrightarrow{\mathbf{G}}^\dagger(\mathbf{r}', \mathbf{x}; \omega) = \mathrm{Im}\,\overleftrightarrow{\mathbf{G}}(\mathbf{r}, \mathbf{r}'; \omega)\,, \tag{D.5c}$$

where the first and second identities reflect, respectively, the analytical properties of the Green's dyadic and reciprocity. Lastly, the relation (D.5c) is, in essence, a statement of the fluctuation-dissipation theorem.

Furthermore, the current $\mathbf{j}(\mathbf{r}, \omega)$—representing a dynamic variable of the system composed by the EM field and the medium—may be written as [3, 4]

---

[6] If spin-orbit coupling is considered, then H_{int} would also have an extra term, $\frac{e\hbar}{2m}\boldsymbol{\sigma} \cdot \mathbf{B}(\mathbf{r})$ (where $\boldsymbol{\sigma}$ denotes the Pauli matrices), responsible for the interaction of the particle's spin with the magnetic field of the EM radiation.

$$j_i(\mathbf{r},\omega) = \omega\sqrt{\frac{\hbar\epsilon_0}{\pi}\,\mathrm{Im}\,\epsilon(\mathbf{r},\omega)}\,f_i(\mathbf{r},\omega)\,, \tag{D.6}$$

where the bosonic field operators obey the commutation relations:

$$\left[f_i(\mathbf{r},\omega),\, f_j^\dagger(\mathbf{r}',\omega')\right] = \delta_{ij}\delta(\mathbf{r}-\mathbf{r}')\delta(\omega-\omega')\,, \tag{D.7a}$$

$$\left[f_i(\mathbf{r},\omega),\, f_j(\mathbf{r}',\omega')\right] = 0\,. \tag{D.7b}$$

Hence, using Eqs. (D.3) and (D.6), one obtains

$$\mathbf{E}(\mathbf{r},\omega) = i\frac{\omega^2}{c^2}\sqrt{\frac{\hbar}{\pi\epsilon_0}}\int d\mathbf{r}'\,\sqrt{\mathrm{Im}\,\epsilon(\mathbf{r}',\omega)}\overleftrightarrow{\mathbf{G}}(\mathbf{r},\mathbf{r}';\omega)\cdot\mathbf{f}(\mathbf{r}',\omega)\,, \tag{D.8}$$

so that, while realizing that in our gauge $\mathbf{E}(\mathbf{r},\omega) = i\omega\mathbf{A}(\mathbf{r},\omega)$, we find

$$\mathbf{A}(\mathbf{r},\omega) = \frac{\omega}{c^2}\sqrt{\frac{\hbar}{\pi\epsilon_0}}\int d\mathbf{r}'\sqrt{\mathrm{Im}\,\epsilon(\mathbf{r}',\omega)}\overleftrightarrow{\mathbf{G}}(\mathbf{r},\mathbf{r}';\omega)\cdot\mathbf{f}(\mathbf{r}',\omega)\,. \tag{D.9}$$

From here, and noting that $\mathbf{A}(\mathbf{r}) = \int d\omega\,\mathbf{A}(\mathbf{r},\omega) + \text{H.c.}$, we now write the components of the vector potential operator figuring in the interaction Hamiltonian (D.1d) as

$$A_i(\mathbf{r}) = \sqrt{\frac{\hbar}{\pi\epsilon_0}}\int d\omega\,\frac{\omega}{c^2}\int d\mathbf{r}'\sqrt{\mathrm{Im}\,\epsilon(\mathbf{r}',\omega)}G_{ij}(\mathbf{r},\mathbf{r}';\omega)f_j(\mathbf{r}',\omega) + \text{H.c.}\,, \tag{D.10}$$

where summation over repeated indices is implicitly assumed (Einstein's summation notation).

Fermi's Golden Rule and first-order processes. In what follows, we consider electronic first-order processes alone. This means, for instance, that we neglect the second term of the minimal-coupling Hamiltonian (D.1d) (i.e., the term proportional to $\mathbf{A}^2 \equiv \mathbf{A}\cdot\mathbf{A}$). Moreover, for definiteness, we assume that the emitter is a Hydrogen atom. Hence, the atom-field interaction Hamiltonian then becomes

$$H_{\text{int}} = \frac{e}{2m}\left[\mathbf{p}\cdot\mathbf{A}(\mathbf{r}) + \mathbf{A}(\mathbf{r})\cdot\mathbf{p}\right]\,. \tag{D.11}$$

We want to determine the transition rate between an initial (excited) state $|i\rangle = |e,0\rangle \equiv |e\rangle\otimes|0\rangle$ and a final (not necessarily the lowest) state $|f\rangle = \left|g, 1_{\mathbf{x}j\omega_0}\right\rangle = f_j^\dagger(\mathbf{x},\omega_0)\,|g,0\rangle$. The transition rate for such a first-order process calculated within first-order perturbation theory follows from Fermi's Golden Rule [5, 6]

$$\Gamma_{i\to f} = \frac{2\pi}{\hbar^2}\,|\langle f|\,H_{\text{int}}\,|i\rangle|^2\,\delta(\omega_i-\omega_f-\omega)\,. \tag{D.12}$$

The matrix element entering in Eq. (D.12) reads[7] [7]

$$
\begin{aligned}
\langle f|\, H_{\rm int}\, |i\rangle &= \frac{e}{m}\sum_k \langle g, 1_{\mathbf{x}k\omega_0}|\, \mathbf{A}(\mathbf{r})\cdot\mathbf{p}\,|e,0\rangle \\
&= \frac{e}{m}\sum_k \langle g,0|\, f_k(\mathbf{x},\omega_0)\mathbf{A}(\mathbf{r})\cdot\mathbf{p}\,|e,0\rangle \\
&= \frac{e}{m}\sqrt{\frac{\hbar}{\pi\epsilon_0}\frac{\omega_0}{c^2}}\sum_k \langle g,0|\,\sqrt{\operatorname{Im}\epsilon(\mathbf{x},\omega_0)}G^{\dagger}_{kn}(\mathbf{r},\mathbf{x};\omega_0)p_n\,|e,0\rangle\ . \qquad \text{(D.13)}
\end{aligned}
$$

where the commutation relations of the bosonic field operators have been used. Carrying out a similar calculation for $\langle i|\, H_{\rm int}\,|f\rangle$, and making use of the identity (D.5c), one may write (taking the wavefunctions as real)

$$
\Gamma = \frac{2\pi}{\hbar^2}\frac{e^2\hbar}{\pi\epsilon_0 m^2c^2}\int d\mathbf{r}\int d\mathbf{r}'\,\psi_e(\mathbf{r}')\psi_e(\mathbf{r})\operatorname{Im}G_{ij}(\mathbf{r},\mathbf{r}';\omega_0)(p_i\psi_g(\mathbf{r}))(p_j^*\psi_g(\mathbf{r}'))\,. \qquad \text{(D.14)}
$$

Now, the electromagnetic Green's function admits an analytical expression of the form (for $z > 0$)

$$
\overleftrightarrow{\mathbf{G}}(\mathbf{r},\mathbf{r}';\omega_0) = \frac{i}{8\pi^2}\int d\mathbf{q}\left[\overleftrightarrow{\mathbf{M}}_s + \overleftrightarrow{\mathbf{M}}_p\right]e^{i\mathbf{q}\cdot\boldsymbol{\rho}+ik_z z}e^{-i\mathbf{q}\cdot\boldsymbol{\rho}'+ik_z z'}\,. \qquad \text{(D.15)}
$$

Since deviations of the LDOS from classicality owing to quantum surface corrections incur at relatively short emitter–metal distances, i.e., at relatively large wavevectors, we here assume the electrostatic limit, i.e., $q \gg k_0$ (except for the reflection coefficient, which is still assumed to be the retarded version in order to obtain the plasmon pole accurately), such that $k_z \to iq$ and $\|\overleftrightarrow{\mathbf{M}}_s\| \to 0$ as $c \to \infty$:

$$
\overleftrightarrow{\mathbf{G}}(\mathbf{r},\mathbf{r}';\omega_0) \overset{c\to\infty}{\simeq} \frac{i}{8\pi^2}\int d\mathbf{q}\,\overleftrightarrow{\mathbf{M}}_p\big|_{c\to\infty} e^{i\mathbf{q}\cdot\boldsymbol{\rho}-qz}e^{-i\mathbf{q}\cdot\boldsymbol{\rho}'-qz'}\,, \qquad \text{(D.16)}
$$

where

$$
\overleftrightarrow{\mathbf{M}}_p\big|_{c\to\infty} = -\frac{qc^2}{\omega_0^2}r_p\begin{pmatrix} i & 0 & 1\\ 0 & 0 & 0\\ -1 & 0 & i\end{pmatrix} = -\frac{qc^2}{\omega_0^2}r_p\begin{pmatrix} i\\ 0\\ -1\end{pmatrix}\otimes\begin{pmatrix}1 & 0 & -i\end{pmatrix} \equiv -2i\frac{qc^2}{\omega_0^2}r_p\hat{\boldsymbol{\epsilon}}_{\hat{\mathbf{q}}}\otimes\hat{\boldsymbol{\epsilon}}^*_{\hat{\mathbf{q}}}, \qquad \text{(D.17)}
$$

with $\hat{\boldsymbol{\epsilon}}_{\hat{\mathbf{q}}}$ denoting the polarization vector defined via $\hat{\boldsymbol{\epsilon}}_{\hat{\mathbf{q}}} = (\hat{\mathbf{q}} + i\hat{\mathbf{z}})/\sqrt{2}$. Therefore, the transition rate (D.14) is now given by

$$
\Gamma = \frac{e^2}{2\pi^2\epsilon_0\hbar m^2\omega_0^2}\int_0^{2\pi} d\varphi\int_0^{\infty} dq\, q^2\,\left|\langle e|\,(\hat{\boldsymbol{\epsilon}}_{\hat{\mathbf{q}}}\cdot\mathbf{p})e^{i\mathbf{q}\cdot\boldsymbol{\rho}-qz}\,|g\rangle\right|^2\operatorname{Im} r_p. \qquad \text{(D.18)}
$$

[7] Neglecting the contribution from $\nabla\cdot\mathbf{A}$ which is approximately zero away from the metal (where $\nabla\cdot\mathbf{A} = 0$).

It it useful to rewrite Eq. (D.18) as

$$\Gamma = \frac{e^2}{2\pi^2\epsilon_0\hbar m^2\omega_0^2}\int_0^{2\pi} d\varphi \int_0^{\infty} dq\, q^2 \left|\langle e|\,(\hat{\boldsymbol{\epsilon}}_{\hat{\mathbf{q}}}\cdot\mathbf{p})e^{i\mathbf{q}\cdot\boldsymbol{\rho}-q(z-z_0)}\,|g\rangle\right|^2 \operatorname{Im} r_p \quad \text{(D.19)}$$

which allows the evaluation of the matrix element with the emitter at the origin.

For simplicity, in what follows we restrict ourselves to electric multipolar transitions whose initial and final states posses $m_f = m_i = 0$. Then, we may choose $\mathbf{q}$ along the x-direction, and, writing the momentum operator as $\mathbf{p} = -i\hbar\nabla$, we obtain

$$\Gamma = \frac{4\alpha\hbar^2\omega_0}{m^2c^2}\int_0^{\infty} du\, u^2 e^{-2uk_0z_0}\left|\langle e|\,e^{iuk_0x-uk_0(z-z_0)}\hat{\boldsymbol{\epsilon}}_{\hat{\mathbf{q}}}\cdot\nabla\,|g\rangle\right|^2 \operatorname{Im} r_p, \quad \text{(D.20)}$$

where we have introduced the dimensionless variable $u \equiv q/k_0$. Since for an arbitrary nth electric transition, En, the matrix element in (D.20) is dominated by only one of the terms coming from the expansion of the exponential,[8] we get

$$\left|\langle e|\,e^{ik_0ux-k_0u(z-z_0)}\hat{\boldsymbol{\epsilon}}_{\hat{\mathbf{q}}}\cdot\nabla\,|g\rangle\right|^2 \propto u^{2(n-1)}. \quad \text{(D.21)}$$

Finally, this leads to the following expression the transition rate associated with a given En transition:

$$\Gamma_{\mathrm{E}n} = 2\alpha^3\omega_0\left[\frac{(k_0a_{\mathrm{B}})^{n-1}}{(n-1)!}\right]^2 \Xi^{l_e,l_g}_{n_e,n_g}\int_0^{\infty} du\, u^{2n} e^{-2uk_0z_0}\operatorname{Im} r_p, \quad \text{(D.22)}$$

where a_{B} denotes the Bohr radius and where the quantity Ξ originates from the matrix element,

$$\Xi^{l_e,l_g}_{n_e,n_g} \equiv \left|\int_0^{2\pi} d\phi \int_0^{\pi} d\theta\, \sin\theta \int_0^{\infty} d\bar{r}\, \bar{r}^2\, \Psi_e(\bar{r},\theta,\phi)\,(i\,\bar{r}\,\sin\theta\cos\phi - \bar{r}\,\cos\theta)^{n-1} \right.$$
$$\left. \times\,(\sin\theta\cos\phi + i\cos\theta)\frac{\partial}{\partial\bar{r}}\Psi_g(\bar{r},\theta,\phi)\right|^2, \quad \text{(D.23)}$$

with $\bar{r} \equiv r/a_{\mathrm{B}}$ (note that here the wavefunctions Ψ are dimensionless).

In passing, we comment that a classical electrodynamics calculation using the Green's dyadics yields the same *qualitative* result as Eq. (D.22) (i.e., up to a proportionality factor). The decay rates *relative to the ones in vacuum*, however, should be *quantitatively* the same (at least within the approximations assumed in this derivation) [8–10].

---

[8] In particular, the dominating term arising from the expansion of the exponential is the one corresponding to the change in the angular momentum of the atom by an integer n, i.e., $|l_f - l_i| = n$; see Supplementary Material of Ref. [7].

References

1. Jackson JD (1998) Classical electrodynamics, 3rd edn. Wiley, New York
2. Novotny L, Hecht B (2012) Principles of nano-optics, 2nd edn. Cambridge University Press, Cambridge
3. Peřina J (ed) (2001) Coherence and statistics of photons and atoms. Wiley, New York
4. Scheel S, Buhmann S (2008) Acta Physica Slovaca 58(5):675
5. Griffiths DJ (2005) Introduction to quantum mechanics, 2nd edn. Prentice-Hall, New Jersey
6. Cohen-Tannoudji C, Diu B, Laloe F (1978) Quantum mechanics, vol 2. Wiley, New York
7. Rivera N, Kaminer I, Zhen B, Joannopoulos JD, Soljačić M (2016) Science 353(6296):263. http://science.sciencemag.org/content/353/6296/263, https://doi.org/10.1017/CBO9780511619915
8. Klimov VV, Ducloy M (2005) Phys Rev A 72:043809. https://doi.org/10.1017/CBO9780511619915; https://doi.org/10.1017/CBO9780511619915
9. Crosse JA, Scheel S (2009) Phys Rev A 79:062902. https://link.aps.org/doi/10.1103/PhysRevA.79.062902, https://doi.org/10.1017/CBO9780511619915
10. Cotrufo M, Fiore A (2015) Phys Rev B 92:125302. https://link.aps.org/doi/10.1103/PhysRevB.92.125302, https://doi.org/10.1017/CBO9780511619915.

The manufacturer's authorised representative in the EU is Springer Nature Customer Service Centre GmbH, Europaplatz 3, 69115 Heidelberg, Germany. If you have any concerns regarding our products, please contact ProductSafety@springernature.com

Printed and bound by CPI Group (UK) Ltd, Croydon, CR0 4YY
30/09/2026
02225462-0004